# The IMA Volumes in Mathematics and its Applications

**Volume 124**

*Series Editor*
Willard Miller, Jr.


Springer
*New York*
*Berlin*
*Heidelberg*
*Barcelona*
*Hong Kong*
*London*
*Milan*
*Paris*
*Singapore*
*Tokyo*

# Institute for Mathematics and its Applications IMA

The **Institute for Mathematics and its Applications** was established by a grant from the National Science Foundation to the University of Minnesota in 1982. The IMA seeks to encourage the development and study of fresh mathematical concepts and questions of concern to the other sciences by bringing together mathematicians and scientists from diverse fields in an atmosphere that will stimulate discussion and collaboration.

The IMA Volumes are intended to involve the broader scientific community in this process.

Willard Miller, Jr., Professor and Director

* * * * * * * * * *

## IMA ANNUAL PROGRAMS

| | |
|---|---|
| 1982–1983 | Statistical and Continuum Approaches to Phase Transition |
| 1983–1984 | Mathematical Models for the Economics of Decentralized Resource Allocation |
| 1984–1985 | Continuum Physics and Partial Differential Equations |
| 1985–1986 | Stochastic Differential Equations and Their Applications |
| 1986–1987 | Scientific Computation |
| 1987–1988 | Applied Combinatorics |
| 1988–1989 | Nonlinear Waves |
| 1989–1990 | Dynamical Systems and Their Applications |
| 1990–1991 | Phase Transitions and Free Boundaries |
| 1991–1992 | Applied Linear Algebra |
| 1992–1993 | Control Theory and its Applications |
| 1993–1994 | Emerging Applications of Probability |
| 1994–1995 | Waves and Scattering |
| 1995–1996 | Mathematical Methods in Material Science |
| 1996–1997 | Mathematics of High Performance Computing |
| 1997–1998 | Emerging Applications of Dynamical Systems |
| 1998–1999 | Mathematics in Biology |
| 1999–2000 | Reactive Flows and Transport Phenomena |
| 2000–2001 | Mathematics in Multimedia |
| 2001–2002 | Mathematics in the Geosciences |
| 2002–2003 | Optimization |
| 2003–2004 | Probability and Statistics in Complex Systems: Genomics, Networks, and Financial Engineering |

Continued at the back

Lisa J. Fauci Shay Gueron
Editors

# Computational Modeling in Biological Fluid Dynamics

With 93 Illustrations

Springer

Lisa J. Fauci
Department of Mathematics
Tulane University
New Orleans, LA 70118, USA
ljf@math.tulane.edu

Shay Gueron
Department of Mathematics
Technion—Israel Institute of Technology
Haifa 32000, Israel
shay@math.technion.ac.il

*Series Editor:*
Willard Miller, Jr.
Institute for Mathematics and its Applications
University of Minnesota
Minneapolis, MN 55455, USA

Mathematics Subject Classification (2000): 76Z10, 76Z05, 92C05, 92C17, 92C35

Library of Congress Cataloging-in-Publication Data
Fauci, Lisa J.
Computational modeling in biological fluid dynamics / Lisa J. Fauci, Shay Gueron.
p. cm. — (The IMA volumes in mathematics and its applications series ; 124)
Based on papers from a workshop held Jan. 1999.
Includes bibliographical references.
ISBN 0-387-95233-0 (alk. paper)
1. Fluid dynamics—Mathematical models—Congresses. 2. Biophysics—Mathematical models—Congresses. I. Fauci, Lisa J. II. Gueron, Shay. III. Institute of Mathematics and Its Applications IV. Title. V. IMA volumes in mathematics and its applications ; v. 124.
QH513.5 .F38 20001
571.4′3—dc21 00-067921

Printed on acid-free paper.



Production managed by Yong-Soon Hwang; manufacturing supervised by Jerome Basma.
Camera-ready copy prepared by the IMA.
Printed and bound by Sheridan Books, Inc., Ann Arbor, MI.
Printed in the United States of America.

9 8 7 6 5 4 3 2 1

ISBN 0-387-95233-0 SPIN 10792730

Springer-Verlag New York Berlin Heidelberg
*A member of BertelsmannSpringer Science+Business Media GmbH*

# FOREWORD

This IMA Volume in Mathematics and its Applications

## COMPUTATIONAL MODELING IN BIOLOGICAL FLUID DYNAMICS

is based on the proceedings of a very successful workshop with the same title. The workshop was an integral part of the September 1998 to June 1999 IMA program on "MATHEMATICS IN BIOLOGY." I would like to thank the organizing committee: Lisa J. Fauci of Tulane University and Shay Gueron of Technion - Israel Institute of Technology for their excellent work as organizers of the meeting and for editing the proceedings.

I also take this opportunity to thank the National Science Foundation (NSF), whose financial support of the IMA made the Mathematics in Biology program possible.

Willard Miller, Jr., Professor and Director
Institute for Mathematics and its Applications
University of Minnesota
400 Lind Hall, 207 Church St. SE
Minneapolis, MN 55455-0436
612-624-6066, FAX 612-626-7370
miller@ima.umn.edu
World Wide Web: http://www.ima.umn.edu

# PREFACE

A unifying theme in biological fluid dynamics is the interaction of moving, elastic boundaries with a surrounding fluid. A complex dynamical system describes the motion of red blood cells through the circulatory system, the movement of spermatazoa in the reproductive tract, cilia of microorganisms, or a heart pumping blood. The revolution in computational technology has allowed tremendous progress in the study of these previously intractable fluid-structure interaction problems.

This volume contains invited and refereed papers based upon presentations given in the IMA workshop on "Computational Modeling in Biological Fluid Dynamics" during January of 1999. This workshop brought together biologists, zoologists, engineers, and mathematicians working on a variety of issues in biological fluid dynamics. It was a lively forum where new developments in numerical methods, modeling, and experimental results were shared. It is our hope that this collection will allow others to become aware of and interested in the exciting accomplishments and challenges uncovered during this workshop.

As the organizers, it is our pleasant duty to thank the IMA for the workshop. It was a success in all respects: friendly atmosphere, high quality talks, fruitful scientific interactions, and extraordinary logistics. We were extremely impressed by the friendliness, efficiency and resourcefulness of the entire IMA staff. Their meticulous and professional organization contributed immensely to the smooth operation during the week of the workshop. In particular, we would like to acknowledge the important contributions of Dr. Fred Dulles to the success of the workshop.

Lisa J. Fauci
Tulane University

Shay Gueron
University of Haifa

## CONTENTS

# FLUID MECHANICS OF CILIARY PROPULSION*

JOHN BLAKE†

**Abstract.** Cilia have many functions in the animal kingdom, some of these being cleansing, feeding, excretion, locomotion and reproduction. They occur in all phyla of the animal kingdom with the possible exception of the class *Nematoda*. This lecture will discuss the development of fluid mechanical models and theories that help with our understanding and interpretation of locomotion of protozoa, mucous transport in the lung, filter feeding in bivalve molluscs and gamete transport.

The theoretical development for the fluid mechanics requires obtaining the fundamental singularities and image systems pertinent to the system under study, the physical interpretation of them and their constructive use to model the flow fields generated by fields of cilia.

These theories allow estimates of the flow fields in the cilia sublayer and for a greater understanding of propulsive mechanisms in both micro-organisms and mucous transport in the lower respiratory tract. The sophisticated models in turn allow us to develop better approximations for simplified models that provide an improved understanding of more complicated flows involving filter feeding in bivalve molluscs and with ovum transport in the oviduct.

Finally, studies of possible filter feeding strategies in the sessile organism, *Vorticella*, which alters the length of its stalk periodically, has led to the development of some interesting non-linear mathematics in a simplified 'blinking stokeslet' model of this filter feeding phenomenon. We shall demonstrate that this can lead to chaotic dynamics, which has been shown to enhance mixing and hence improve the efficiency of feeding currents. The continuous system is reduced to an area-preserving map, which allows for greater analytical progress to be made in this inertia-free system. Poincaré sections and Lyapunov exponents are used alongside other chaotic measures to determine the nature and extent of the chaos. Effects of molecular diffusion are mimicked via the incorporation of white noise in the map and enhanced feeding levels are predicted.

The author of this paper acknowledges, with gratitude, the enormous influence that Sir James Lighthill has had on his life and academic career. This paper is dedicated to his memory and is based on research work conducted by the author of this review over the last 30 years with the material in this article being taken from papers over this period.

## Contents



*Dedicated to the memory of Sir James Lighthill.

†School of Mathematics and Statistics, The University of Birmingham, Edgbaston, Birmingham B15 2TT, U.K.

**1. Introduction.** In his lecture to the International Congress of Theoretical and Applied Mechanics in Moscow in 1972, Sir James Lighthill made the following observation:

> *Generally speaking, progress in Zoologically Applied Mechanics can be made only through close collaboration between a zoologist and a mechanician. The animal kingdom is vast, and the problem of discerning and describing what types of animal motion have in different parts of it been significant, above all in the evolutionary sense, is one for the zoologist alone. At the same time, study of the complicated interactions between those motions and a surrounding aqueous medium is one for the hydrodynamicist alone. It is therefore when a zoologist and a hydrodynamicist have got to know each other well enough to be able to talk together about the problems, and gradually to learn enough of each other's language so as to be able to communicate effectively, that collaborative progress involving hydrodynamically sound analysis of zoologically significant motions becomes possible.*

Those present at this IMA Workshop on 'Computational Modeling in Biological Fluid Dynamics' and indeed throughout this year-long programme on 'Mathematics in Biology' would be in strong agreement with these sentiments although in the other programmes we would replace hydrodynamics with mathematics.

In this paper we are concerned with the principles of the fluid mechanics of ciliary transport in a wide range of environments including locomotion, feeding, digestion, respiration, excretion, reproduction, sensory reception and cleansing of surfaces. Micro-organisms are very small, typically 10–1000$\mu$m in length, and move with low velocities, measured in $\mu$m/sec, so the non-dimensional Reynolds number $R_b$ of the body and $R_c$ for the cilium defined by

$$R_b = \frac{\rho U \ell}{\mu}, \quad R_c = \frac{\rho \omega L^2}{\mu} \tag{1}$$

are naturally very small. Here $U$ is the velocity of propulsion, $\ell$ the length of the micro-organism, $\omega$ the angular frequency, $L$ the length of the cilium while $\mu$ is the dynamic viscosity and $\rho$ the density. This means that the propulsion of micro-organisms is due mainly to the viscous stresses rather than the inertial forces we associate with the propulsion of larger and faster objects such as fish, ships and aeroplanes.

**2. Structure of a cilium.** Cilia, flagella and sperm tails are three types of cellular organelle which have similar structure, the internal mechanism consisting of a 9 + 2 fibril complex with each fibril of constant length being able to slide over each other by interaction between the peripheral doublet - the contraction hypothesis (Sleigh 1962). In Figure 1 details of the structure of a cilium are shown both in length and in cross-section (Sleigh *et al.* 1988). For a summary of more recent research in this area, further information may be found in Satir *et al.* (1998).

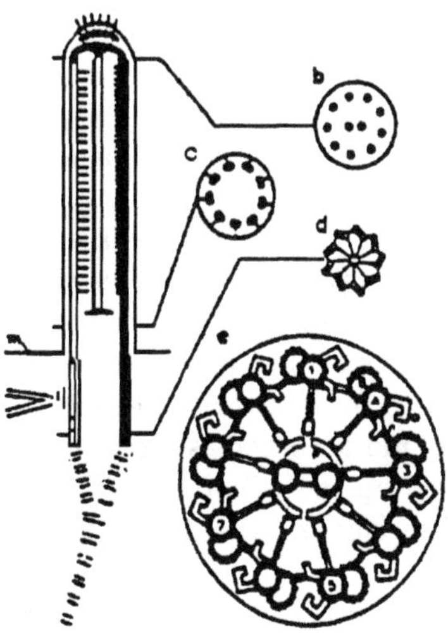


FIG. 1. *Structure of a cilium.*

**3. Ciliary propulsion — descriptive.** Ciliated protozoa are propelled by a large field of cilia that beat in a well-coordinated manner. The organisms may vary in length from 20–1000$\mu$m although some may be larger. Some representative organisms are shown in Figure 2 (from Sleigh and Blake 1975).

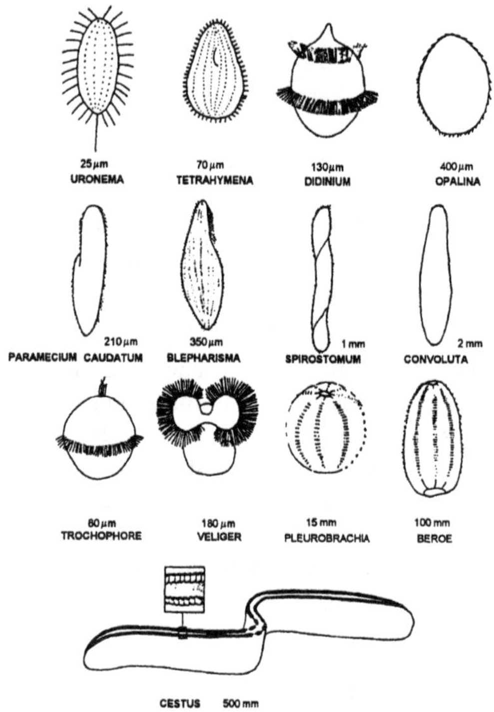


FIG. 2. *Some representative organisms propelled by cilia.*

The density of cilia on the cell surface is usually in the range 0.1–10 cilia/$\mu$m$^2$ and the cilia length is normally in the range 5–20 $\mu$m. Cilia have a characteristic beat pattern that is illustrated in Figure 3. The beat pattern can conveniently be split into two phases known universally as the (a) effective and (b, c) recovery stroke. The effective stroke consists of a fast rotating rigid body motion whereas the recovery stroke is much slower as a wave is propagated up the cilium from the base bringing the cilium much closer to the cell surface as it drags itself back to begin a new effective stroke.

There are several important fluid mechanical features associated with this propulsive mechanism which make it more effective.

1. The higher velocity in the effective stroke leads to a greater force being exerted on the fluid,

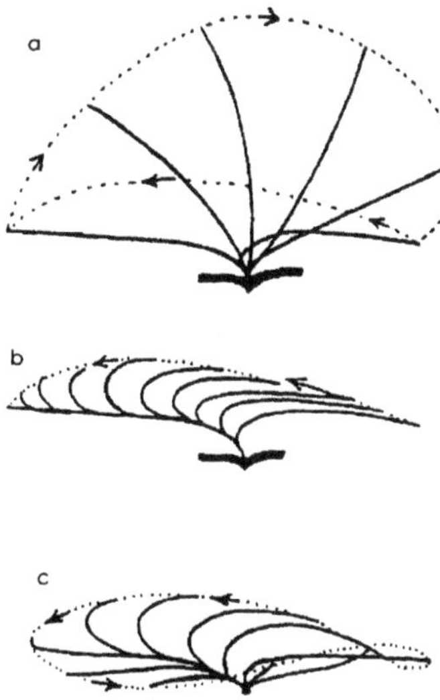


FIG. 3. *The beat cycle of a typical water-propelling cilium consists of 2 parts, an effective stroke (a) and a recovery stroke (b), both seen from the side, with profiles at equal time intervals; the dotted line shows the path taken by the ciliary tip. When the cilium is seen from above (c), the sideways swing of the shaft during the recovery stroke is evident.*

2. *the orientation effect* for a slender body means that a body having a predominant motion normal to its centre-line will exert almost twice the force for a tangential mode of motion at the same velocity, and
3. *the wall effect* which shields the interaction of the cilium with the liquid medium. In fluid mechanical terms the far-field velocity is either stresslet or Stokes-quadrupole in character rather than the stokeslet behaviour near the slender body. The strength of the stresslet field depends linearly upon the height above the wall.

Another important observable feature of ciliary propulsion is the metachronal waves — the coordination wave created by the field of cilia — often compared to wind blowing over a wheatfield. Knight-Jones (1954) labelled the different types of metachronal wave as illustrated in Figure 4. The direction of the coordination wave is taken in relation to the beat direction during the effective stroke. The type of metachronal wave has an important bearing on the freedom of movement of cilium during its effective stroke. For example the cilia are very close together in symplectic metachronism whereas in antiplectic metachronism, the cilia are relatively wide apart.

The previous discussion on the beat pattern and the metachronal wave have further implications on the fluid mechanics as is illustrated in Figure 5.

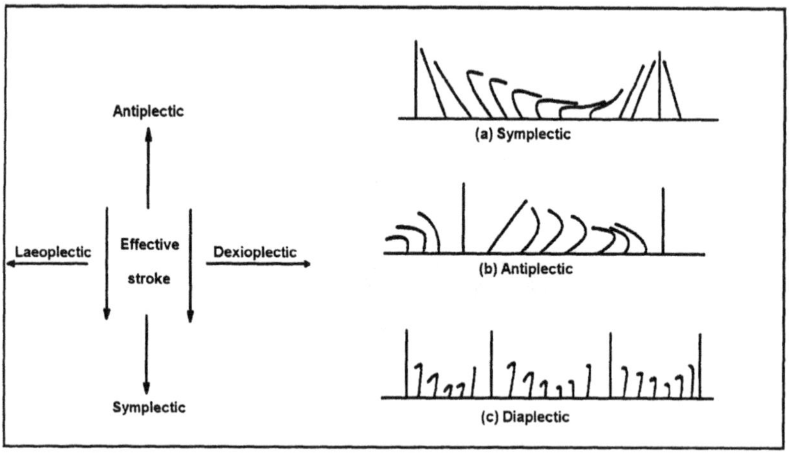


FIG. 4. *(a) The four basic types of metachronism, (b) Illustration of the various types of metachronal wave patterns.*

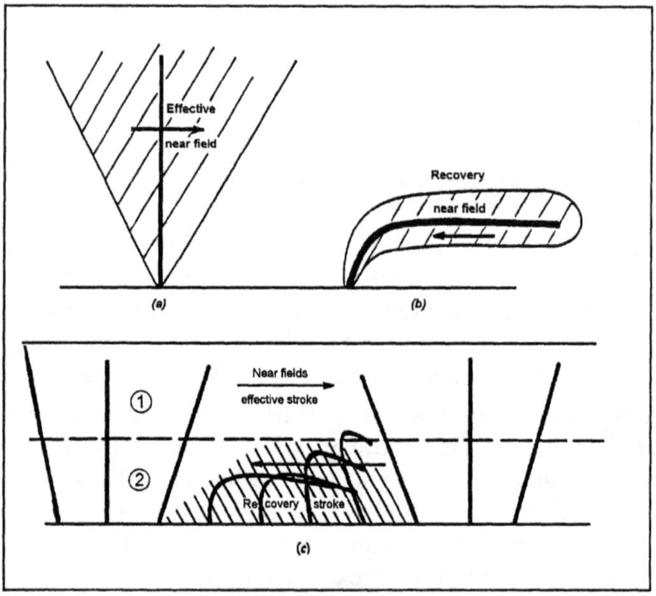


FIG. 5. *The 'zones of influence' during the (a) effective and (b) recovery stroke. In (c) the combined influence of both the effective and recovery strokes is incorporated into a metachronal wave. This leads to a relatively steady flow in the upper part of the ciliary sub-layer and a highly oscillatory motion in the lower part where both the effective and recovery strokes are equally important.*

## 4. Ciliary propulsion — modelling.

**4.1. Introduction.** It is clear that to model the fluid mechanics of ciliary propulsion will be extremely complicated and that various assumptions will need to be made if progress is to be achieved. The enormous number of cilia on the cell surface and the relative flatness of ciliated organisms allows us to approximate the surface as a flat sheet. Schematic diagrams illustrating the geometric description of both the cilia and the metachronal wave representation are shown in 6 and 7 respectively. The mathematical detail will be provided in the following sections.

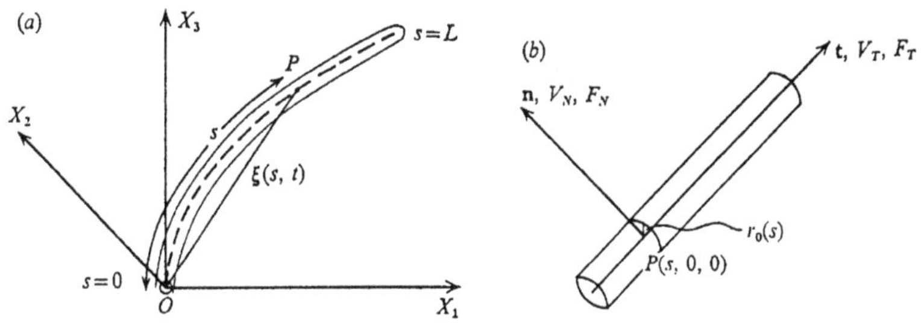


FIG. 6. *Diagrams illustrating the co-ordinate representations for a cilium: (a) Cartesian co-ordinates for $\xi(s,t)$ where s is the arc-length; (b) the local 'inner' co-ordinates, where* **t** *is the tangent and* **n** *is the normal to the centre-line. The radius of the cilium at P is $r_0(s)$.*

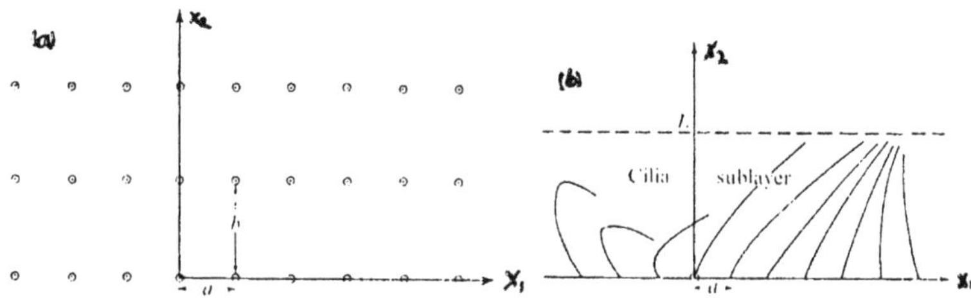


FIG. 7. *Illustrations of regular array of cilia: (a) from above and (b) side-on view for symplectic metachronism (i.e. wave in same direction as effective stroke). The spacing in the $x_1$ direction is a, while in the $x_2$ direction it is b.*

**4.2. Cilia beat cycle.** A realistic model of the ciliary transport system must include an accurate representation of the cilia beat cycle since it is through the beat of the cilia that energy is transmitted to the fluid. The representation should incorporate the periodic nature of the beat and should embody the physical advantages of the observed beat cycle.

The method employed is that described in Blake (1972) where the cartesian position of any given point on the cilium $\xi$ at time $t$ is represented by the truncated Fourier series

$$\xi(s,t) = \frac{1}{2}\mathbf{a}_0(s) + \sum_{n=1}^{N_0}(\mathbf{a}_n(s)\cos n\sigma t + \mathbf{b}_n(s)\sin n\sigma t) \tag{2}$$

where $s$ measures arc length from the base of the cilium and $\sigma$ is the angular beat frequency. The Fourier coefficients $\mathbf{a}_n$ and $\mathbf{b}_n$ are vector quantities which are approximated by,

$$\mathbf{a}_n = \sum_{m=1}^{M_0} \mathbf{A}_{mn}s^m \text{ and } \mathbf{b}_n = \sum_{m=1}^{M_0} \mathbf{B}_{mn}s^m \tag{3}$$

where $\mathbf{A}_{mn}$ and $\mathbf{B}_{mn}$ are vector quantities.

**4.3. Metachronal wave.** We suppose that a regular array of cilia bases is distributed over the plane $x_3 = 0$ with spacing $a$ in the $x_1$-direction and $b$ in the $x_2$-direction. The length of the cilium is $L$ with characteristic radius $r_0$, with $L/r_0 \gg 1$. The cilium centre-line motion is defined

$$\mathbf{x} = \xi(s,t) \tag{4}$$

where $s$ is the arc length and $t$ is time, an illustration is shown in Figure 6(a). To represent a general ciliary motion in a metachronal wave we generalise (2) to

$$\left.\begin{aligned} \xi_1' &= ma + \xi_1(s,\tau) \\ \xi_2' &= nb + \xi_2(s,\tau) \\ \xi_3' &= \xi_3(s,\tau) \end{aligned}\right\} \tag{5}$$

where $\tau = t + \kappa.\mathbf{x}^*$, $\mathbf{x}^* = (x_1, x_2)$ and $\kappa = (\kappa_1, \kappa_2)$. $\kappa$ indicates the direction the wave propagates over the cell surface. We also suppose that exactly $N$ cilia may be found in one wavelength. For convenience, in this study, we suppose that the metachronal wave propagates in the $x_1$-direction.

**4.4. Cilia sublayer model.** It is now possible to write down formally the leading-order term for the velocity field by summing up all the contributions from the doubly infinite array of slender bodies.

$$u_i(\mathbf{x},t) = \sum_{m=-\infty}^{\infty}\sum_{n=-\infty}^{\infty}\int_0^L G_{ij}(\mathbf{x},\xi')F_j[\xi']ds \tag{6}$$

where $G_{ij}(\mathbf{x},\xi')$ is the Stokes flow Green's function near a single plane rigid boundary,

$$\begin{aligned} G_{ij}(\mathbf{x},\xi') = \frac{1}{8\pi\mu}\Bigg[\left(\frac{\delta_{ij}}{r} + \frac{r_ir_j}{r^3}\right) - \left(\frac{\delta_{ij}}{R} + \frac{R_iR_j}{R^3}\right) \\ +2h\left(\delta_{j\alpha}\delta_{\alpha k} - \delta_{j3}\delta_{3k}\right)\frac{\partial}{\partial Rk}\left\{\frac{hR_i}{R^3} - \left(\frac{\delta_{i3}}{R} + \frac{R_iR_3}{R^3}\right)\right\}\Bigg] \end{aligned} \tag{7}$$

where

$$r = [(x_1 - \xi_1')^2 + (x_2 - \xi_2')^2 + (x_3 - \xi_3')^2]^{\frac{1}{2}}$$

$$\text{and} \quad R = [(x_1 - \xi_1')^2 + (x_2 - \xi_2')^2 + (x_3 + \xi_3')^2]^{\frac{1}{2}}$$

The summation and integral signs can be reversed in (6) which together with periodicity associated with exactly $N$ cilia in each wavelength allows us now to write this equation as,

$$u_i(\mathbf{x}, t) = \sum_{n=1}^{N} \int_0^L F_j[\xi'] \left\{ \sum_{m=-\infty}^{\infty} \sum_{n=-\infty}^{\infty} G_{ij}(\mathbf{x}, \xi') \right\} ds. \tag{8}$$

On using the Poisson summation formula, the leading order term in (8) becomes,

$$\sum_{m=-\infty}^{\infty} \sum_{n=-\infty}^{\infty} G_{ij} = \frac{A_{ij}}{N\mu ab} K(x_3, \xi_3') + O\left(\frac{Nab}{L^2}\right), \tag{9a}$$

where $A_{ij} = \delta_{i1}\delta_{j1} + \delta_{i2}\delta_{j2}$

$$K(x_3, \xi_3') = \begin{cases} x_3 : x_3 < \xi_3' \\ \xi_3' : x_3 > \xi_3' \end{cases} \tag{9b}$$

and the metachronal wavelength $\lambda = Na$. The interesting feature about $A_{ij}$ is that there is no vertical velocity $u_3$, as would be expected on mass conservation grounds because of the zero normal velocity component of the rigid no-slip boundary.

Thus the leading-order term yields a linear velocity field $U_\alpha(x_3, t)$ due to the field of cilia, which is only a function of height and time as follows,

$$U_\alpha(x_3, t) = \frac{1}{\mu ab} \frac{1}{N} \sum_{n=1}^{N} \int_0^L K(x_3, \xi_3') A_{\alpha j} F_j[\xi_3'] ds, \qquad \alpha = 1, 2. \tag{10}$$

If we use Gray and Hancock's (1955) resistive force theory for $F_j[\xi_3']$ we obtain

$$F_j[\xi'] = C_T \left( \gamma \delta_{jk} - (\gamma - 1) \frac{\partial \xi_j'}{\partial s} \frac{\partial \xi_k'}{\partial s} \right) \left( \frac{\partial \xi_k'}{\partial t} - U_k(\xi_3') \right) \tag{11}$$

where the summation convention over $k$ has been exploited, $C_T$ is the tangential resistance coefficient and $\gamma$ is the ratio of the normal to tangential resistance coefficients.

We therefore obtain an integral equation for the "mean velocity field" $U_\alpha(x_3, t)$ where $\alpha = 1$ or 2. Further details of this theory may be found in Blake (1972) and Mochon and Liron (1976).

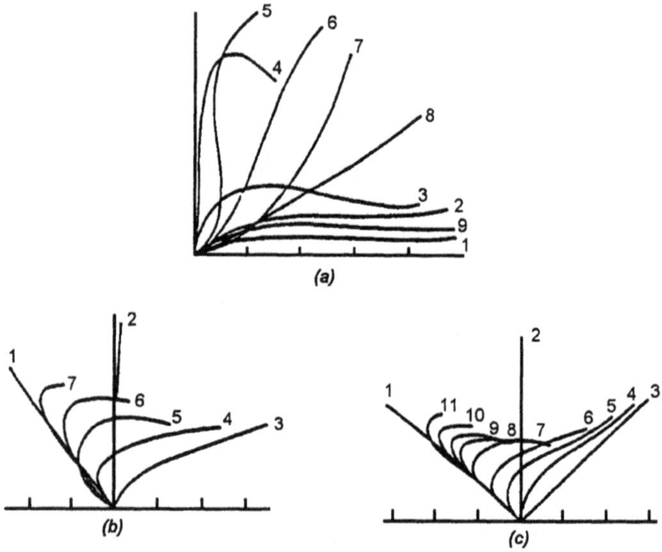


FIG. 8. *The Fourier-series-least squares models for (a) Opalina, (b) Paramecium and (c) Pleurobrachia. Numbers indicate successive stages of the bear. The effective stroke in Opalina is from 5 to 9, in both Paramecium and Pleurobrachia it is from 1 to 3.*

The input data required for these calculations is the ciliary beat cycle data. Examples of the beat patterns of *Opalina*, *Paramecium* and *Pleurobrachia* are shown in Figure 8.

The mean velocity profiles are shown in Figure 9 for the three examples. The common features of all three cases is the relatively low velocities in the lower part of the ciliary sublayer before a rapid increase up to the maximum velocity at the top of the sublayer. In the *Paramecium* and *Pleurobrachia* examples 'backflow' is predicted although the calculated values in this region are probably comparable to the error. Certainly one would expect backflow for that part of the cycle when the cilia are in their recovery stroke.

## 5. The propulsion of mucus by cilia.

**5.1. Introduction.** The mechanics of muco-ciliary transport is one of the fundamental problems in respiratory mechanics and, yet, very little is understood about the basic interaction mechanisms between cilia and mucus. The main function of the mucous layer is in cleaning the inspired air of unwanted particles, cellular debris and secretions from the airways of the lung. In this section the theoretical model of the previous section is further developed to predict velocity profiles in the periciliary and mucous layers.

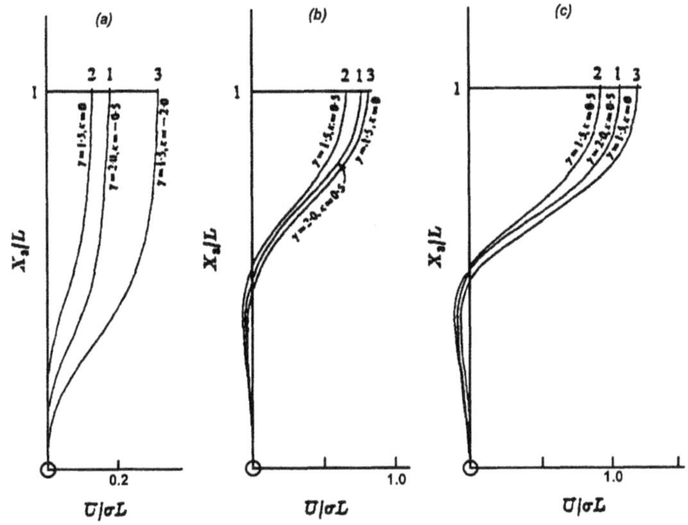

FIG. 9. *Mean velocity profiles for (a) Opalina, (b) Paramecium and (c) Pleurobrachia.*

Cilia are found on the outer surface of the epithelial cells which line most airways of the human respiratory tract, a complex branched network of tubes terminating in the alveoli (Figure 10(a)). Above this lies the periciliary layer of fluid and above this again, a mucous layer, the size and quantity of mucus depending on the location of the airway and the state of respiratory health of the subject (Figure 10(b)).

As stated earlier, cilia are elongated cylindrical projections emerging from the epithelial cell surface. The typical three-dimensional beat cycle of cilia in the lung consists of a fast effective stroke in which the cilium protrudes at full length from the epithelium (see Figure 3). At the end of the effective stroke the cilium is observed to "rest" (Sanderson and Sleigh 1981, Sleigh 1981, 1982). This may be an adaption to the beat cycle of cilia from water-propelling organisms to overcome the difficulty of withdrawing the cilium from the highly viscous mucus. During the slow recovery stroke, the cilia rotate outwards to the right side (from above) keeping close to the epithelium and taking advantage of the shielding of the no-slip boundary.

In discussing the fluid mechanical principles of muco-ciliary transport, it should be noted that there are several length scales of particular relevance to this study. They are (1) molecular length scales relative to the biochemical structure of mucus, (2) length scales associated with the cilium tip (0.1–1.0 $\mu$m), (3) a length scale associated with the cilium length, cell size, ciliary wavelength and coherence of a mucous plaque (5–50 $\mu$m) and

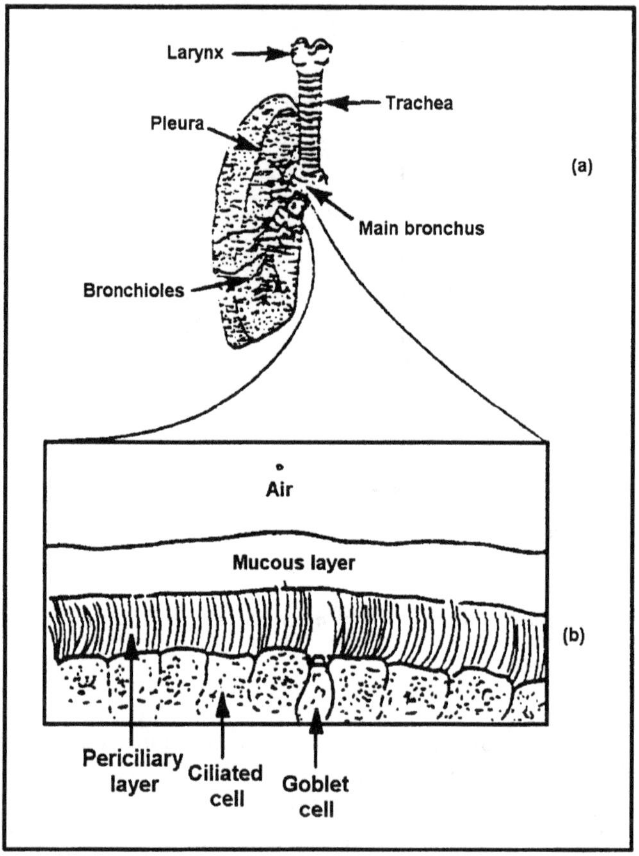


FIG. 10. *Schematic illustration of the muco-ciliary system in the lung: (a) a human lung, (b) cross-section of a major bronchiole.*

(4) a length scale associated with the length of an airway (5–10 mm). Of particular importance with regard to cilia-mucus interaction is the scale of the cilium tip with respect to the entangled network of molecules constituting the mucus. Does the cilium "see" a continuous viscous liquid or a loosely coiled network of molecules?

A qualitative understanding of the ciliary beat cycle was discussed in the previous section. However, for the shorter cilia in mucus-propelling environments where there is not such a clear distinction between effective and recovery strokes, further mechanisms must be employed to produce a continuous steady motion instead of the oscillatory motion that would result. Clearly, an elastic, almost rigid, structure on the macroscale, but a more liquid-like response on the microscale of the cilium tip would be the most desirable attributes for the surface layer of mucus. Visco-elastic properties such as occur in coiled macromolecular networks provide the consistency and integrity that is required for transport in muco-ciliary systems. Clearly then, the rheological properties of the liquid in the lung are central to any scientific discussion of muco-ciliary transport; an illustration of these ideas being found in Figure 11.

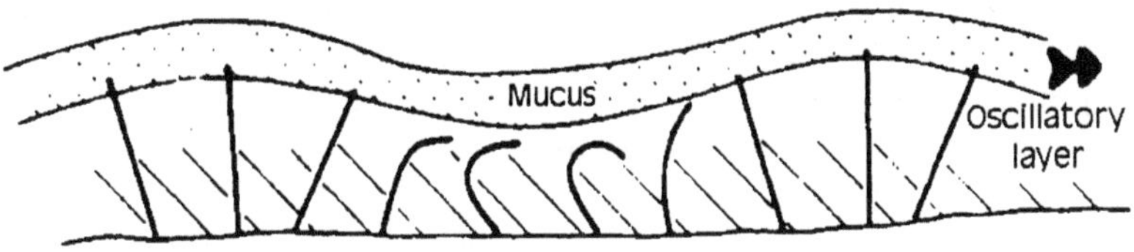


FIG. 11. *The physical interpretation of Figure 10 may be extended by including penetration of the upper mucous layer by cilia in their effective stroke.*

Over the last decade extensive studies have been directed towards the mucus' rheological properties (see e.g. Silberberg 1982) but far fewer attempts have been made to study the mechanical interaction between the mucus and cilia. Litt (1970) discussed the visco-elastic properties of the mucous layer and the advantage of having relatively large relaxation times for the mucus in comparison with the beat period of the cilium. Sade, Eliezer, Silberberg and Nero (1970) argued that the mere size of the glycoprotein molecules would physically inhibit the ciliary motion if present in the cilia sublayer. In pathological circumstances hypersecretion may lead to a deepening of either the periciliary fluid layer, or the mucous layer, or both. If the periciliary layer deepens, mucous transport rates will be substantially diminished because of the difficulty of transmitting the tangential shear stress generated by the cilia across the wide low-viscosity periciliary layer. On the other hand, a thick viscous mucous layer may become entangled in the cilia restricting their beat and hence reducing transport rates. Recent studies have focussed on liquid transport across the epithelia. Apart from altering the thickness of the periciliary layer, it will also influence the rheological properties of the mucus.

The effects of airflow on mucous transport have usually been neglected. However, in cases of hypersecretion, high frequency ventilation or coughing, the influence of surface air stresses cannot be entirely neglected, especially in the first few generations of the lung. Airflow patterns in the lung are extremely complicated as a result both of the strong secondary flows produced by bifurcations and airway curvature and of the oscillatory nature of the flow (see Pedley, Schroter and Sudlow, 1971). Generally, we would expect the boundary-layer structure to be quasi-steady except at extremely high frequencies when the structure will be that of the familiar Stokes unsteady boundary layer. Although the surface air stresses increase in these examples, we would also need a change in properties and depth of the mucous and periciliary layers for there to be effective mucous transport. As discussed by Blake (1975, 1984), the important factor is the depth and viscosity of the periciliary layer: if this layer increases in depth, the mucus becomes disengaged from the cilia and hence considerably eases the movement of mucus.

**5.2. Ciliary beat cycle.** The representation for a cilium associated with muco-ciliary transport follows on from the previous section. Sleigh (1977) proposed a planar cilia beat cycle for the lung where it was required that the minimum radius of curvature along the length is 1.25 $\mu$m. This beat pattern is reconstructed using the above Fourier least squares procedure with $N_0 = 6$ and $M_0 = 3$. The coefficients $\mathbf{A}_{mn}$ and $\mathbf{B}_{mn}$ for this beat pattern are presented in Table 1(a) and the corresponding cilia beat pattern in Figure 12. Likewise the beat pattern of Sanderson and Sleigh (1981) leads to the coefficients given in Table 1(b) and Figure 12. Further details on this procedure can be found in Fulford and Blake (1986). In the Sanderson and Sleigh (1981) study the cilia were observed to penetrate the mucous layer at some stage during their effective stroke. The calculations record that the tip speed of the cilium varies, depending on the location of the cilium. There is a noticeable slowing down of the cilium as it enters the very viscous mucous layer. When the cilium exits from the mucous layer it speeds up considerably due to the elastic recoil energy stored in the cilium.

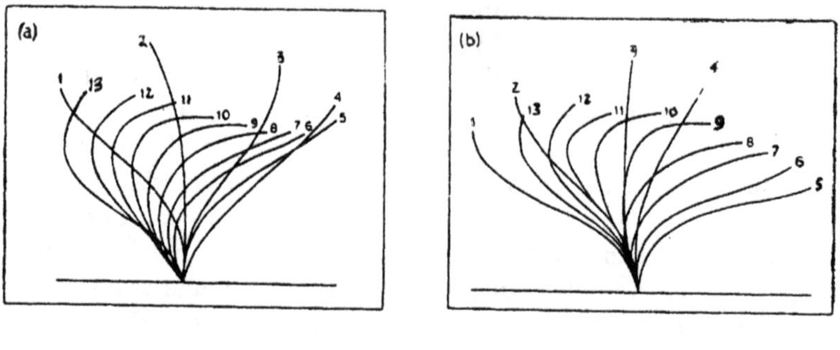


FIG. 12. *Reconstruction of the cilia beat cycle from (a) Sleigh (1977) and (b) Sanderson and Sleigh (1981).*

**5.3. The metachronal wave.** Using the same representation as in Section 5, we use $\tau = kna+\sigma t$ to represent the antiplectic metachronal wave (which propagates in the opposite direction to the effective stroke) which is so common in mucous transport. The frequency of the wave is $\dfrac{\sigma}{2\pi}$ and the wavelength is $\lambda = \dfrac{2\pi}{k}$. In the lung, under normal, healthy conditions, the cilia spacing is such that one wavelength comprises approximately 100 cilia.

TABLE 1

*(a) Fourier least squares coefficients for the cilia beat pattern from Sleigh (1977).*
*(b) Fourier least squares coefficients for the cilia beat pattern from Sanderson and Sleigh (1981).*

$A_{mn}$

| $n=$ | 0 | 1 | 2 | 3 | 4 | 5 | 6 |
|---|---|---|---|---|---|---|---|
| $m=1$ | −0·654 | 0·393 | −0·097 | 0·079 | 0·119 | 0·119 | 0·009 |
| | 1·895 | −0·018 | 0·158 | 0·010 | 0·003 | 0·013 | 0·040 |
| $m=2$ | 0·787 | −1·516 | 0·032 | −0·302 | −0·252 | −0·015 | 0·035 |
| | −0·552 | −0·126 | −0·341 | 0·035 | 0·006 | −0·029 | −0·068 |
| $m=3$ | 0·202 | 0·716 | −0·118 | 0·142 | 0·110 | −0·013 | −0·043 |
| | 0·096 | 0·263 | 0·186 | −0·067 | −0·032 | −0·002 | 0·015 |

$B_{mn}$

| $n=$ | 1 | 2 | 3 | 4 | 5 | 6 |
|---|---|---|---|---|---|---|
| $m=1$ | 0·284 | 0·006 | −0.059 | 0·018 | 0·053 | 0·009 |
| | 0·192 | −0·050 | 0·012 | −0·007 | −0·014 | −0·017 |
| $m=2$ | 1·045 | 0·317 | 0·226 | 0·004 | −0·082 | −0·040 |
| | −0·499 | 0·423 | 0·138 | 0·125 | 0·075 | 0·067 |
| $m=3$ | −1·017 | −0·276 | −0·196 | −0·037 | 0·025 | 0·023 |
| | 0·339 | −0·327 | −0·114 | −0·105 | −0·057 | −0·055 |

(a)

$A_{mn}$

| $n=$ | 0 | 1 | 2 | 3 | 4 | 5 | 6 |
|---|---|---|---|---|---|---|---|
| $m=1$ | −0·449 | 0·130 | −0·169 | 0·063 | −0·050 | −0·040 | −0·068 |
| | 2·076 | −0·003 | 0·054 | 0·007 | 0·026 | 0·022 | 0·010 |
| $m=2$ | −0·072 | −1·502 | 0·260 | −0·123 | 0·011 | −0·009 | 0·196 |
| | −1·074 | −0·230 | −0·305 | −0·180 | −0·069 | 0·001 | −0·080 |
| $m=3$ | 0·658 | 0·793 | −0·251 | 0·049 | 0·009 | 0·023 | −0·111 |
| | 0·381 | 0·331 | 0·193 | 0·082 | 0·029 | 0·002 | 0·048 |

$B_{mn}$

| $n=$ | 1 | 2 | 3 | 4 | 5 | 6 |
|---|---|---|---|---|---|---|
| $m=1$ | −0·030 | −0·093 | 0·037 | 0·062 | 0·016 | −0·065 |
| | 0·080 | −0·044 | −0·017 | 0·052 | 0·007 | 0·051 |
| $m=2$ | 1·285 | −0·036 | −0·244 | −0·093 | −0·137 | 0·095 |
| | −0·298 | 0·513 | 0·004 | −0·222 | 0·035 | −0·128 |
| $m=3$ | −1·034 | 0·050 | 0·143 | 0·043 | 0·098 | −0·054 |
| | 0·210 | −0·367 | 0·009 | 0·120 | −0·024 | 0·102 |

The upper and lower numbers in each entry of the above tables correspond to the $x_1$ and $x_3$ components respectively.

(b)

In Figure 13, the computer-generated profile of a wave of ciliary activity to be used in our model of muco-ciliary transport is illustrated for the ciliary beat cycle in Figure 12(a). The metachronal wave is clearly seen moving to the left.

**5.4. Cilia sublayer model.** Like the studies reported in the previous section, the cilia sublayer is modelled by an infinite array of flexing slender bodies attached to a plane boundary. Coordination of the cilia will be

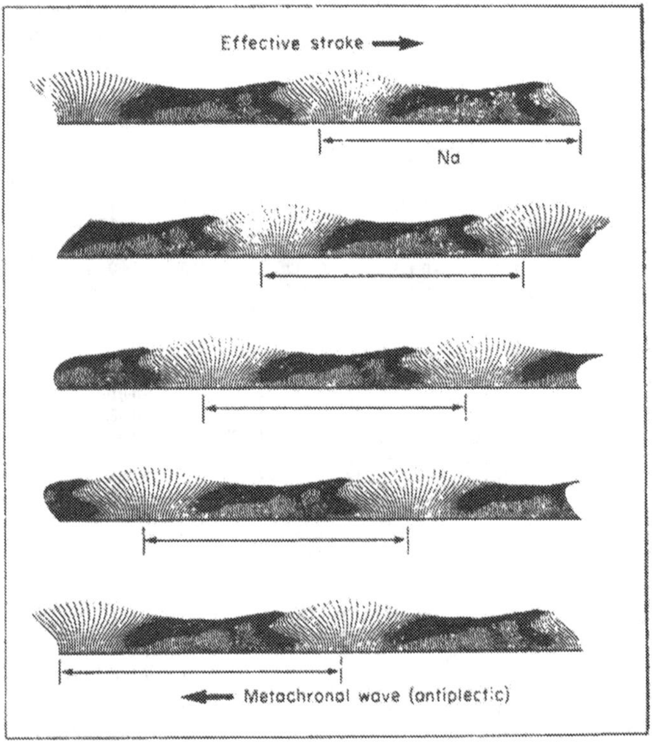


FIG. 13. *Antiplectic metachronal wave with the cilia beat from Figure 12(a) at equal time intervals. The direction of the effective stroke and the metachronal wave are indicated on the diagram.*

assumed to be antiplectic and exactly $N$ cilia comprise one wavelength of the metachronal wave. Thus lines of cilia in the $x_2$-direction and every $N$th row in the $x_1$-direction will beat in synchronisation.

Figure 14 illustrates the periciliary layer has depth $h$ and the mucous layer depth $(H - h)$ giving a total thickness of the two liquid layers of $H$ with $L$ the depth of the cilia sublayer. The viscosities of the periciliary and mucous layers are $\mu_1$ and $\mu_2$ respectively. The imposed boundary conditions at the interface are continuity of tangential stress and velocity with the normal velocity being zero.

Following the approach developed in the previous section, the velocity and pressure field may be represented by a doubly infinite sum overall the cilia of a distribution of 'stokeslets' and associated images along the centre-line of every cilium as follows,

$$u_i^{(k)} = \sum_{n=-\infty}^{\infty} \sum_{m=-\infty}^{\infty} \int_0^L G_{ij}^{(k,\ell)}(\mathbf{x}, \xi') F_j^{(\ell)}(\xi') ds, \tag{12}$$

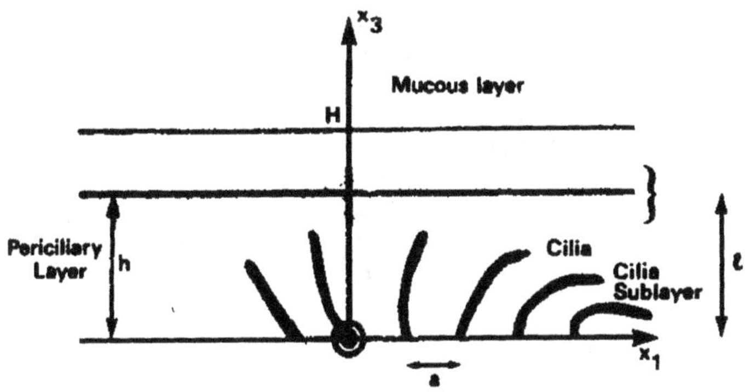


FIG. 14. *Schematic side-on view of geometry used in model.*

where $k$ and $\ell$ are either 1 or 2 and refer to whether we are considering the upper or lower fluid, $i$ and $j$ are either 1, 2 or 3 and refer to the cartesian coordinate system, and where $\xi'$ is summed over $m$ and $n$ given the definition in (5). The force per unit length is given by $F^{(\ell)}$ while, in principle, the Green's function $G_{ij}^{(k,\ell)}$ may be obtained by Fourier transform methods. Since we are assuming every $N$ cilia are in phase, we can rearrange, as previously, (12) to yield,

$$u_i^{(k)} = \sum_{n=1}^{N} \int_0^L F_j^{(\ell)}(\underline{\xi}') \left\{ \sum_{-\infty}^{\infty} \sum_{-\infty}^{\infty} G_{ij}^{(k,\ell)}(\mathbf{x}, \xi') \right\} ds. \tag{13}$$

The quantity in brackets corresponds to the flow field due to all the cilia in phase. This doubly infinite sum over a periodic array may be converted to the Fourier transformed version by using the Poisson summation formula. The leading order terms correspond to a plane of constant force per unit area $(F/Nab)$ at $x_3 = \xi_3'$ yielding the following 'mean-field' Green's function in the case when the interface is at $x_3 = h$.

(i) $0 < \xi_3' < h$

$$K^{(1)}(x_3, \xi_3') = \begin{cases} \dfrac{x_3}{\mu_1}; & 0 \le x_3 < \xi_3' \\[2ex] \dfrac{\xi_3'}{\mu_1}; & \xi_3' < x_3 \le h \\[2ex] \dfrac{\xi_3'}{\mu_1}; & h \le x_3 \le H \end{cases}$$

(14)

(ii) $h < \xi_3' < H$

$$K^{(2)}(x_3, \xi_3') = \begin{cases} \dfrac{x_3}{\mu_1}; & 0 \le x_3 \le h \\[2ex] \dfrac{1}{\mu_2}[x_3 + (\theta - 1)h]; & h \le x_3 \le \xi_3' \\[2ex] \dfrac{1}{\mu_2}[\xi_3' + (\theta - 1)h]; & \xi_3' < x_3 \le H \end{cases}$$

where $\theta = \dfrac{\mu_2}{\mu_1}$, the viscosity ratio. The other terms in the doubly infinite sum decay exponentially with distance from the cilium. This automatically leads to the 'mean field' equations for the average velocity,

$$\begin{aligned} U_\alpha(x_3, t) &= \frac{1}{Nab} \sum_{\lambda=1}^{N} \int_0^L F_\alpha^{(\ell)} K^{(\ell)}(x_3, \xi_3') ds \\ U_3(x_3, t) &= 0, \end{aligned} \tag{15}$$

where $\alpha = 1$ or 2.

To complete the computation we now require an approximate expression for the force/unit length $F_\alpha^{(\ell)}$. By generalising the Gray and Hancock (1955) theory, used in (11), we can approximate the force per unit length as follows,

$$F_j^{(\ell)}(\xi^n) = C_T^{(\ell)} \left( \gamma^{(\ell)} \delta_{jk} - (\gamma^{(\ell)} - 1) \frac{\partial \xi_j^n}{\partial s} \frac{\partial \xi_k^n}{\partial s} \right) \left( \frac{\partial \xi_k^n}{\partial t} - U_k(\xi_3^n, t) \right), \tag{16}$$

where $C_T^{(\ell)}$ is the tangential resistance coefficient, $\gamma^{(\ell)}$ is the ratio of the normal to tangential resistance coefficients, the superscript $n$ refers to the $n$th cilium in the metachronal wave and $\ell$, the layer. Using the first order solutions from Fulford & Blake (1986), we obtain

$$\left. \begin{aligned} &\gamma^{(1)} = \gamma^{(2)} = 2 \\ \text{and} \quad &C_T^{(1)} = \frac{2\pi\mu_1}{\log \dfrac{L_1}{r_0}}, \qquad C_T^{(2)} = \frac{2\pi\mu_2}{\log \dfrac{L_2}{r_0}} \end{aligned} \right\} \tag{17}$$

where $L_1$ is the length of the cilium in the periciliary layer of viscosity $\mu_1$ and $L_2$ the length in the mucous layer of viscosity $\mu_2$. It should be emphasised at this stage that only for part of the cycle will the cilium be embedded in the mucous layer and that on some occasions the cilium will not penetrate the mucous layer at all.

The 'mean field' equations of (15) now reduce to

$$
\begin{aligned}
U_\alpha(x_3,t)+\sum_{n=1}^{N}\frac{C_T^{(\ell)}}{Nab}\int_0^L\left[\gamma^{(\ell)}\delta_{\alpha\beta}-(\gamma^{(\ell)}-1)\frac{\partial\xi_\alpha^n}{\partial s}\frac{\partial\xi_\beta^n}{\partial s}\right]U_\beta(x_3,t)K^{(\ell)}(x_3,\xi_3^n)ds \\
=\sum_{n=1}^{N}\frac{\ell_T^{(\ell)}}{Nab}\int_0^L\left[\gamma^{(\ell)}\delta_{\alpha j}-(\gamma-1)\frac{\partial\xi_\alpha^n}{\partial s}\frac{\partial\xi_j^n}{\partial s}\right]\frac{\partial\xi_j^n}{\partial t}K^{(\ell)}(x_3,\xi_3^n)ds.
\end{aligned}
\tag{18}
$$

On scaling lengths with respect to $\ell$ and velocities with respect to $\sigma\ell$, the following dimensionless cilium density parameters are obtained,

$$
\left.
\begin{aligned}
\tau_1 &= \frac{2\pi L^2}{Nab\log\dfrac{L_1}{r_0}}=\frac{2\pi L^2}{\lambda b\log\dfrac{L_1}{r_0}}\\
\tau_2 &= \frac{2\pi L^2}{Nab\log\dfrac{L_2}{r_0}}=\frac{2\pi L^2}{\lambda b\log\dfrac{L_2}{r_0}}
\end{aligned}
\right\}
\tag{19}
$$

where $\lambda$ is the wavelength of the metachronal wave. Typically in the lung this parameter is in the range 5–10.

In Figure 15 the mean velocity profile for different values of the parameter $\tau_1$ are illustrated. As $\tau_1$ increases from 0.1 to 6 the mean velocity in the upper part of the cilia sublayer steadily increases, associated with the increase in the force/unit volume acting on the fluid due to the increasing cilia density. For larger values of $\tau_1$ there is only a marginal increase in the mean flow velocity. This limiting velocity was referred to by Blake and Winet (1980) as the 'active porous medium limit' where there is extremely strong viscous coupling between adjacent cilia.

Using the three cilia beat cycles illustrated in Figure 12 we calculate the mean velocity profiles in both the periciliary layer and the mucous layer with cilia density parameter $\tau_1 = 6$ and with 100 cilia per wavelength. This viscosity of the mucus is 1000 times that of the fluid in the periciliary layer. The results are illustrated in Figure 16 for various degrees of penetration of the cilia, in their effective stroke, into the mucous layer. the maximum mucus velocity occurs at an optimal penetration depth of about 10% ($h/L = 0.9$) for the cilia beat of Sleigh (1977) and the idealised beat, whereas for the cilia beat of Sanderson and Sleigh (1981) it occurs when $h/L = 0.8$. The cilia beat cycle observed by Sanderson and Sleigh (1981) is a realistic model when the cilia penetrate the mucus because the

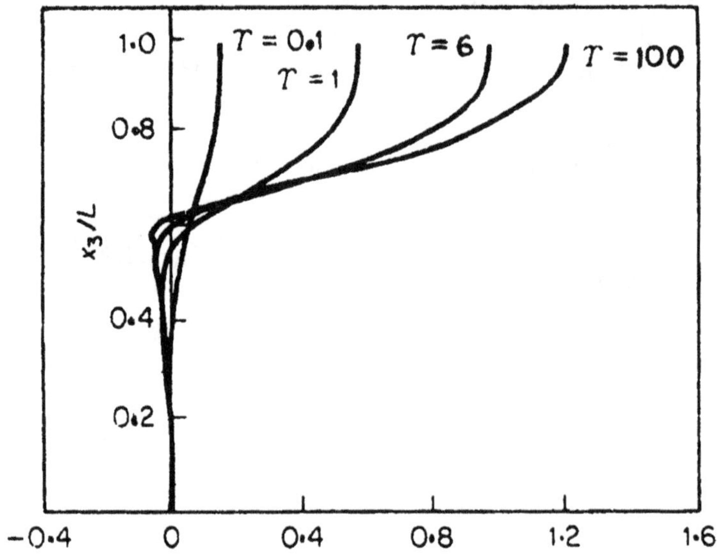


FIG. 15. *Velocity profiles for mucous transport when the cilia do not penetrate the mucous layer using the cilia beat pattern in Figure 12. Various values of the dimensionless parameter* $\tau(=\tau_1)$ *are indicated on the diagram.*

effective stroke is slowed down due to the higher resistance to movement in the mucous layer.

In Figure 16(b) the calculations are repeated with only 10 cilia per wavelength. If the wavelength of the metachronal wave remains the same, then the value of the density parameter $\tau_1$ will remain at 6, however, if we consider an identical reduction of the cilia density in the $x_2$-direction, then $\tau_1 = 0.6$. This lower value leads to larger velocities in the periciliary layer. In summary the calculations record that, while it is the cilia density which determines the mean velocity in the periciliary layer, in the mucous layer the extremely high viscosity requires only a small number of cilia per wavelength to penetrate the mucous layer to propel the mucus at near the tip speed of the cilia.

**5.5. Future developments.** The theory reported in this section is drawn mainly from the papers of Blake (1984), Blake and Fulford (1984) and Fulford and Blake (1986). This fundamental problem of muco-ciliary transport concerns the complex interaction between the cilia and the overlying mucus, involving the penetration and withdrawal of the cilia from the mucous layer. The two layer nature of bronchial mucus, consisting of a "watery" periciliary layer and a much more viscous and non-Newtonian layer, are modelled by two layers of Newtonian fluid of differing viscosity. Predictions show that very little mean flow occurs in the periciliary layer

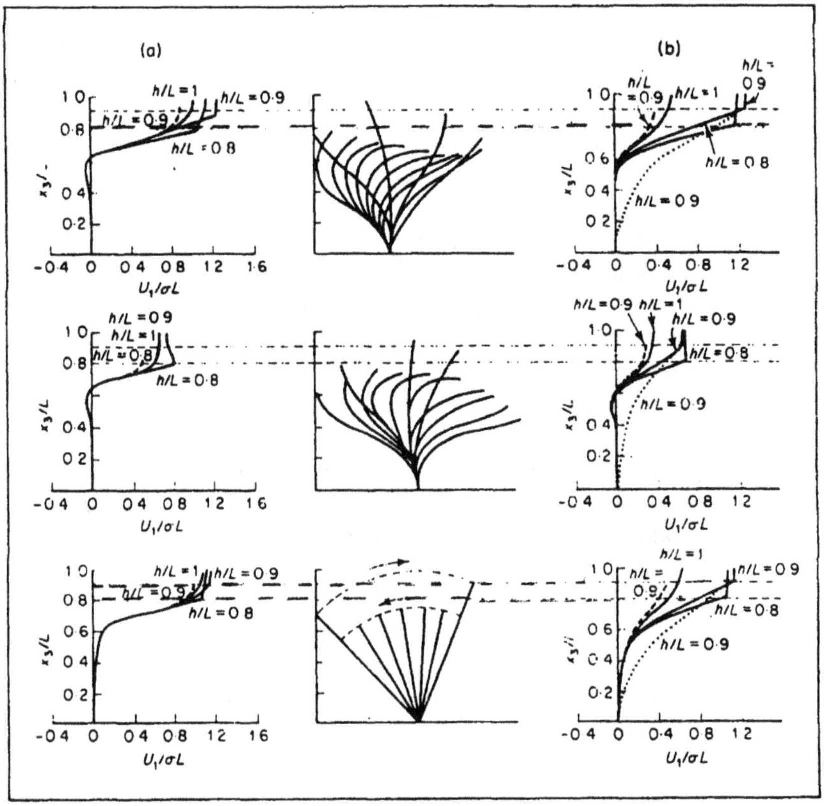

FIG. 16. *Velocity profiles with cilia penetration into the mucous layer: (a) (100 cilia/wavelength) and (b) (10 cilia/wavelength). The depth of penetration of the cilia into the mucous layer is indicated by the value of $h/L$ ($h/L = 1$ corresponds to no penetration). The cilia beat used is indicated alongside each set of profiles. [- - -$\tau_1 = 6$, $\theta = 1000$, — $\tau_1 = 6$, $\theta = 0$ and $\cdots \tau_1 = 0.6$, $\theta = 1000$].*

except near the mucus interface where high shear rates occur to produce the high observed rates in the mucous layer. However, recent studies by Matsui, Randell, Peretti, Davis and Boucher (1998) suggest that marker particles in both the periciliary layer and mucous layer move with similar speeds, even though the theoretical mean flows yield a very different picture. At present these observations have no satisfactory explanation. Perhaps mixing processes, identified in the final section, may be important for liquid transport in the lung as particles from the mucous layer are drawn down into the periciliary layer. It would be interesting to know the concentration dependence as a function of length transported in both the layers. There may also be significant transport across the epithelia driven by osmotic gradients. Research is continuing on this topic with particular reference to cystic fibrosis.

While substantial progress has been made over the last decade in the study of muco-ciliary transport both from the experimental and theoretical side, considerable additional study needs to be done before we have a clear understanding of the underlying principles. Much of the experimental work has been conducted on the frog palate which is under neural control whereas it is thought this is not the case in the mammalian trachea. Studies of the ciliary beat pattern have been made in a specially cultured environment that may not be entirely representative of the respiratory tract (e.g. less damaged cilia, smoother epithelia, no airflow).

Clearly on the theoretical side a more realistic model that includes a non-Newtonian mucous layer is required, and which also allows for the non-linear deformation and penetration of an initially flat interface by a slender body would yield valuable information on the interaction of the cilium with the mucus during the cilium's effective stroke. Theory also needs to be developed to account for the non-uniform distribution of cilia over the epithelia, the problem of a mucous layer becoming partially attached to the epithelia and the importance of airflow including coughing on mucous transport rates.

## 6. Ovum transport.

**6.1. Introduction.** Ovum transport along the oviduct is a vital link in the procreation of man, yet our knowledge of the mechanics of ovum transport is limited. There exists a vast amount of biochemical and physiological research into ovum transport, not all of it in agreement, and even now it is difficult to obtain a consensus on the relative importance of muscular and ciliary activity. In this section a fluid dynamical model is developed that incorporates the principle transport mechanisms of ciliary and muscular activity and pressure differences across the oviduct.

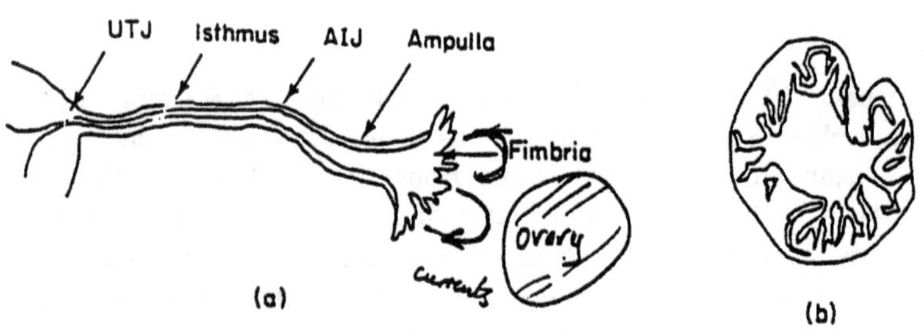


FIG. 17. *(a) The human oviduct, (b) cross-section of the isthmus in rabbit oviduct.*

Although not conclusive, experimental data does suggest that cilia are necessary for ovum transport, particularly in the ampulla. There appears to be a correlation between the number of ciliated cells on the fimbria and fertility in humans (Broseus and Vasquez, 1976). Pauerstein and Eddy (1979) showed that reversal of 1cm segments of ampulla in the rabbit prevents pregnancy, whereas reversal of similar lengths of the isthmus did not interfere with normal fertility. Thus, the present trend of evidence favours cilia as a necessary condition for ovum transport. Unfortunately, the abundance of smooth muscle in the isthmus and its occurrence in varying amounts in the ampulla have thus far prevented direct observation of cilia-ovum interaction in the intact tube. The intact tube is a necessary condition for meaningful analysis of the fluid dynamical mechanism of ovum transport since any pressure gradients along the axis which are present *in vivo* are removed by opening the duct.

The effect of muscular activity on net ovum transport was an unresolved question. For example, Halbert, Tam and Blandau (1976) obtained experimental evidence that ovum transport in the ampulla of the mammalian oviduct is not affected when muscular activity is suppressed. Similarly Croxatto *et al.* (1979) concluded that compounds which stimulate tubal contractility do not necessarily accelerate ovum transport. In contrast, Hodgson, Talo and Pauerstein (1977) and Talo (1974) have obtained evidence that muscular activity in the isthmus of the oviduct is clearly related to ovum transport.

Little experimental work exists on the effect of pressure differences across the oviduct on ovum transport. Local pressure gradients in the oviduct (e.g. as induced by ciliary and muscular activity) are obviously necessary for ovum transport. Thus, applied pressure differences from one end of the oviduct to the other will play an important rôle in ovum transport.

For fluid mechanical purposes, the oviduct may be thought of as a long trumpet-shaped tube, about 10 cm in length with an internal diameter varying from 100–200 $\mu$m at the AIJ (Ampulla-Isthmus junction) through to around 0.5–1.0 cm at the fimbria (see Figure 17(a)). A human ovum is typically around 150–200 $\mu$m in diameter. Normally an ovum is transported to the AIJ in a few minutes where it remains for 24–72 hours before proceeding along the isthmus to the uterus. There is often a much higher density of cilia in the ampulla than in the isthmus while cilia lengths are in the range 5–7 $\mu$m and they beat at frequencies of 15–30 Hz.

As far as the fluid mechanics of ovum transport is concerned, there are several comments that can be made immediately. To gain insight, let us consider the simplest case in the absence of muscular activity and secretion, when the conservation of mass implies that the volume flow rate across each section must be constant. Thus, if we define $A_1$, $V_1$ and $A_2$, $V_2$ to be respectively the cross-sectional area and average velocities at two cross-sections, then their products are equal to the same constant value, say $Q$, i.e.

$$A_1V_1 = A_2V_2 = Q. \tag{20}$$

Thus if $A_2$ is greater than $A_1$, then $V_2$ is less than $V_1$. Of course these velocities are averages, but often, in practice, the velocity across a section will change sign. Thus, the change in cross-sectional area along the oviduct will lead to changes in the velocity profile and hence will affect the rate of ovum transport. Because there are very narrow sections of the oviduct such as the AIJ or the UTJ, or due to the presence of mucus (Jansen, 1980), the overall volume flow rate $Q$ may be extremely small. If this is the case both $V_1$ and $V_2$ will be close to zero indicating that at each section there could be almost equal flow backwards and forwards. In addition, the presence of an ovum in the oviduct will restrict fluid motion and may also bring about the zero volume flow rate condition.

We can also make further important qualitative statements about aspects of the flow field due to changes in the cross-sectional area. If in a segment of an open tube the wall contracted due to muscular activity, most of the fluid would be moved in the direction of the larger cross-sectional area (lower resistance). Conversely if the tube expands, the fluid would flow into this region from the larger cross-sectional area. Thus the uncoordinated contractions and expansions of the oviduct will produce a highly oscillatory ovum movement. It is clear from the above discussion that one of the important aspects of fluid flow within the oviduct will be the relative cross-sectional areas.

An additional morphological feature that complicates the development of a theoretical model is the highly convoluted nature of cross-sections of the oviduct (see Figure 17(b)). Cross-sectional views of the oviduct do not display the smooth features found in other conduits in the body, for example, the circular cross-sections of blood vessels. The existence of this complex geometry could allow some interesting flow patterns, some of which are beneficial to pro-uterine ovum transport. For example, pro-uterine ovum transport could be accomplished while fluid moves in an opposite direction in some other folds. However, in a wide tube, pro-uterine transport of the ovum in the presence of a net pro-ovarian fluid flow would only be possible if the ovum is adjacent to the cilia tips.

In this section we concentrate on modelling the motion of the ovum in the narrow isthmus by adopting a lubrication layer approach and by using a force balance on the ovum to calculate its motion. In the ampulla we expect the ovum and surrounding cumulus to be swept along close to the cilia tips.

**6.2. The fluid dynamical model.** To develop a fluid dynamical model we need to incorporate two dominant morphological features of the oviduct, firstly the taper from the ovarian to the uterine end and secondly the complex cross-sectional profiles. These features have led to the development of a model which consists of two layers, a schematic illustration being shown in Figure 18.

1. An inner tapered region through which the ovum moves. In our model we suppose cilia are found around the periphery of this region.
2. An outer region, through which only fluid flows representing the various channels between the folds of the lumen in the oviduct. Flow in this region is due entirely to gradients in the pressure.

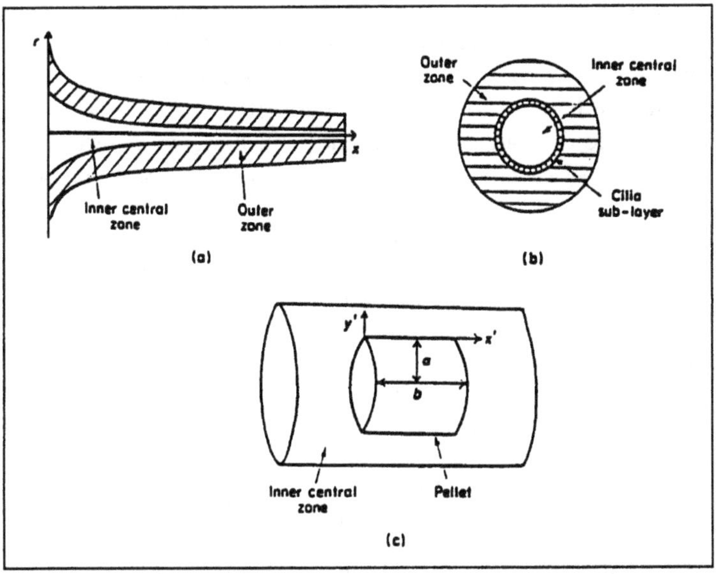


FIG. 18. *(a) The trumpet-shaped cylindrical geometry which is used to model the oviduct, (b) Cross-sectional view of the geometry used in the model, (c) the geometry of the "pellet" in the narrow sections of the oviduct.*

The radius of the inner core is defined by $R(x,t)$, where $x$ is the distance along the oviduct from the ovarian end and $t$ is time. The dependence on time allows for the existence of muscular activity which produces time dependent variations of the radius. Thus we define $R(x,t)$ to be

$$R(x,t) = R_0(x)H_1(x,t), \tag{21a}$$

where

$$R_0(x) = a(1+ke^{-\epsilon x}), \tag{21b}$$

$$H_1(x,t) = 1 - C_1(t)e^{-0.75\left(\dfrac{x-x_1}{S_1}\right)^2}, \tag{21c}$$

and

$$C_1(t) = \frac{R_1}{2}(1-\cos 2\pi f t). \tag{21d}$$

Here $R_0(x)$ corresponds to the undisturbed shape of the channel and the negative exponential models the taper in the oviduct. In the calculations we have taken $a = 0.01$ cm, $k = 49$ and $\epsilon = 0.6\text{cm}^{-1}$ giving values of the diameter equal to 0.0224 cm and 1.0 cm at the uterine and oviduct ends respectively. $H_1(x,t)$ has been introduced to simulate a muscular contraction around $x_1$ with a frequency of $f$ Hz. $R_1$, which is in the range of 0 to 1, corresponds to the extent the contraction closes the lumen and $S_1$ is a measure of the length of oviduct affected by the contraction. Additional sites of muscular activity can be included by multiplying equation (21a) by functions similar to $H_1(x,t)$. It should be noted that a sequence of contractions can be used to simulate a peristaltic wave.

To facilitate the calculations of the fluid flow, we separate the total volume flow rate at each section into 3 time dependent components $Q_p$, $Q_c$ and $Q_0$ due to (a) the local pressure gradient, (b) the cilia activity and (c) flow in the outer porous regions respectively

$$Q(t) = Q_p(t) + Q_c(t) + Q_0(t). \tag{22}$$

Muscular activity will result in the flow rate having a dependence on $x$. Mass conservation for incompressible flow in a deformable tube requires that,

$$\frac{\partial Q(x,t)}{\partial x} + \frac{\partial A(x,t)}{\partial t} = 0, \tag{23}$$

where $A(x,t)$ is the cross-sectional area. Integration of (23) yields an expression for $Q(x,t)$,

$$Q(x,t) = Q(0,t) - \int_0^x \frac{\partial A}{\partial t} dx. \tag{24}$$

The exceedingly small Reynolds numbers and the lubrication theory approximation reduces the Stokes flow equations to the axial flow component

$$\frac{dp}{dx} = \frac{\mu}{r}\frac{\partial}{\partial r}\left(r\frac{\partial u}{\partial r}\right) + F(r,t), \tag{25}$$

where $\dfrac{dp}{dx}$ is the pressure gradient along the tube, $u$, the axial velocity, $\mu$, the viscosity and $F(r,t)$ is the volume force distribution in the axial direction due to the cilia. To model the cilia sublayer we will suppose we have a distribution of volume force $\mathbf{F}(r,t)$ which has the function form,

$$\mathbf{F}(r,t) = \begin{cases} 0; & 0 \le r \le R - L \\ F_0 \sin \dfrac{\pi(R-r)}{L}; & R - L \le r \le R \end{cases} \tag{26}$$

where $R$ is the radius of the tube and $L$ the length of the cilia.

The theory used here is based on the approach of Keller *et al.* (1975) and Wu (1977).

The resulting velocity in the "inner" region is,

$$U(r,t) = U_p(r,t) + U_c(r,t) \tag{27a}$$

where the velocity field $U_p$ due to the pressure gradient is,

$$U_p(r,t) = \frac{1}{4\mu}\frac{dp}{dx}(r^2 - R^2). \tag{27b}$$

In obtaining these velocities we are ignoring the radial velocities due to exchange of fluid between the "inner" core and the outer porous zone. We obtain the cilia sublayer velocity profile $U_c$ by assuming that the layer is locally two-dimensional. This approximation requires $L \ll R$ and gives the following profile,

$$U_c(r,t) = \begin{cases} \dfrac{F_0 L^2}{\mu\pi}; & 0 \le r \le R - L \\ \dfrac{F_0 L^2}{\mu\pi^2}\left[\dfrac{\pi}{L}(R-r) + \sin\left\{\dfrac{\pi(R-r)}{L}\right\}\right]; & R-L \le r \le R. \end{cases} \tag{27c}$$

The volume flow rate due to the pressure gradient, $Q_p$, and the cilia, $Q_c$, in the "inner" channel are respectively,

$$Q_p = 2\pi \int_0^R rU_p(r)dr = -\frac{\pi}{8\mu}R^4\frac{dp}{dx} \tag{28}$$

and

$$Q_c = 2\pi \int_0^R U_c(r)dr = \frac{F_0 R^4}{\mu}\alpha^2\left[1 + \left(\frac{4}{\pi^2} - 1\right)\alpha + 0(\alpha^2)\right] \tag{29}$$

where $\alpha = L/R$.

In the "outer" porous region the volume flow rate $Q_0$ is proportional to the pressure gradient while we will for convenience write as,

$$Q_0 = -\beta\frac{\mu}{8\mu}R^4\frac{dp}{dx}; \quad \beta > 0. \tag{30}$$

The dimensionless quantity $\beta$ is a function of the geometry and the cross-sectional area at each section. In general $\beta$ is a function of $x$ and $t$, although here it will be taken as a constant. Physically $\beta$ is the ratio of the volume flow rate due to the pressure gradient in the "outer" porous region to that in the "inner" central region. The magnitude of this "$\beta$-factor" will be of major importance in determining the movement of an ovum in the oviduct.

The treatment of the flow in the outer region is a reasonable first approximation in view of the lack of morphological data to support a more detailed model. However we have retained the essential feature of the "leakback" in the outer region which may allow forward movement of the ovum in the inner region, particularly in the isthmus.

In the case of no muscular activity (23) provides the solution of a constant $Q$ with respect to $x$ as one would expect. We may invoke either of two physical conditions,

**I** The total volume flow rate $Q(0, t)$ is known or specified. This allows us to calculate the pressure rise $\Delta P$ down the oviduct in the absence of an ovum.

**II** The pressure rise $\Delta P$ down the oviduct in the absence of an ovum is specified and this allows us to calculate $Q$.

For a given flow rate $Q$, we can obtain the pressure gradient by a rearrangement of (22), (28) and (30) to obtain

$$\frac{dp}{dx} = \frac{8\mu}{\pi(1+\beta)R^4}\left[\frac{F_0 R^4}{\mu} f(\alpha) - Q\right] \tag{31}$$

where $f(\alpha) = \alpha^2\left[1 + \left(\frac{4}{\pi^2} - 1\right)\alpha\right]$.

By integrating $\frac{dp}{dx}$ along the length of the tube we obtain $\Delta p$

$$\Delta P = \frac{8\mu}{\pi}\int_0^L \left[\frac{F_0 f(\alpha)}{\mu} - \frac{Q}{R^4}\right]\frac{dx}{1+\beta(x)}. \tag{32}$$

This provides us with the relation between $Q$ and $\Delta p$ in the absence of an ovum, or a good approximation in those regions where the ovum's dimensions are much less than the width of the oviduct. The presence of an ovum in sections of comparable dimensions will alter the pressure change across the complete oviduct, which is discussed later. It is apparent from (32) that the magnitude of the pressure change along the oviduct will depend on the sign of the term in the square brackets. The bulk of experimental evidence supports the case where the maximum pressure is located at the uterus end of the oviduct, i.e. where $\Delta p \geq 0$. The value of $\frac{dp}{dx}$ may, of course, also be altered by mucosal secretion or absorption. It is clear from (32) that the total pressure rise is particularly sensitive to the radius of the "inner" region and also the magnitude of $\beta(x)$.

Before we can carry out the calculations we have to give $F_0$ a realistic value. This can be achieved if we can observe the velocity of the fluid at the cilia tips. In this case we have the following relation,

$$F_0 = \frac{\mu\pi V_c}{L^2}. \tag{33}$$

In the calculations we use $V_c = 0.03$ cm/sec (see Winet and Blake, 1980) with the viscosity of water at 20°C, $\mu = 0.01$ poise.

It is very likely that the ovum will often be in close proximity to the oviduct walls, this being particularly the case in the isthmus region. We will approximate an ovum by considering a finite length pellet of radius $a$ and length $b$. Thus, in this model, we will have a finite cylinder which has a variable radius $R(x,t)$. On the very outside we have the "outer" region. The geometry of the problem is illustrated in Figure 18(c). In the analysis that follows we define

$$g(x',t) = R(x,t) - a \text{ where } x' = x - x_0$$

and $x_0$ is the location of the trailing edge of the pellet in the tapered cylinder coordinate system. The pellet is moving with a velocity $U$. Since the problem is linear we can divide the velocity profile into three parts corresponding to the pressure gradient contribution, the shear flow due to the pellet translation and the cilia induced contribution. A local two-dimensional coordinate system $(x', y')$ is used in the gap between the ovum and tube wall. The resulting velocity field is,

$$U'(y') = \frac{1}{2\mu}\frac{dp}{dx}y'(y'-g) + U\left(1 - \frac{y'}{g}\right) + U_c'(y') \tag{34a}$$

where

$$U_c'(y') = \begin{cases} \dfrac{F_0 L^2 y'}{\mu g \pi}; & 0 \le y' \le g - L \\ \dfrac{F_0}{\mu}\left[\dfrac{(g-y')L(g-L)}{g\pi} + \dfrac{L^2}{\pi^2}\sin\dfrac{\pi(g-y')}{L}\right]; & g-L \le y' < g. \end{cases} \tag{34b}$$

The flux $Q'$ through the annular gap is approximately

$$Q' = 2\pi R \int_0^g u(y')dy' \tag{35a}$$

$$= 2\pi R\left[-\frac{1}{12\mu}\frac{dp}{dx}g^3 + \frac{Ug}{2} + \frac{F_0 L^2}{2\mu\pi}\left\{g + \left(\frac{4}{\pi^2} - 1\right)L\right\}\right]. \tag{35b}$$

Mass conservation requires that

$$Q = Q_0 + Q' + \pi a^2 U, \tag{36}$$

where $Q_0$ is defined in (30).

An additional equation is needed to balance the number of unknowns; this being obtained from the force balance equation. For simplicity it is assumed there are only two contributions to the force balance (i) the tangential shear stress on the pellet wall and (ii) the pressure difference acting on the ends of the pellet. The shear stress $\tau(x')$ on the pellet ($y' = 0$), at location $x'$, is,

$$\tau(x') = \mu \left[\frac{du'}{dy'}\right]_{y'=0} \tag{37a}$$

$$= -\frac{1}{2}\frac{dp}{dx}g - \frac{\mu U}{g} + \frac{F_0 L^2}{g\pi}. \tag{37b}$$

The force balance on the pellet yields,

$$2\pi a \int_0^h \tau(x)dx - \pi a^2 \int_0^b \frac{dp}{dx}dx = 0, \tag{38}$$

which on substitution for $\tau(x)$ and rearrangement yields,

$$\int_0^b \left(\frac{2F_0L^2}{g\pi} - \frac{2\mu U}{g} - R\frac{dp}{dx}\right) dx = 0. \tag{39}$$

From (36b) and (36) it is possible to obtain an expression for $\dfrac{dp}{dx}$

$$\frac{dp}{dx} = \frac{16\mu}{R^3\left(\beta + \frac{4}{3}\left(\frac{g}{k}\right)^3\right)} \left[\frac{Ug}{2} + \frac{F_0L^2}{2\mu\pi}\left\{g + \left(\frac{4}{\pi^2} - 1\right)L\right\} + \frac{\pi a^2 U - Q}{2\pi R}\right]. \tag{40}$$

The total pressure drop $\Delta P_t$ across the oviduct including the ovum in the isthmus is,

$$\Delta P_t = \Delta P + \int_0^b \left\{\left[\frac{dp}{dx'}\right]_{ovum} - \left[\frac{dp}{dx}\right]_{tube}\right\} dx', \tag{41}$$

where $\Delta P$ is the pressure calculated from (31), $\left[\dfrac{dp}{dx'}\right]_{ovum}$ from (40) and $\left[\dfrac{dp}{dx}\right]_{tube}$ from (30). Equation (41) provides a relation between $\Delta P_t$, $Q$ and $U$.

As indicated earlier two cases I and II can occur. In case I with $Q$ prescribed we obtain the velocity of the pellet as,

$$V = \frac{A}{B} \tag{42a}$$

where

$$A = \int_0^b \frac{F_0L^2}{\mu\pi g} - \frac{4\left[\frac{F_0L^2}{\mu\pi}\left\{g + L\left(\frac{4}{\pi^2} - 1\right) - \frac{Q}{\pi R}\right\}\right]}{R^2\left[\beta + \frac{4}{3}(\frac{g}{R})^3\right]} dx\,, \tag{42b}$$

$$B = \int_0^b \left[ g^{-1} + \frac{4(g + \frac{a^2}{R})}{R^2[\beta + \frac{4}{3}(\frac{g}{R})^3]} \right] dx'. \tag{42c}$$

These integrals can be easily evaluated provided we know $F_0$, $L$, $g$, $R$, $\beta$ and $a$. Once we have obtained $U$, we can easily obtain the pressure rise along the oviduct.

Although one site of muscular activity produces no net ovum displacement in the absence of other transport mechanisms, muscular activity may possibly interact with other transport mechanisms and change transport times. To consider possible interactions, we have calculated ovum movement through the isthmus under the influence of muscular activity ($x_1 = 7.5$, $f = 0.1$) and beating cilia, assuming there is no pressure difference across the oviduct. Typical ovum movements can be observed in Figure 19 where transport due to cilia only and the movement due to muscular activity alone are also shown for comparison.

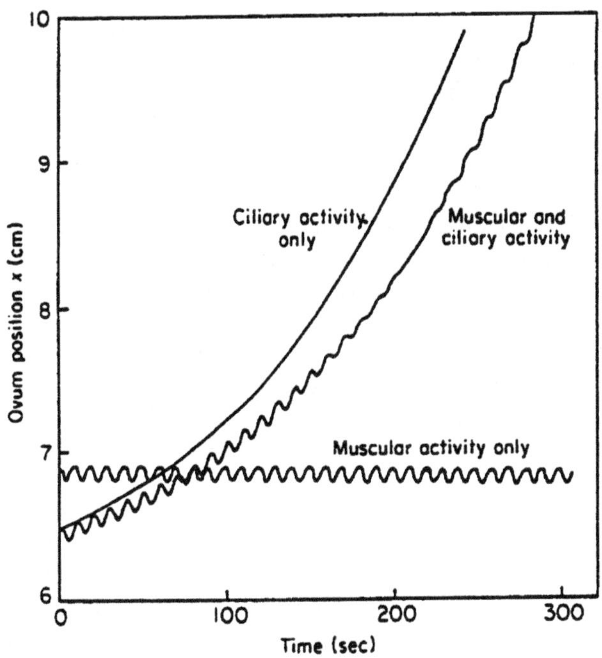


FIG. 19. *Ovum movement through the isthmus due to muscular and/or ciliary activity.*

The back and forth motion, due to muscular activity, that is superimposed on the net forward movement provided by the cilia is clearly seen. As the ovum moves through the constriction, the oscillations due to the muscular activity are damped out since volume flow rates through the region of the constriction do not vary as those outside the constriction. This model suggests that one site of muscular activity does not significantly alter the average progress of ova due to the cilia. It would therefore be beneficial if there was some form of coordination, such as a peristaltic wave.

## 7. Filter feeding.

**7.1. Introduction.** The filtering mechanism in bivalve molluscs, such as *Mytilus edulis*, is a complex fluid mechanical problem that is based on three different sets of cilia which have differing functions. The most commonly observed cilia are the latero-frontal cilia (or cirri) which have a fern-like appearance. These are thought to be primarily responsible for filtering out the particles in the incoming stream. There are two sets of these identical cilia: one set covers the opening to the gill filament, the other set are upright for part of the cycle, apparently out of the way not causing any interference to the general flow field. The two sets of cirri interchange positions in a regular cycle. The second set of cilia are the lateral cilia; the "pump" cilia that provide the motive power to develop the necessary filtering currents. These cilia have dexioplectic metachronal coordination — i.e. the coordinating wave amongst the cilia progresses along the band of cilia to the right (Figure 4; Silvester 1987). The third set of cilia are the short mucus-carrying cilia that transport the captured particles along the muco-ciliary escalator to the mouth.

However the capture mechanism for the particles in the incoming stream has continued to fascinate marine biological scientists. It has usually been assumed that the particles are mechanically filtered out by the side branches on the compound latero-frontal cilia and then transferred to the frontal cilia when they open up. However Jørgenson (1982) proposed the idea of hydrodynamic shear forces that would push the particles sideways in the direction of the mucus-carrying frontal cilia. In this section we will develop a qualitative theory that will assist in our understanding of some of the observed phenomena.

**7.2. Biological background and data.** In Figure 20(a) a schematic diagram of the water flow pattern in *Mytilus* illustrates the principal features that may be observed (adapted from Silvester and Sleigh 1984). Inflow into *Mytilus* occurs through the relatively wide inhalant aperture into the mantle cavity. From here the incoming sea water is "pumped" through the filtering filaments to the suprabronchial cavity from which the water is ejected to outside in a relatively narrow high speed jet (2 cm/s). The size of the external openings is under muscular control which is reflected in the spacing of the filtering filaments inside.

Figure 20(b) is a schematic illustration of the three flow zones across 2 gill filaments where the 3 sets of cilia are clearly identified. The distance

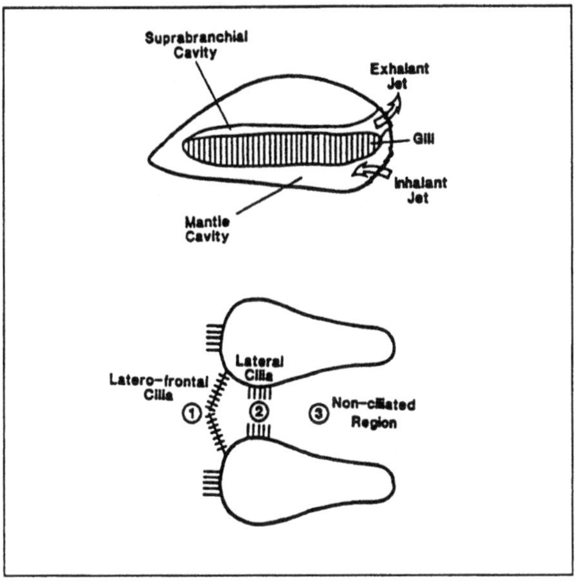


FIG. 20. *(a) Schematic diagram of the inhalant and exhalant jets in Mytilus, (b) The location of the 3 sets of cilia on a filament of the gill.*

between the filaments is usually around 40 $\mu$m, the length of the latero-frontal cilia 20–30 $\mu$m, while the lateral cilia are about 15 $\mu$m and the frontal cilia near 6 $\mu$m in length. The frontal cilia transport the captured food particles and mucus towards the ventral marginal grooves of the gills. The latero-frontal cilia consist of two alternating sets that cover the opening to the inter-filament canal.

The organisation and coordination of the lateral cilia (the "pump" cilia) are of particular interest. Silvester (1988) has outlined the principal features of the cilia organisation into a metachronal wave. A schematic illustration of the metachronal wave is shown in Figure 21(a). The interfilament spacing is 40 $\mu$m, the length of the cilia 15 $\mu$m while the metachronal wave is 12 $\mu$m. The metachronal wave is dexioplectic and since the cilia are facing each other the metachronal waves move in opposite directions. In Figure 21(b) further details are provided on the metachronal wave. The band of lateral cilia has only a total width of 10 $\mu$m, while the cilia are "fanned-out" over a 2.4 $\mu$m section in the effective stroke. Further data on the dimensions and frequencies may be found in Table 2.

Table 2 represents a collection of data on the dimensions, velocities and pressures that have been reported in the literature, principally by Jørgenson (1989) and Sleigh and Silvester (1984). Jørgenson (1987, 1989) also reports a number of additional interesting features of the flow patterns. If the latero-frontal cilia are inactive, currents produced by the recovery stroke

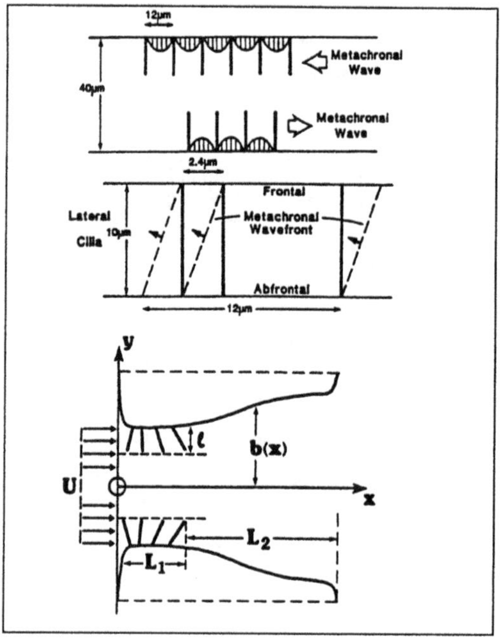


FIG. 21. *(a) Schematic illustration of the metachronal wave, (b) Oblique propagation of the metachronal wave along the band of lateral cilia, (c) Details of the geometry and dimensions for the model of the filament.*

TABLE 2

| | |
|---|---|
| Length of ciliated section | $L_1 = 10\ \mu$m. |
| Length of non-ciliated section | $L_2 = 190\ \mu$m. |
| Viscosity of water | $\mu = 10^{-2}$ gm/cms |
| Cilium length | $L = 15\ \mu$m. |
| Beat frequency | $f = 15 - 20$ Hz. |
| Mean flow velocity | $U = 2.36$ mm/s |
| Half width of "inner core" | $a_0 = 5\ \mu$m. |
| Spacing of cirri branches | $d = 1.3\ \mu$m |
| Radius of cirri branches | $r_0 = 0.1\ \mu$m. |
| Radius of pump cilia | $r_p = 0.1\mu$m |
| Spacing of pump cilia | $d_p = 3\mu$m |

can extend 50 $\mu$m or more above the gill surface whereas when they are functioning normally, these currents barely extend beyond the gill surface.

Later we will see that these currents or eddies need not be associated with the recovery stroke but may be due to the pressure generation by the lateral cilia. Another interesting feature is the high observed flow velocities in the inter-filamentary canal, often thought to be as much as 2–3 times the ciliary tip speeds.

**7.3. Pressure estimates: theory.** The objective of the theoretical model is to obtain estimates of the pressure changes through the gills due to (i) the latero-frontal cilia (the "filter" cilia) (ii) the lateral cilia (the "pump" cilia) and (iii) the non-ciliated region. These regions are clearly indicated in Figure 20(b). The total pressure drop across the gill is

$$\Delta P = \Delta P_1 + \Delta P_2 + \Delta P_3 \tag{43}$$

where

$\Delta P_1$ = pressure drop across the latero-frontal cilia
$\Delta P_2$ = pressure change across the lateral cilia, and
$\Delta P_3$ = pressure drop along the non-ciliated region.

We now proceed to obtain estimates for these three quantities.

(i) *Pressure drop across the latero-frontal cilia*

The latero-frontal cilia have a well developed beat pattern that appears to have a rest period when covering the lumen to the gill entrance. In our analysis we estimate the pressure drop as that over a single row of stationary parallel cylinders as has been done by previous authors (see e.g. Silvester and Sleigh 1984, Jørgenson 1982). The contribution due to the beat pattern of the latero-frontal cilia is ignored. Thus a common formula for estimates of this type is,

$$\Delta P_1 = \frac{8\pi\mu U}{d(1 - 2\ell n\tau + 1/6\tau^2 + O(\tau^4))} \tag{44}$$

where $U$ is the incident velocity, $\mu$ the viscosity, $d$ the cilia branch spacing, $r_0$ the branch radius and $\tau = \dfrac{2\pi r_0}{d}$.

(ii) *Pressure change across the lateral cilia*

For the purposes of the mathematical model, we suppose that the edge of the channel is given by $y = \pm b(x)$ for $0 \le x \le L_1 + L_2$ (see Figure 21(c)). The few lateral cilia of length $\ell$ are spread out over the section $0 \le x \le L_1$. To simplify the calculations we will suppose that

$$b(x) = \begin{cases} b_0 & 0 \le x \le L_1 \\ b_0 + \dfrac{b_1 - b_0}{L_2} L_1 \left(1 - \dfrac{x}{L_1}\right)^{\gamma}, & L_1 \le x \le L_1 + L_2 \end{cases} \tag{45}$$

where $\gamma$ is a parameter to be specified. Thus the half width of the "central core" is $a_0 = b_0 - L$, the appropriate half width dimension that is needed in calculating the pressure gradient.

We consider separately the flow due to the cilia and the flow due to a pressure gradient.

The velocity field due to the cilia in the absence of a pressure gradient will be approximated by,

$$u_c(y) = \begin{cases} U_c; & 0 \le |y| \le a_0 \\ \dfrac{U_c(b_0 - y)}{L}; & a_0 \le |y| \le b_0 \end{cases} \tag{46}$$

where $U_c$ is the average velocity of the liquid at the tips of the cilia. This is a further simplification from the model used in Section 6 but the models used here will enable a qualitative picture to emerge. In many organisms this is found to be usefully approximated by,

$$U_c \approx 2\pi f L \tag{47}$$

where $f$ is the beat frequency and $L$, the length of the cilium (see e.g. Blake 1972, Blake and Sleigh 1974, Fulford and Blake 1986). Thus the volume flow rate due to the cilia is,

$$Q_c = U_c(2a_0 + L). \tag{48}$$

Now consider the problem of uni-directional flow in a channel of width $2b_0$ under the action of a uniform pressure gradient $\frac{dp}{dx}$. Attached to the inner surfaces of the channel is a doubly infinite parallel array of cylindrical rods, height $L = b_0 - a_0$, radius $r_0$ and spacing of the rods $d$ (see Figure 21(c)). We will suppose that in the "inner channel" the flow of the viscous fluid obeys the Stokes flow equations, i.e.

$$\frac{dp}{dx} = \mu \frac{\partial^2 u}{\partial y^2} \tag{49}$$

where $p$ is the pressure, $\mu$ the viscosity and $u_p$ the velocity in the $x$-direction, while Darcy's law for flow in a porous medium is applicable for flow through the rods, i.e.

$$u_p(y) = \frac{-k}{\mu} \frac{dp}{dx} \tag{50}$$

where $k$ is the permeability.

The boundary condition on the plane surface between the two regions is a matter of some controversy but typically two are commonly used: (i) continuity of velocity between the two regions or (ii) a mixed relation connecting the shear stress and velocities. In case (i), the velocity fields is as follows,

$$u_p(x,y) = \begin{cases} \dfrac{1}{2\mu}\dfrac{dp}{dx}(y^2 - a^2) - \dfrac{k}{\mu}\dfrac{dp}{dx}; & 0 < |y| \le a_0, \\ -\dfrac{k}{\mu}\dfrac{dp}{dx}; & a_0 < |y| < b_0. \end{cases} \tag{51}$$

The volume flow rate per unit length $Q_p$ in the channel is defined as,

$$Q_p = \int_{-b_0}^{b_0} u_p(y)dy, \tag{52}$$

yielding

$$Q_p = -\frac{2}{\mu}\frac{dp}{dx}\left(\frac{1}{3}a^3 + kb\right). \tag{53}$$

From Happel and Brenner (1964), it is possible to obtain an estimate for $k$ for normally oriented rods,

$$k = \frac{bd_p^2}{4}\left(\log\left(\frac{d_p}{r_p}\right) - \frac{1}{2} + \left[\frac{r_p^4}{(d_p^4 + r_p^4)}\right]\right) \tag{54}$$

where $d_p$ is the distance between the "spacing" cilia and $r_p$ is the radius of these cilia.

Thus the total flow rate $Q_0$, which is equal to the flow through the latero-frontal cilia is given by,

$$\begin{aligned} Q_0 &= Q_c + Q_p \\ &= U_c(2a_0 + L) - \frac{2}{\mu}\frac{dp}{dx}\left(\frac{1}{3}a_0^3 + kb_0\right) \end{aligned} \tag{55}$$

Now the pressure change across the lateral cilia may be obtained by exploiting the above formula. Since the gradient is constant, the pressure change is simply obtained by multiplying it by the length of lateral cilia.

(iii) *Pressure change along the non-ciliated region*

The theory for calculating the pressure gradient in the non-ciliated region has effectively been developed in the previous section where we can set $U_c = 0$, but with the additional complication that the width of the channel is now varying with distance. Thus the pressure gradient is given by,

$$\frac{dp}{dx} = -\frac{3\mu}{2}\frac{Q_0}{b^3(x)}. \tag{56}$$

While the pressure drop is given by,

$$\Delta P_3 = -\frac{3Q_0\mu}{2}\int_{L_1}^{L_1+L_2} \frac{1}{b^3(x)}dx. \tag{57}$$

**7.4. Capture mechanism.** The filtrate passes through one of about 20 parallel gill filaments. We may therefore employ symmetry arguments and consider the incident flow as occurring through the periodic structure indicated in Figure 20(a).

Thus, we wish to consider flow in a semi-infinite ($x \geq 0$) channel of width $2h$ ($|y| \leq h$). Far upstream ($x \to -\infty$) the velocity is uniform ($U$) across the width of the channel. Because the gill filament system is periodic we require symmetry conditions on the "boundaries". In terms of the velocities we have that

$$\frac{\partial u}{\partial y} = 0 \text{ and } v = 0 \text{ at } y = \pm h \tag{58}$$

where $u$ is the $x$-velocity and $v$ in the $y$-direction. On the end boundary $x = 0$ we have mixed conditions, the no-slip condition on the solid boundary and a prescribed velocity profile over the remainder. Thus on $x = 0$ we have the following conditions,

$$\begin{aligned} u(0,y) &= \begin{cases} 0 & \text{for } b_0 \leq |y| \leq h \\ \dfrac{U}{b_0} - \beta^* \left( y^2 - \dfrac{b_0^2}{3} \right) & \text{for } 0 < |y| \leq b_0, \end{cases} \\ v(0,y) &= 0, \end{aligned} \tag{59}$$

where $\beta^*$ is constant and directly proportional to the pressure gradient.

The Reynolds number of this flow is very small because of the microscopic length scales and velocities. Thus we require the fluid to satisfy the Stokes flow equations,

$$\left. \begin{array}{rcl} \nabla p & = & \mu \nabla^2 \mathbf{u} \\ \nabla . \mathbf{u} & = & 0 \end{array} \right\}. \tag{60}$$

To solve these equations subject to the above boundary conditions we first non-dimensionalise with respect to the far-field velocity $U$ and the half width of the channel $h$.

It is convenient to solve the Stokes flow equations in terms of the stream function $\Psi$ which is related to the velocities as follows,

$$u = \frac{\partial \Psi}{\partial y} \text{ and } v = -\frac{\partial \Psi}{\partial x}, \tag{61}$$

which when substituted into the vorticity equation yields the biharmonic equation

$$\nabla^4 \Psi = 0, \tag{62}$$

for the stream function. In terms of $\Psi$ the boundary conditions become

$$\Psi(x,y) \to y \text{ as } x \to \infty \tag{63}$$

while on $y = h$

$$\Psi(x,h) = -1 \text{ and } \frac{\partial^2 \Psi}{\partial y^2}(x,h) = 0 \tag{64}$$

and on $y = -h$

$$\Psi(x,-h) = 1 \text{ and } \frac{\partial^2\Psi}{\partial y^2}(x,-h) = 0. \tag{65}$$

Thus the flux through the system is 2. On $x = 0$ we have,

$$\frac{\partial\Psi}{\partial x}(0,y) = 0$$

and

$$\Psi(0,y) = \begin{cases} -1 & b_0 \le y \le h, \\ 1 & -h \le y \le -b_0, \\ \dfrac{y}{a_0} - \dfrac{\beta^*}{3}(y^3 - a^2 y); & |y| \le a_0. \end{cases} \tag{66}$$

A solution to this equation is given by the series,

$$\Psi(x,y) = -y + \sum_{n=1}^{\infty} A_n(1 + n\pi x)\sin n\pi e^{-n\pi x} \tag{67}$$

where

$$A_n = 2\left\{\sin n\pi a\left(\frac{-1}{n^2\pi^2 a} + \frac{\frac{2}{3\alpha a^2}}{n^2\pi^2} - \frac{2\alpha}{n^4\pi^4}\right) + \frac{2a\alpha}{n^3\pi^3}\cos n\pi a\right\} \tag{68}$$

If we define $Q' = \dfrac{Q_0}{Q_c}$, the ratio of the inflow rate into the filament to that due to the cilia, we find that eddies are found for $Q' < 0.733$ for the parameters used in this study (see Blake & Fulford 1995 for further details). It is interesting that in the case of $Q' = 0.25$ the eddies extend outwards for a distance similar to the spacing between the filaments, as observed by Jorgenson.

An illustration of the streamline patterns may be found in Figure 22.

## 8. Chaotic filtration, dispersion and a blinking stokeslet.

**8.1. Introduction.** In *Vorticella*, the flexibility and contractile nature of the stalks (see 23) may play an important rôle in increasing the efficiency of filter feeding in the organism (see Blake & Otto, 1996; Blake, Otto & Blake, 1998) because the organism is effectively exerting a net force on the liquid at different heights above the boundary.

Filter feeding in the sessile organisms is therefore modelled by a point force above a rigid boundary. The point force changes its point of application according to a given protocol (known as a blinking stokeslet). The resulting flow can show a much enhanced mixing and particle dispersion.

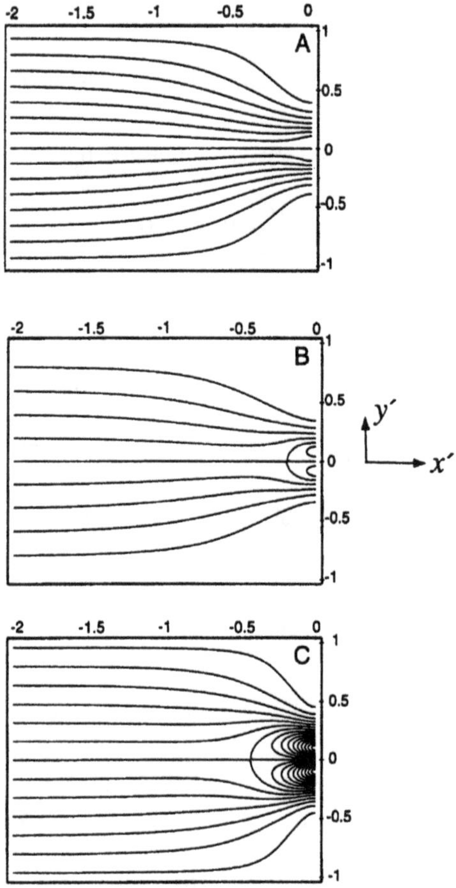


FIG. 22. *Examples of streamline patterns for $Q'$ equal to (A) 0.75, (B) 0.5 and (C) 0.25.*

**8.2. Eddy pattern due to a stationary cell.** Higdon (1979a) developed a model for a sessile organism with a single flagellum providing the motive force to produce a toroidal eddy pattern in the liquid around the cell body. This flow pattern may readily be modelled by the stream function equation for a single stokeslet in the case of a single, separate organism or by a line distribution of stokeslets for a linear arrangement of organelles as occur, for example, with the latero-frontal cilia in *Mytilus*. We present here the stream function equations and streamlines for both the axisymmetric and two-dimensional flows due to a stokeslet above a rigid boundary and directed away from it.

1. **Axisymmetric.** In this case the flow field may be represented in terms of a Stokes stream function $\Psi(r, z)$, defined in terms of the velocities as follows,

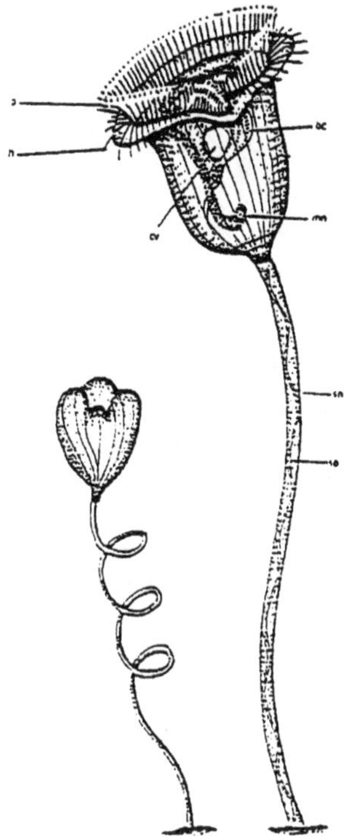

FIG. 23. *The sessile organism Vorticella illustrating the stalk shortening and currents (Sleigh, 1973).*

(69a) $$u_z = \frac{1}{r}\frac{\partial\Psi}{\partial r}, \quad \text{and} \quad u_r = -\frac{1}{r}\frac{\partial\Psi}{\partial z},$$

where $r$ is the radial and $z$ the vertical coordinate. The stream function (Aderogba and Blake, 1978) has a simple form and may be written as,

(69b) $$\Psi(r,z) = \frac{F}{8\pi\mu}\left[\frac{r^2}{(r^2+(z-h)^2)^{\frac{1}{2}}} - \frac{r^2}{(r^2+(z+h)^2)^{\frac{1}{2}}} - \frac{2hr^2z}{(r^2+(z+h)^2)^{\frac{3}{2}}}\right],$$

where $\mu$ is the viscosity of the fluid, and $h$ and $F$ are the height and force associated with the stokeslet, respectively. The streamlines are shown in Figure 24. Aderogba and Blake (1978) showed that this toroidal eddy pattern will also exist at a two fluid interface but not at a free surface.

2. **Two-dimensional stokeslet.** The equivalent two-dimensional stokeslet has been studied by Liron and Blake (1981) and, in terms of the two-dimensional stream function, may be written as

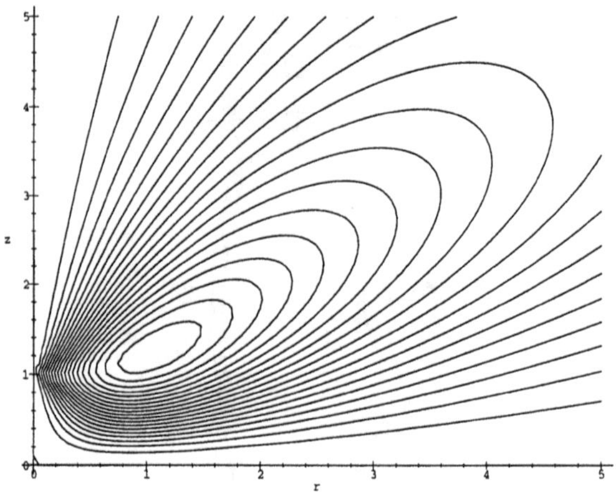


FIG. 24. *Stream function for axisymmetric stokeslet, with one toroidal eddy.*

$$\Psi(x,y) = \frac{F}{8\pi\mu}x\left[\frac{1}{2}\ln\left\{\frac{x^2+(y+h)^2}{x^2+(y-h)^2}\right\} - \frac{2hy}{x^2+(y-h)^2}\right], \tag{70}$$

where $F$ is located at $y = h$ and is directed away from the rigid boundary given by $y = 0$. The velocity components are given by

$$u = \frac{\partial\Psi}{\partial y} \quad \text{and} \quad v = -\frac{\partial\Psi}{\partial x},$$

where $u$ and $v$ are the velocities in the $x$ and $y$ directions respectively. The streamlines can be seen in Figure 25 where it should be noted that two eddies are formed instead of the single toroidal eddy depicted in Figure 24.

In both cases it is clear that this steady eddy structure is not a particularly desirable flow pattern to filter the water in the near vicinity of the organism and that perhaps it might be advantageous to alter the length of the stalk in some periodic manner. This forms the conceptual basis for the following part of the study.

**8.3. Chaotic behaviour associated with the periodic motion of a cell body.** The concept of a 'blinking stokeslet' is based on the ideas developed by Aref (1984) for a blinking vortex. This approach has also been recently exploited in the work of Meleshko & Aref (1996) concerning a blinking rotlet in a problem of chaotic advection. In earlier work, Blake & Otto (1996), we discussed the flow associated with a periodically moving

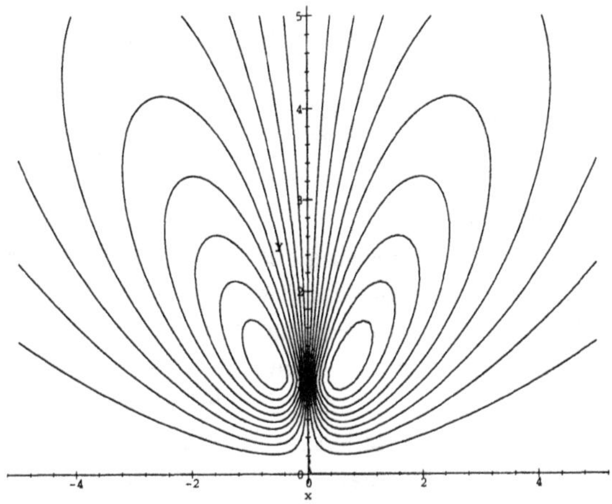


FIG. 25. *Stream function for two-dimensional stokeslet, with two infinite eddies.*

stokeslet; thus in some way replicating the motion of the organism. A summary of those results is given here.

We shall now endeavour to describe the basic algorithm which was used to study the flow. We shall assume that the two configurations can be represented by the stream functions $\psi_+$ and $\psi_-$, where at this stage we shall not further specify the flows. We presume that the configuration is not biased towards either flow and consider the superposition of the stream functions, so that

$$\Psi(x, y; t) = \Omega_+(t)\psi_+(x, y) + \Omega_-(t)\psi_-(x, y).$$

We use the switching protocol exploited by Aref & Balachandar (1986), that is

$$\Omega_\pm(t) = 2\Omega_\pm^0 \sin\left(\frac{t\pi}{\tau}\right)\left\{\sin\left(\frac{t\pi}{\tau}\right) \pm \left|\sin\left(\frac{t\pi}{\tau}\right)\right|\right\},$$

where the configuration switches after $T$ with $\tau = 2T$. This continuous protocol has the advantage of being differentiable. The system can now be written as

$$\text{(71)} \qquad \frac{dx}{dt} = \frac{\partial \Psi}{\partial y} \quad \text{and} \quad \frac{dy}{dt} = -\frac{\partial \Psi}{\partial x}.$$

To illustrate the dispersive behaviour associated with the system, we consider the motion of $N$ particles located at $(x_n^0, y_n^0)$ for $n = 1, \cdots, N$ subject to (71). The equations now constitute a Hamiltonian system. This system is integrated using a fourth order symplectic technique as described in De Frutos & Sanz–Serna (1992).

We shall provide diagrams in the form of Poincaré sections to allow some discussion of the integrability of the systems. The basic algorithm being that we shall integrate the system through one period of the switching protocol, and then draw a point. If the system is integrable a solid line will result, else the diagram will become blurred and it is then that we postulate the system is chaotic.

We note the initially surprising result that the slower one switches between configurations the more chaotic the system becomes. This phenomenon was observed in Aref (1984) in chaotic advection; two point vortices were considered much in the same way as we consider our two flows here. Since we are dealing with Stokes flow here, and inertia does not play a rôle, it is perhaps understandable that as $\tau$ increases the system becomes more chaotic. When $\tau$ is very small, which means the time spent with each configuration is very small, it is as if both of the features are turned on concurrently. As $\tau$ increases there is more chance for the particles to dance between the different flow configurations. Indeed this can be seen clearly in the 'still' frames produced in Blake & Otto (1996).

For $\tau = 0.1$ we present a Poincaré section (Figure 26) which clearly shows that the majority of the region is integrable and as $\tau$ is increased to 0.5 (Figure 27) more of the region becomes chaotic, although an inner island remains. The reader is referred to Hénon (1969) for discussion of these invariant curves, although they are intrinsically different due to the unsteadiness of the underlying flow. If $\tau$ were to be increased further (Figure 28) virtually all of the region would become chaotic, although an island still remains, and further satellite islands are observed. The reader is referred to Blake & Otto (1996) for further detail.

Recently in Otto, Yannacopolous & Blake (2000) the structure of the flow in the neighbourhood of a blinking stokeslet has been investigated using the techniques extracted from dynamical systems studies: Poincaré sections, Lyapunov exponents and the construction of the stable foliation lines. The switching protocol is taken to be a train of delta functions, and consequently we are able to determine a map. Higher order island structures are found, and by the use of Lyapunov exponents we are able to determine regions from which the fluid is likely to become well mixed, Figure 29. For a map the local Lyapunov exponents are related to the logarithms of the eigenvalues of the Jacobian.

In Figure 29 we show the LLE for one hundred periods. The grid contains 100 by 100 points and demonstrates the dominant structures. For instance around the central fixed point there is very little stretching, as the flow here is main rotational. Similarly in the neighbourhood of $y = 0$ there is very little stretching. Notice also the 'stripes' above the vortex core, in which there is relatively low stretching. We include the LLE's for

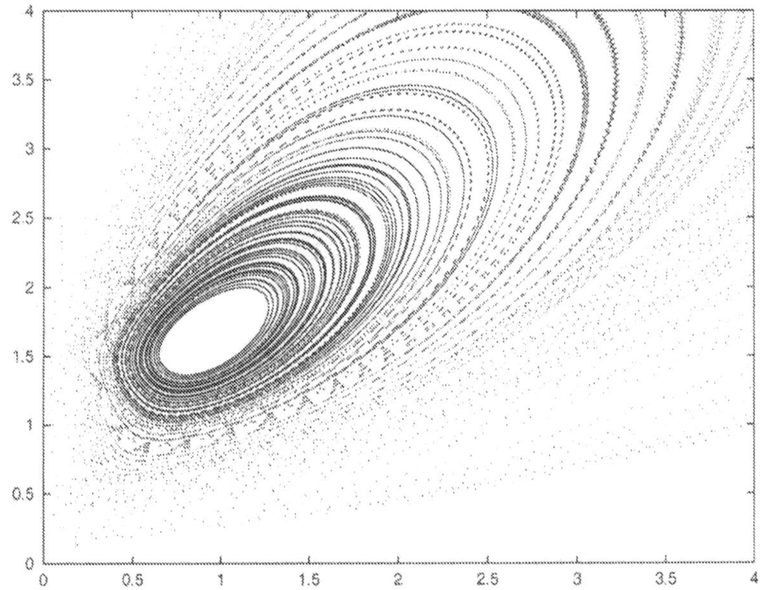


FIG. 26. *Poincaré Section for a single stokeslet, with* $\tau = 0.1$.

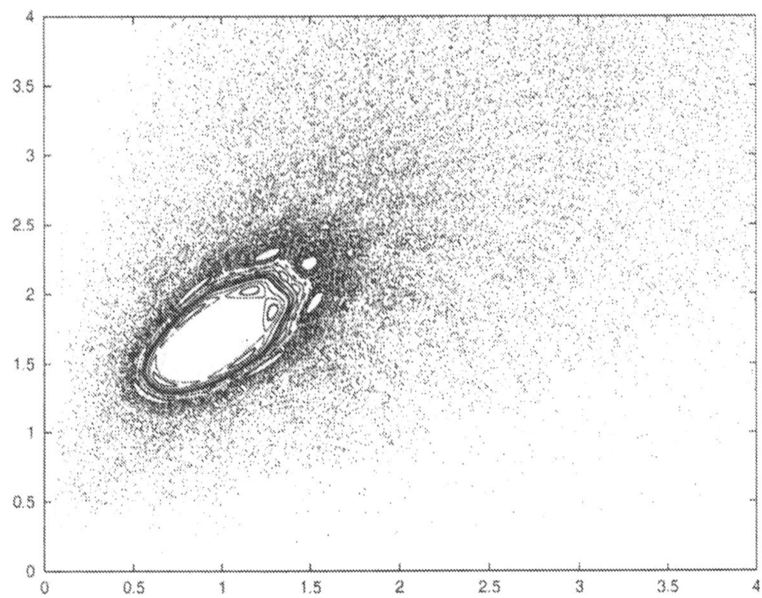


FIG. 27. *Poincaré section for a single stokeslet with* $\tau = 0.5$.

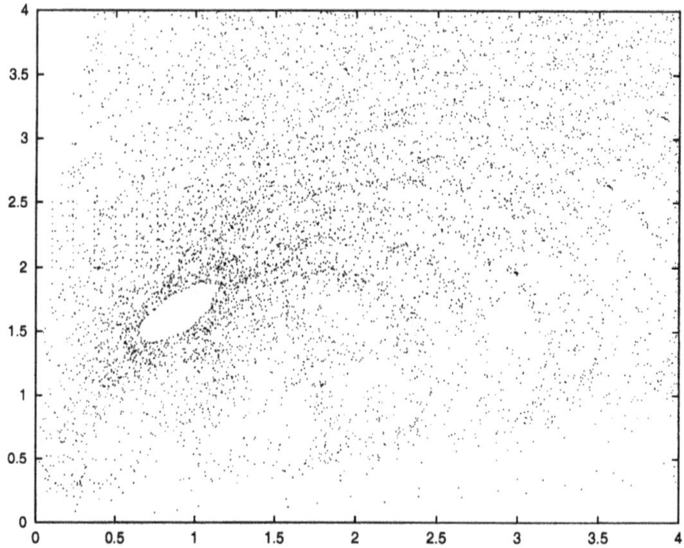


FIG. 28. *Poincaré section for a single stokeslet with* $\tau = 1.0$.

three different values of $\alpha^{(1\pm\epsilon)}$, the strength of the stokeslets. We note that for low values the particles, which are well mixed, originate above the stokeslets' locations. For intermediate values we see that the mixing has been extended although a central core exists around the fixed point in which the action of the map is largely rotational. In the final figure we show a case where most of the region will be mixed. In these figures we also give the values of $\max(\lambda_{LLE})$ over the region (since the shading is relative). We now clearly see that an increase in the period of the stokeslet corresponds to additional mixing, refer to Otto, Yannacopolous & Blake (2000) for more details.

The chaotic nature of the dynamics can lead to enhanced transport in the flow. In a flow with regular streamlines, the tracer is restricted onto tori (defined by the streamlines) and so a great part of phase space (which coincides with physical space in problems of this kind) is inaccessible to it. In this case the tracer can visit certain parts of phase space only under the effect of molecular diffusion. However, in the case where the motion of the tracer is chaotic, the tori onto which the streamlines are confined are broken and this leads to a connection of certain parts of phase space otherwise inaccessible in the absence of molecular diffusion.

Although this study is only qualitative, it clearly indicates the opportunity of efficient mixing in Stokes flow that was previously thought not possible. This is likely not only to have important application to mixing processes in microscopic biological reasons, but also in new areas such as micro-machinery.

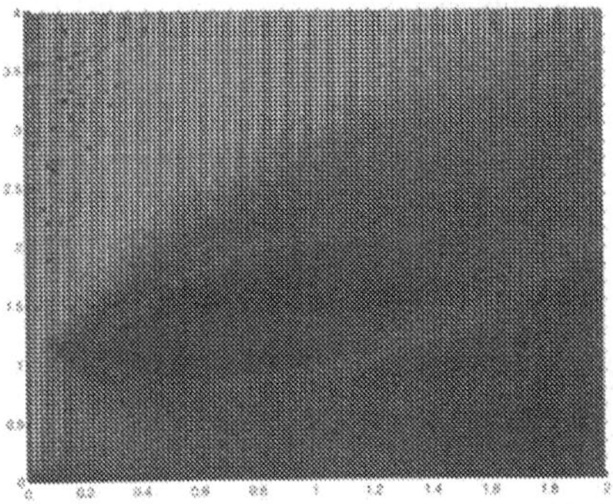

(a) $\alpha^{(1\pm\epsilon)} = 1/10$, $\max(\lambda_{LLE})$=0.1183

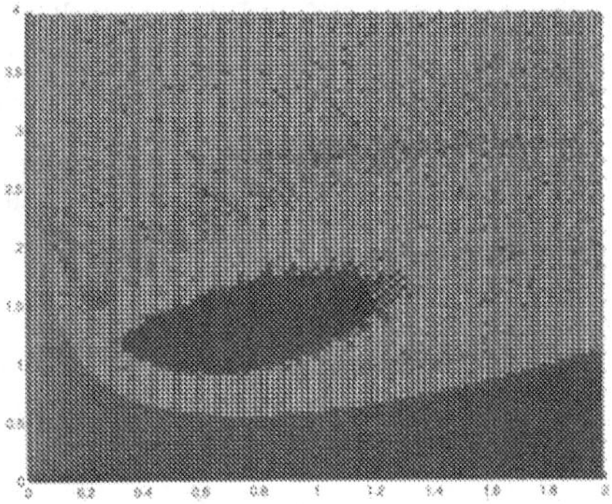

(b) $\alpha^{(1\pm\epsilon)} = 3/4$, $\max(\lambda_{LLE})$=0.2268

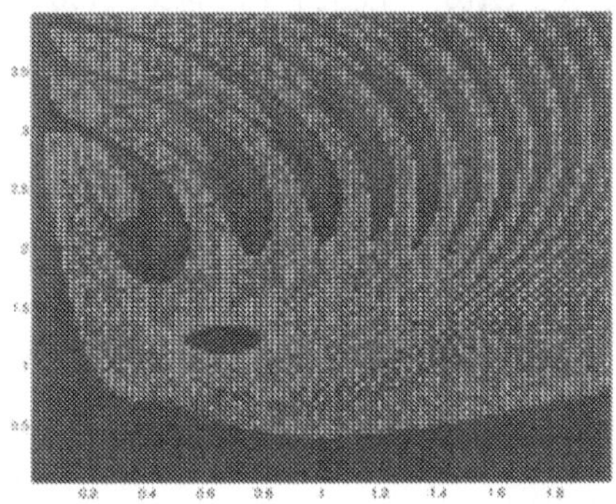

(c) $\alpha^{(1\pm\epsilon)} = 3/2$, $\max(\lambda_{LLE})$=0.3827

FIG. 29. *Local Lyapunov exponents (LLE) for* $\alpha^{(1\pm\epsilon)} = 1/10$, $3/4$ *and* $3/2$. *The values have been averaged over 100 periods of the map. The grey-scale maps are relative, and the maximum values of the LLE's are given in the individual captions.*

**9. Conclusions.** In Lighthill's 1969 *Annual Review of Fluid Mechanics* article on the 'Hydromechanics of Aquatic Animal Locomotion' he identified ciliary propulsion as a research area that needs further development. This paper has reviewed the major developments in the theory of ciliary propulsion over the last 30 years, together with a range of applications to locomotion, feeding, excretion and procreation.

From the theoretical biological fluid mechanics side it has required the development of appropriate singular solutions of the Stokes flow equations, the interactions of slender bodies with both themselves and viscous interfaces, the development of sophisticated models for the fields of cilia and the exploitation of this knowledge base to develop simplified models of complex flow structures in the areas of feeding, reproduction and respiratory physiology. Finally, this area of study has led to some interesting and exciting studies in non-linear fluid mechanics involving the concept of the blinking stokeslet which may be used as a paradigm for mixing processes at very low Reynolds number.

**Acknowledgements.** The material for this presentation is drawn from printed notes for a lecture course in Biological and Physiological Fluid Mechanics which, in turn, is based on published and recently conducted research in this area. Although Sir James Lighthill had a profound influence on my research, I would also like to acknowledge the significant contributions made by Michael Sleigh, Nadav Liron, Glenn Fulford, Peter Vann, Ernie Tuck, Ted Wu, Allen Chwang, Howard Winet, Chris Brennen, David Katz, Steve Otto, Thanos Yannacopolous, Barry Leadbeater, Belinda Orme and Michala Pettitt.

## REFERENCES

[1] Aderogba, K. and Blake, J.R., 1978. Action of a force near the planar surface between semi-infinite immiscible liquids at very low Reynolds numbers: Addendum. *Bull. Aust. Math. Soc.* **19**, 309–318.

[2] Aiello, E. and Sleigh, M.A., 1972. The metachronal wave of *Mytilus edulis*, *J. Cell. Biol.* **54**, 493–506.

[3] Aref, H., 1984. Stirring by chaotic advection, *J. Fluid Mech.* **143**, 1–21.

[4] Aref, H. and Balachandar, S., 1986. Chaotic advection in a Stokes flow. *Phys. Fluids* **29**, 3515–3521.

[5] Blake, J.R., 1972. A model for the microstructure in ciliated organisms, *J. Fluid Mech.* **555**, 1–23.

[6] Blake, J.R., 1975. On the movement of mucus in the lung, *J. Biomech.* **8**, 179–190.

[7] Blake, J.R., 1975. Fluid flows in fields of resistance. *Bull. Aust. Math. Soc.* **13**, 129–145.

[8] Blake, J.R., 1984. Mechanics of muco-ciliary transport, *IMA J. Appl. Math.* **32**, 69–87.

[9] Blake, J.R. and Fulford, G.R., 1984. Mechanics of muco-ciliary transport, *Physico. Chem. Hydrodynamics* **5**, 401–411.

[10] Blake, J.R. and Fulford, G.R., 1995. Hydrodynamics of filter feeding. In: *Biological Fluid dynamics*, Editors: C.P. Ellington and T.J. Pedley. *Soc. Exp. Biol.*, pp. 183–197.

[11] Blake, J.R., Liron, N., and Aldis, G.K., 1982. Flow patterns around microorganisms and in ciliated ducts. *J. Theor. Biol.* **98**, 127–141.

[12] Blake, J.R. and Otto, S.R., 1996 Ciliary propulsion, chaotic filtration and a blinking stokeslet. *J. Eng. Math.* **30**, 151–168.

[13] Blake, J.R., Otto, S.R., and Blake, D.A., 1998. Filter feeding, chaotic filtration and a blinking stokeslet. *Theor. Comp. Fluid Dyn.* **10** 23–36.

[14] Blake, J.R. and Sleigh, M.A., 1974. Mechanics of ciliary locomotion, *Biol. Rev.* **49**, 85–125.

[15] Blake, J.R., Vann, P., and Winet, H., 1983. A model for ovum transport, *J. Theor. Biol.* **102**, 145–166.

[16] Blake, J.R. and Winet, H., 1980. On the mechanics of muco-ciliary transport, *Biorheol.* **17**, 125.

[17] BRENNEN, C. AND WINET, H., 1977. Fluid mechanics of propulsion by cilia and flagella, *Ann. Rev. Fluid Mech.* **9**, 339–398.

[18] BROSENS, I.A. AND VASQUEZ, G., 1976. *J. Reprod. Med.* **16**, 17.

[19] CHEER, A.Y.L. AND KOEHL, M.A.R., 1987. Paddles and Rakes: Fluid flow through bristled appendages of small organisms. *J. Theor. Biol.* **129**, 17–39.

[20] CROXATTO, H.B., ORITZ, M.E., DIAZ, S., AND HESS, R., 1979. *J. Reprod. Fert.* **55**, p. 231.

[21] DRESDNER, R.D., KATZ, D.F., AND BERGER, S.A., 1980. The propulsion by large amplitude waves of uniflagellar micro-organisms of finite length, *J. Fluid Mech.* **97**, 591–621.

[22] DE FRUTOS, J. AND SANZ-SERNA, J.M., 1992. An easily implementable fourth-order method for the time integration of wave problems. *J. Comp. Phys.* **103**, 160–168.

[23] FULFORD, G.R. AND BLAKE, J.R., 1986. Muco-ciliary transport in the lung, *J. Theor. Biol.* **121**, 381–402.

[24] GRAY, J. AND HANCOCK, G.J., 1955. The propulsion of sea urchin spermatozoa, *J. Exp. Biol.* **32**, 802–814.

[25] HALBERT, S.A., TAM, P.Y., AND LANDAU, R.J., 1976. *Science* **191**, 1052.

[26] HANCOCK, G.J., 1953. The self propulsion of microscopic organisms through liquids, *Proc. Roy. Soc.* **A219**, 96–121.

[27] HAPPEL, J. AND BRENNER, H., 1965. *Low Reynolds Number Hydrodynamics*, Prentice-Hall, Englewood Cliffs, N.J.

[28] HARPER, J.F. AND WAKE, G.C., 1983. Stokes flow between parallel plates due to a transversely moving end wall. *IMA J. Appl. Math.* **30**, 141–149.

[29] HÉNON, M., 1969. Numerical Study of Quadratic Area-Preserving Mappings. *Quart. Appl. Maths* **27**, 291–312.

[30] HIGDON, J.J.L., 1979. A hydrodynamic analysis of flagellar propulsion, *J. Fluid Mech.* **90**, 685–711.

[31] HIGDON, J.J.L., 1979. The generation of feeding currents by flagellar motions *J. Fluid Mech.* **94**, 305–330.

[32] HIGDON, J.J.L., 1979. The hydrodynamics of flagellar propulsion: helical waves, *J. Fluid Mech.* **94**, 331–351.

[33] HODGSON, B.J., TALO, A., AND PAUERSTEIN, C.J., 1977. *Biol. Reprod.* **16**, p. 394.

[34] JANSEN, R.P.S., 1980. *Am. J. Obstet. Gynaecol.* **136**, p. 292.

[35] JOHNSON, R.E. AND BROKAW, C.J., 1979. Flagellar hydrodynamics. A comparison between resistive force theory and slender body theory, *Biophys. J.* **25**, 113–127.

[36] JONES, H.D., RICHARDS, O.G., AND SOUTHERN, T.A., 1992. Gill dimensions, water pumping rate and body size in the mussel *Mytilus edulis. L. J. Exp. Mar. Biol. Ecol.* **155**, 213–237.

[37] JØRGENSON, C.B., 1981. A hydromechanical principle for particle retention in *Mytilus edulis* and other ciliary suspension feeders. *Marine Biol.* **61**, 277–282.

[38] JØRGENSON, C.B., 1981. Feeding and cleaning mechanisms in the suspension feeding bivalve *Mytilus edulis. Marine Biol.* **65**, 159–163.

[39] JØRGENSON, C.B., 1982. Fluid mechanics of the mussel gill: The lateral cilia. *Marine Biol.* **70**, 275–281.

[40] JØRGENSON, C.B., 1991. *Bivalve Filter Feeding: Hydrodynamics, Bioenergetics, Physiology and Ecology*, Denmark: Olsen and Olsen.

[41] JØRGENSON, C.B., LARSEN, P.S., MOHLENBERG, F., AND RIISGARD, H.V., 1988. The mussel pump: properties and modelling, *Mar. Ecol. Prog. Ser.* **45**, 205–216.

[42] JØRGENSON, C.B., MOHLENBERG, F., AND STEN-KNUDSON, O., 1986. Nature of relation between ventilation and oxygen consumption in filter feeders. *Mar. Ecol. Prog. Ser.* **29**, 73–88.

[43] JØRGENSON, C.B., FAMME, P., KRISTENSEN, H.S., LARSEN, P.S., MOHLENBERG, F., AND RIISGARD, H.U., 1986. The bivalve pump. *Mar. Ecol. Prog. Ser.* **34**, 69–77.

[44] JOSEPH, D.D. AND STURGES, L., 1978. The convergence of biorthogonal series for biharmonic and Stokes' flow edge problems. Part II. *SIAM J. Appl. Math.* **34**, 7–26.

[45] KATZ, D.F. AND BLAKE, J.R., 1975. Flagellar motions near walls, In: *Swimming and Flying in Nature*, Eds. Wu, T.Y., Brokaw, C.J. and Brennen, C. Plenum, New York, pp. 173–184.

[46] KNIGHT-JONES, E.W., 1954. Relations between metachronism and the direction of ciliary beat ion Metazoa, *Quart. J. Microsc. Sci.* **95**, 503–521.

[47] LIGHTHILL, M.J., 1970. Aquatic animal propulsion of high hydromechanical efficiency. *J. Fluid Mech.* **44**, 265–301.

[48] LIGHTHILL, M.J., 1976. Flagellar hydrodynamics, *SIAM Rev.* **18**, 161–230.

[49] LIRON, N. AND BLAKE, J.R., 1981. Existence of viscous eddies near boundaries. *J. Fluid Mech.* **107**, 109–129.

[50] LIRON, N. AND MOCHON, S., 1976. The discrete cilia approach to propulsion of ciliated micro-organisms, *J. Fluid Mech.* **75**, 593–607.

[51] LITT, M., 1970. Mucus rheology, *Archs. Intern. Med.* **126**, 417–423.

[52] MATSUI, H., RANDELL, S.H., PERETTI, S.W., DAVIS, C.W., AND BOUCHER, R.C., 1998. Coordinated clearance of periciliary liquid and mucus from airway surfaces. *J. Clin. Invest.* **102**, 1125–1131.

[53] MELESHKO, V.V. AND AREF, H., 1996. A blinking rotlet model for chaotic advection. *Phys. Fluids A* **8**, 3215–3217.

[54] OTTO, S.R., YANNACOPOLOUS, A.N. AND BLAKE, J.R., 2000. Transport and mixing in Stokes flow: the effect of chaotic dynamics on the blinking stokeslet. *J. Fluid Mech.* (in press).

[55] PAUERSTEIN, C.J. AND EDDY, C.A., 1979. *J. Reprod. Fert.* **55**, 223.

[56] PEDLEY, T.J., SCHROTER, R.C., AND SUDLOW, M.F., 1971. Flow and pressure drop in systems of repeatedly branching tubes, *J. Fluid Mech.* **46**, 365–383.

[57] SADE, J., ELIEZER, N., SILBERBERG, A., AND NERO, A.C., 1970. The rôle of mucus in transport by cilia, *Am. Rev. Resp. Dis.* **102**, 48–52.

[58] SANDERSON, M.J. AND SLEIGH, M.A., 1981. Ciliary activity of cultured rabbit tracheal epithelium: Beat pattern and metachrony, *J. Cell. Sci.* **47**, 331–347.

[59] SATIR, P., HAMASAKI, T., AND HOLWILL, M., 1998. Modeling outer dynein arm activity and its relation to the ciliary beat cycle. In: *Cilia, Mucus and Mucociliary Interactions*, Eds. G.L. Baum, Z. Priel, Y. Roth, N. Liron and E.J. Ostfeld. Marcel Dekker.

[60] SHACK, W.J., FRAY, C.S., AND LARDNER, T.J., 1971. Observations of the hydrodynamics and swimming motions of mammalian spermatozoa, MIT Preprint.

[61] SILBERBERG, A., 1982. Rheology of mucus, muco-ciliary interaction, and ciliary activity, *Cell. Not. (Suppl.)* **1**, 25–28.

[62] SILVESTER, N.R., 1988. Hydrodynamics of flow in *Mytilus gills. J. Exp. Mar. Biol. Ecol.* **120** 171–182

[63] SILVESTER, N.R. AND SLEIGH, M.A., 1984. Hydrodynamic aspects of particle capture by *Mytilus*, *J. Mar. Biol. Assoc. U.K.* **64**, 859–879.

[64] SLEIGH, M.A., 1962. *The Biology of Cilia and Flagella*, Pergamon.

[65] SLEIGH, M.A., 1973. *The Biology of Protozoa*, London, Edward Arnold, p. 315.

[66] SLEIGH, M.A., 1977. The nature and action of respiratory tract cilia, In: *Respiratory Defence Mechanisms*, (Brain, J.D., Proctor, D.F. and Reid, L. Eds.) pp. 247–288, New York: Marcel Dekker.

[67] SLEIGH, M.A., 1981. Ciliary function in mucus transport, *Chest* **805**, 791–795.

[68] SLEIGH, M.A., 1982. Movement and coordination of tracheal cilia and the relation of these to mucus transport, *Cell. Not. (Suppl.)* **1**, 19–24.

[69] SLEIGH, M.A. AND BLAKE, J.R., 1975. Hydromechanical aspects of ciliary propulsion. In: *Swimming and Flying in Nature*, Eds. Wu, T.Y., Brokaw, C.J. and Brennen, C. Plenum, New York, pp. 185–210.

[70] SLEIGH, M.A., BLAKE, J.R., AND LIRON, N., 1988. The propulsion of mucus by cilia. *Am. Rev. Resp. Dis.* **137**, 726–741.

[71] SPENCE, D.A., 1982. A note on the eigenfunction expansion for the elastic strip. *SIAM J. Appl. Math.* **42**, 155–174.

[72] SPENCE, D.A., 1983. A class of biharmonic end-strip problems arising in elasticity and Stokes flow. *IMA J. Appl. Math.* **30**, 107–139.

[73] TALO, A., 1974. *Biol. Reprod.* **11**, p. 335.

[74] TAYLOR, G.I., 1951. Analysis of the swimming of microscopic organisms, *Proc. Roy. Soc.* **A209**, 447–461.

[75] TEODORESCU, P.P., 1960. Sur le probleme de la demibande plane elastique. *Arch. Mech. Stos.* **12**, 313–331.

# THE ROLE OF CYCLIC NUCLEOTIDE PATHWAYS AND CALMODULIN IN CILIARY STIMULATION

ALEX BRAIMAN*, NATALYA UZLANER*, AND ZVI PRIEL*†

**1. Introduction.** Cilia are tiny hairlike protrusions enveloped by a membrane contiguous with the cell membrane, which beat in a cooperative pseudo-periodic, spatial and temporal pattern called the metachronal wave. They are thin (0.25–0.3 $\mu$m), relatively long (6–50 $\mu$m) and densely packed on the cell surface (100–200 cilia per cell). In the mucociliary system, their primary function is transport of a mucus layer over the cell together with various objects that may be trapped in this layer. Highly cooperative beating of cilia at high frequencies enables the mucociliary system to carry relatively large objects, at remarkable velocities. Furthermore, high frequency of ciliary beating results in increased energy expenditure. Therefore, under normal conditions, cilia beat with either low frequency or may even be at rest. However, they can dramatically change their activity in response to a variety of receptor-mediated stimuli. For example, ciliary cells possess purinergic $P1$ and $P2$ [7, 12, 32, 33], cholinergic [1, 6, 24], adrenergic [19, 31, 34] receptors.

It is widely accepted that changes in intracellular levels of either calcium ions [4, 13, 15, 16, 23, 28, 30, 32, 34] or cAMP [3, 15, 26, 31, 34] or cGMP [5, 21, 22, 34], lead to CBF enhancement. Although these second messengers are well known and widely investigated modulators of ciliary activity, it is still unclear whether they act independently, each of them representing a separate pathway of ciliary stimulation, or whether their action is cooperative. Moreover, the pathways of ciliary stimulation, represented by these second messengers, co-exist in the same cell types. For example, in rabbit airway epithelium, ciliary activity is regulated by $[Ca^{2+}]_i$ [12, 14, 29], cAMP [15, 26] and cGMP [25, 27, 29]. Thus, a question can be raised about the physiological significance of this apparent multitude of ways to enhance ciliary activity in the same cell.

**2. Materials and methods.**

*Tissue culture.* Experiments were carried out on monolayer tissue cultures grown from rabbit trachea using the procedure described previously [12]. Briefly, adult white rabbits (at least 1.5 kg in mass) were killed according to the guidelines laid down by the animal welfare committee of Ben-Gurion University by gradual exposure to carbon dioxide followed by exsanguination. Care was taken to slowly increase the gas flow over several minutes to prevent any visual signs of distress. Tracheas were removed,

*Department of Chemistry, Ben-Gurion University of the Negev, P.O. Box 635, Beer - Sheva 84105, Israel.

†E-mail: alon@bgumail.bgu.ac.il.

and the ciliary epithelium was peeled off the cartilage rings and cut into small pieces. Two or three pieces of epithelium were placed on a glass coverslip and were incubated in RPMI-1640 growth medium supplemented with 7.5% fetal calf serum, 20 units $ml^{-1}$ penicillin, 2.5 units $ml^{-1}$ nystatin and 20 $\mu g\ ml^{-1}$ streptomycin at 37°C with 5% $CO_2$. Prior to use, the glass coverslips were sterilized with ethanol, placed in plastic Petri dishes (Nunk, Denmark) and incubated with a small amount of medium for 24 hours. The medium was replaced every two days. Measurements were performed on 8–20 day-old tissue cultures; this was the time period for which the epithelial monolayers were big enough to be used in the experiments.

*Chemicals and solutions.* Fetal calf serum, RPMI-1640 culture medium, and antibiotics were from Biological Industries (Bet-Haemek, Israel). Fura-2/AM was from either Molecular Probes (Eugene, OR, USA) or from Teflabs (Austin, TX, USA). The dye was stored in solid form at −20°C and fresh solutions were made before each experiment in dry dimethyl sulfoxide (DMSO). 6-anilino-5, 8-quinolinedione (LY-83583) was from Biomol (Plymouth Meeting, PA, USA). All other materials were obtained from Sigma Chemical Co. (St. Louis, MO, USA). The standard Ringer solution was composed of (mM): 150 NaCl, 2.5 KCl, 1.5 $CaCl_2$, 1.5 $MgCl_2$, 5 d-glucose, 5 Hepes. The pH of all the solutions was set to 7.4 with NaOH and HCl . LY-83583 was dissolved in $CH_3OH/C_2H_5OH$ (1:4). Stock solutions of ATP were prepared in 0.05M Hepes buffer (pH=7.4) and diluted into the Ringer solutions immediately prior to use. All other chemicals were dissolved in water. The final concentration of the alcohol was not more than 0.5%.

*Simultaneous measurement of $[Ca^{2+}]_i$ and ciliary beating.* We used the method of simultaneous measurement of $[Ca^{2+}]_i$ and ciliary beating that was extensively described previously [13]. Briefly, $[Ca^{2+}]_i$ was measured with the fluorescent indicator Fura-2. The dye-loaded cells were epi-illuminated with light from a 150 W xenon lamp (Oriel Corp. Stamford, CT, USA) filtered by 340 and 380 nm interference filters (Oriel) mounted on a four position rotating filter wheel. The fluorescence, emitted at 510 nm, was detected by a photon-counting photomultiplier (H3460-53, Hamamatsu, Japan). The 340/380 fluorescence ratio averaged over a period of 1 s, was stored in a computer (Pentium®). CBF was measured by trans-illuminating the same cells with light at 600 nm (so as not to interfere with the Fura-2 fluorescence at 510 nm). The light scattering from the beating cilia created amplitude modulations of the 600 nm light that were detected by a photomultiplier (R2014, Hamamatsu, Japan).

The $[Ca^{2+}]_i$ was calculated using a calibration curve that correlates Fura-2 fluorescence ratio to calcium concentration. The calibration curve was constructed from measurements of the fluorescence ratio obtained from calibration solutions composed of 115 mM KCl, 20 mM NaCl, 5 mM $MgCl_2$, 5 mM D-glucose, 5 mM HEPES, 10 mM EGTA, 1$\mu$M $K_2$Fura-2 and $CaCl_2$ at a range of concentrations [8]. The $[Ca^{2+}]_i$ concentration was calculated directly from the calibration curve using a table look-up algorithm. The frequency response was calibrated by:1) oscillating lamp in the range of 2 to 40 Hz. 2) vibrating needle between 2 to 40 Hz. A linear response was revealed in these dynamic ranges.

*Experimental procedure and data presentation.* Cells were pre-loaded with Fura-2 by incubating the tissue in growth medium containing 8.5$\mu$M Fura-2/AM for 60 min at 37$^o$C in a rotating water bath. The cells were then rinsed with Ringer solution and maintained at room temperature for 30 min to allow full de-esterification and equilibration of the dye in the cytoplasm. Only evenly fluorescent cells which did not display bright spots and had a steady CBF were used. In each experiment, basal CBF (Fo) and basal $[Ca^{2+}]_i$ ($[Ca^{2+}]_{io}$) were measured for 2–5 min in the appropriate solution. These were taken as reference values. Then, the test chemical was introduced by exchanging the bath solution, and the frequency (F) and $[Ca^{2+}]_i$ were monitored in the same ciliary area for 10–15 min. Each experimental condition was tested on naive cultures to avoid possible bias arising from previous manipulations. Beat frequencies are presented as the fold increase in CBF (F/ Fo). Changes in $[Ca^{2+}]_i$ are presented as the difference between the observed $[Ca^{2+}]_i$ and the basal $[Ca^{2+}]_i$ ($[Ca^{2+}]_i$-$[Ca^{2+}]_{io}$). Each experiment was carried out on 3–15 tissue cultures from at least 2–3 animals. Results were averaged and are displayed as mean ± SE, with $n$ (number of experiments) shown in parentheses.

**3. Results and discussion.** Intracellular calcium concentration ($[Ca^{2+}]_i$) plays a crucial role in regulation of ciliary activity, as first demonstrated by Verdugo [30]. A rise in $[Ca^{2+}]_i$ correlates with a rise in ciliary beat frequency (CBF) [3, 12, 14, 16–18, 23]. Furthermore, an elevation in $[Ca^{2+}]_i$ is essential for initiation of a response to many receptor-mediated stimuli. Thus, depletion of intracellular calcium stores, using calcium ionophores or the sarcoplasmic/endoplasmic $Ca^{2+}$- ATPase inhibitor thapsigargin, abolishes the ciliary stimulation normally induced by extracellular ATP or by acetylcholine [12, 23, 28, 29].

Calmodulin is a ubiquitous $Ca^{2+}$-binding protein. It has been identified in all animal and plant cells and appears to be an intracellular $Ca^{2+}$ receptor, which as a $Ca^{2+}$-calmodulin complex participates in a majority of $Ca^{2+}$-regulated processes [2, 11, 20]. Therefore, it is reasonable to suggest that $Ca^{2+}$ delivers its stimulatory effect on cilia, at least partially, via the $Ca^{2+}$-calmodulin complex. However, the available data on the involvement of calmodulin in regulation of ciliary activity is contradictory. For example, application of a calmodulin inhibitor blocked the increase in CBF normally produced by ionomycin in human respiratory epithelium [4]. However, application of a calmodulin inhibitor was ineffective in blocking the CBF rise induced by acetylcholine in cultured ovine trachea [24]. Therefore, a possible involvement of calmodulin in the CBF stimulation has been examined. Fig. 1A, B shows a normal response of ciliary cells from rabbit trachea to extracellular ATP. Fig. 1C, D shows the response obtained in presence of the calmodulin inhibitor W-7 (100 $\mu$M). The calmodulin inhibitor produced only a moderate attenuation of the response to ATP in $[Ca^{2+}]_i$. However, despite a significant rise in $[Ca^{2+}]_i$, the elevation in CBF has been almost completely abolished. The average responses to 100 $\mu$M ATP were $\delta[Ca^{2+}]_i = 373 \pm 7$ nM ($n$=40) and $F/Fo = 2.2 \pm 0.1$ ($n$=40). The average responses to 100 $\mu$M ATP in the presence of 100 $\mu$M W-7 were $\delta[Ca^{2+}]_i = 255 \pm 30$ nM ($n$=10) and $F/Fo = 1.3 \pm 0.1 (n = 10)$. This effect of a calmodulin inhibitor is not restricted to ATP or to the rabbit trachea. Similar inhibition of CBF enhancement inflicted by calmodulin inhibitors (TFP or W-7) was obtained when $[Ca^{2+}]_i$ was elevated artificially by application of the calcium ionophore ionomycin in both rabbit trachea and frog esophagus (unpublished results). Furthermore, an essentially identical effect of the calmodulin inhibitor W-7 has been demonstrated in frog esophagus stimulated by acetylcholine [35].

These results suggest that the $Ca^{2+}$-calmodulin complex plays an important role in the process of ciliary stimulation. Furthermore, since the effect of the calmodulin inhibitors is limited neither to a single type of stimulation, nor to a single type of ciliary tissue, it is reasonable to suggest that the critical role of calmodulin in the process of CBF enhancement is a general phenomenon.

As was mentioned above, the $Ca^{2+}$- calmodulin complex can influence activity of many cellular enzymes. Activation of the nitric oxide synthase (NOS) is of special interest in ciliary cells, since it initiates another important pathway of ciliary stimulation - the cGMP pathway. The increased production of nitric oxide (NO) leads to activation of guanylyl cyclase (GC). This results in an increase in a cGMP level and activation of the cGMP-dependent protein kinase (PKG). We have recently reported that inhibition of either NOS, or GC, or PKG blocks the CBF enhancement normally evoked by ATP, even though the $[Ca^{2+}]_i$ elevation is not significantly impaired by these procedures [29]. Fig. 2 shows a typical response to 100 $\mu$M ATP obtained in the presence of 100 $\mu$M LY-83583, an inhibitor of GC. In other words, the effect produced by these manipulations is analogous to the effect produced by the calmodulin inhibitors. The essence of this effect is uncoupling between the rise in $[Ca^{2+}]_i$ and the rise in CBF. These results suggest that at least one of the important functions performed by the $Ca^{2+}$-calmodulin complex in the ciliary stimulation is activation of the cGMP pathway. Moreover, it appears that any manipulation, which in the end prevents activation of PKG, nullifies the stimulatory effect of a high $[Ca^{2+}]_i$ on CBF.

To verify that this obstruction of the CBF enhancement is caused by a lack of PKG activity and is not a side effect of the inhibitors, the following experiment has been performed. Dibutyryl-cGMP (db-cGMP), a cell-permeable analog of cGMP, has been added to the cells treated by the GC inhibitor LY-83583. Then, 100 $\mu$M ATP have been applied. As can be

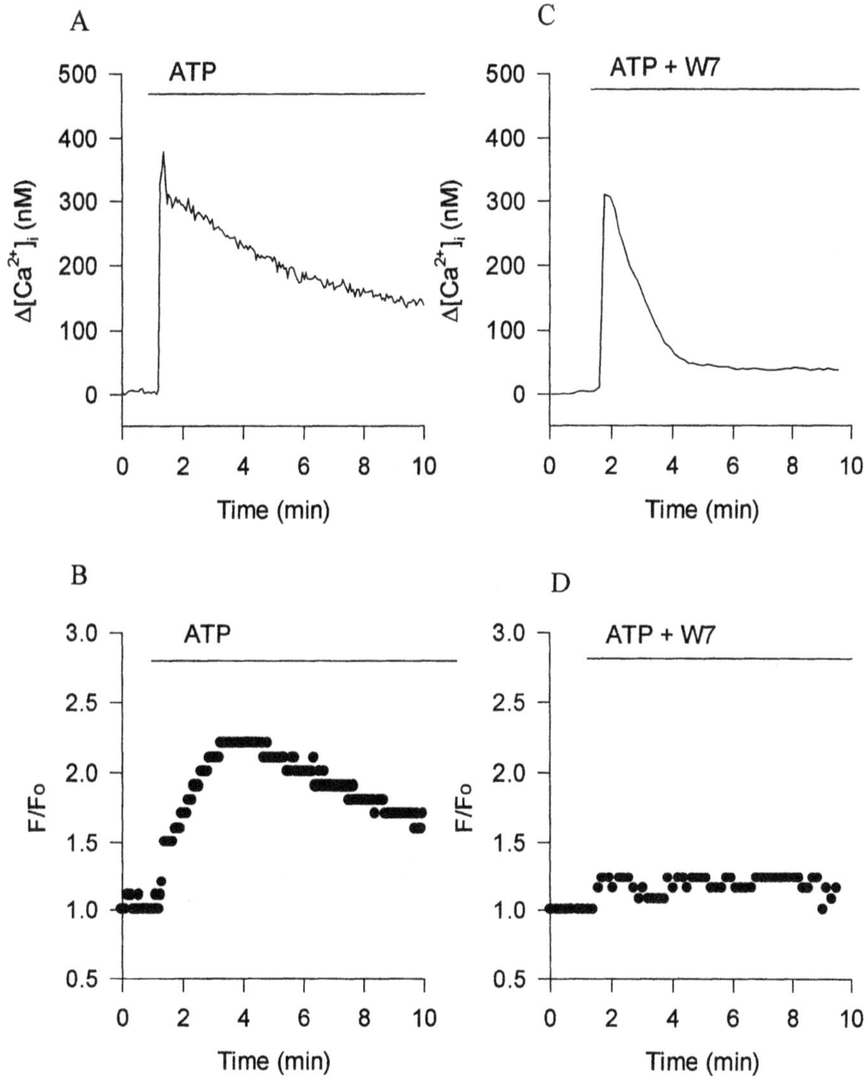


FIG. 1. *Involvement of calmodulin in ciliary activation. Time course of the effect of extracellular ATP (100 μM) on $[Ca^{2+}]_i$ (A, C) and CBF enhancement (B, D) in the presence (C, D) and the absence(A, B) of W-7 (100 μM). Inhibition of calmodulin decouples between rise of $[Ca^{2+}]_i$ and CBF enhancement induced by ATP.*

seen in Fig. 3, bypassing of the GC inhibition, achieved by the addition of 100 μM db-cGMP, restores the ATP-induced CBF enhancement.

It deserves mentioning that addition of 100 μM db-cGMP alone to untreated cells produced a negligible effect on CBF [29]. The elevation in the db-cGMP concentration to as high as 1 mM resulted in a moderate

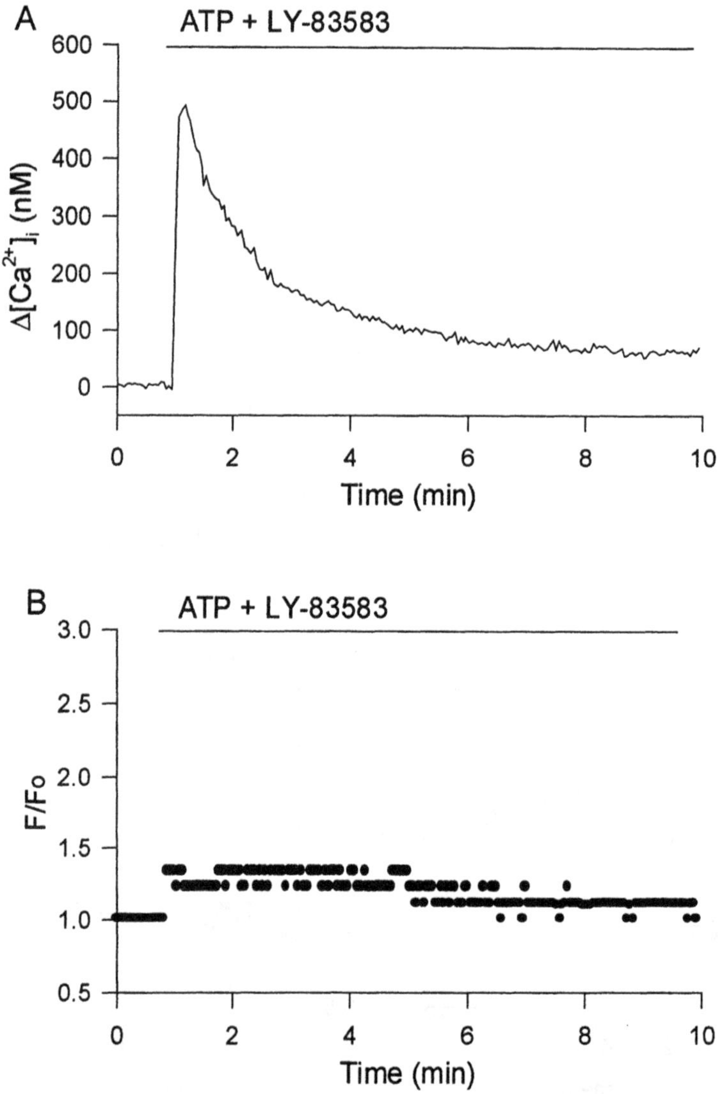


FIG. 2. *Role of GC in ciliary stimulation. Time course of the effect of extracellular ATP (100 μM) on $[Ca^{2+}]_i$ (A) and CBF (B) in the presence of LY-83583, a GC inhibitor . LY-83583 (100 μM) did not affected significantly the magnitude of the response in $[Ca^{2+}]_i$ but strongly inhibited CBF enhancement caused by ATP.*

increase in CBF (not shown). The CBF enhancement has never exceeded 30% under these conditions. Similar small elevations in CBF ranging from 10% to 50% in response to an increase in either NO or cGMP levels have been observed in human airway epithelium [5, 21, 22]. Thus, the cGMP

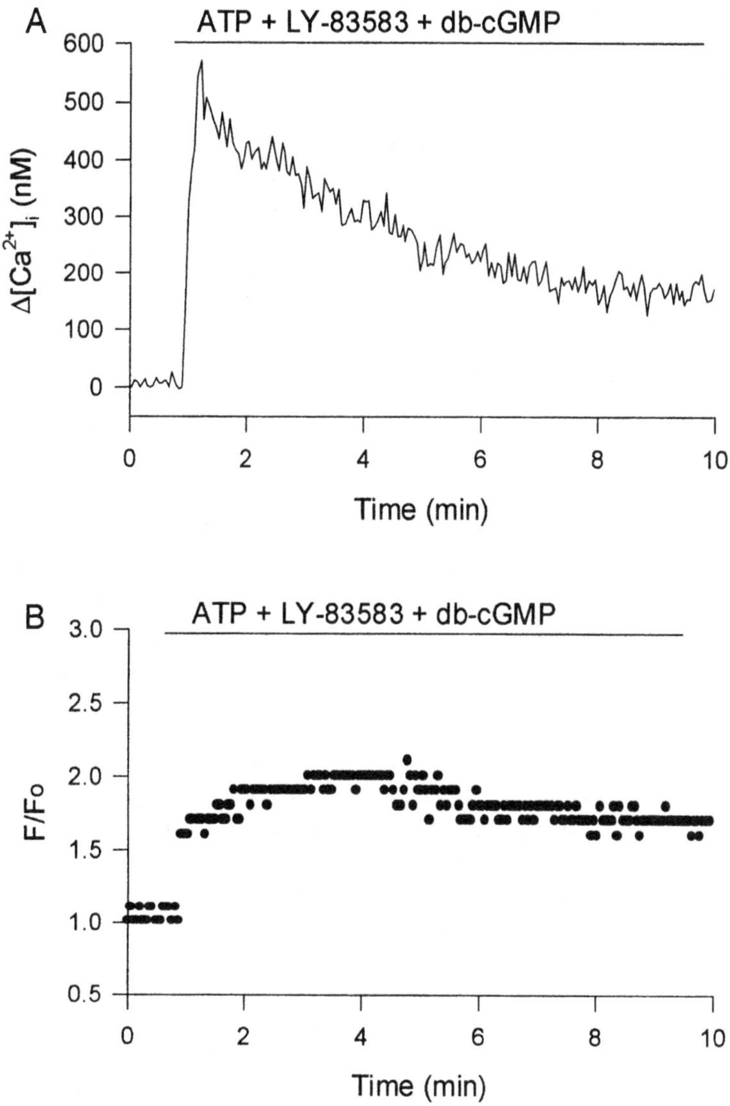


FIG. 3. *The by-pass of GC inhibition by supplying the cells with cGMP. Time course of the effect of extracellular ATP (100 μM) on* $[Ca^{2+}]_i$ *(A) and CBF (B) in the presence of 100 μM LY-83583 and 100 μM db-cGMP. Addition of db-cGMP prevents the inhibitory effect of LY-83583 on CBF enhancement (Fig. 2).*

pathway alone can probably elevate CBF in a calcium-independent manner, although the magnitude of this elevation is relatively small.

Cyclic-AMP is another well known modulator of ciliary activity [3, 4, 26, 34]. Moreover, production of cAMP can also be regulated by the $Ca^{2+}$-

calmodulin complex [2]. Cyclic-AMP-dependent protein kinase (PKA) has been shown to phosphorylate specific axonemal targets that increase the forward swimming speed in Paramecium [9, 10]. We have recently reported that in tissue cultures from frog esophagus, the increase in cAMP concentration induced CBF enhancement, even when a rise in $[Ca^{2+}]_i$ was deterred [3]. It should be emphasized that, just as in the case of cGMP, the magnitude of the CBF enhancement, produced by the cAMP elevation in a calcium-independent manner, was relatively small (50%–70%). To examine the effect of an increase in cAMP levels on the ciliary activity in cultured rabbit trachea, we treated the cells with forskolin, a powerful activator of adenylyl cyclase. Addition of 25$\mu$M forskolin produced a negligible effect on $[Ca^{2+}]_i$ and caused moderate enhancement of CBF (Fig. 4). The average responses to 25$\mu$M forskolin were $\delta[Ca^{2+}]_i = 34 \pm 5$ nM ($n$=10) and $F/Fo = 1.3 \pm 0.1 (n = 10)$.Thus, these results conform to the results obtained in the frog esophagus [3], once again implying a general trend.

Taken together, these results suggest that the elevation in $[Ca^{2+}]_i$ is neither necessary nor sufficient a condition for CBF enhancement. It is not a sufficient condition for CBF enhancement, because blockage of PKG activation renders the stimulatory effect of $[Ca^{2+}]_i$ void. It is not a necessary condition, because an increase in the levels of the cyclic nucleotides can induce enhancement of CBF in a calcium-independent manner. This enhancement is presumably achieved through a direct phosphorylation of axonemal proteins by PKA and/or PKG. It is important to emphasize, however, that although the ciliary activation induced in this fashion can be sustained for a long time [3], its magnitude is relatively small.

Apparently, phosphorylation(s) performed by the protein kinases is/are a necessary condition for the CBF enhancement, while a calcium elevation is not. On the other hand, impeding of $[Ca^{2+}]_i$ elevation blocks ciliary response to the stimuli, which are conveyed by receptors coupled to phospholipase $C$ [12, 23, 28, 29]. An elevation in $[Ca^{2+}]_i$ cannot be a single stimulatory effect of these receptors, since such an elevation alone is not sufficient for the CBF enhancement. Therefore, it appears that calmodulin plays a key role in the process of ciliary stimulation providing a link between the $[Ca^{2+}]_i$ elevation and the protein kinases (Fig. 5). Following the receptor-induced $[Ca^{2+}]_i$ elevation, the $Ca^{2+}$ - calmodulin complex stimulates production of the cyclic nucleotides, leading to the PKA- and/or PKG-dependent phosphorylations. Moreover, these phosphorylations are not only necessary for the CBF enhancement, but can sustain ciliary activity at a moderately enhanced level during the later stages of the stimulation, when $[Ca^{2+}]_i$ subsides to its resting level [35].

To summarize, the $[Ca^{2+}]_i$ elevation alone is insufficient to induce CBF enhancement. The cyclic nucleotide-dependent phosphorylations can produce a moderate enhancement of CBF in a calcium-independent manner. However, the combined effect of the calcium elevation and of the cyclic nucleotide-dependent kinases activated by the $Ca^{2+}$- calmodulin complex,

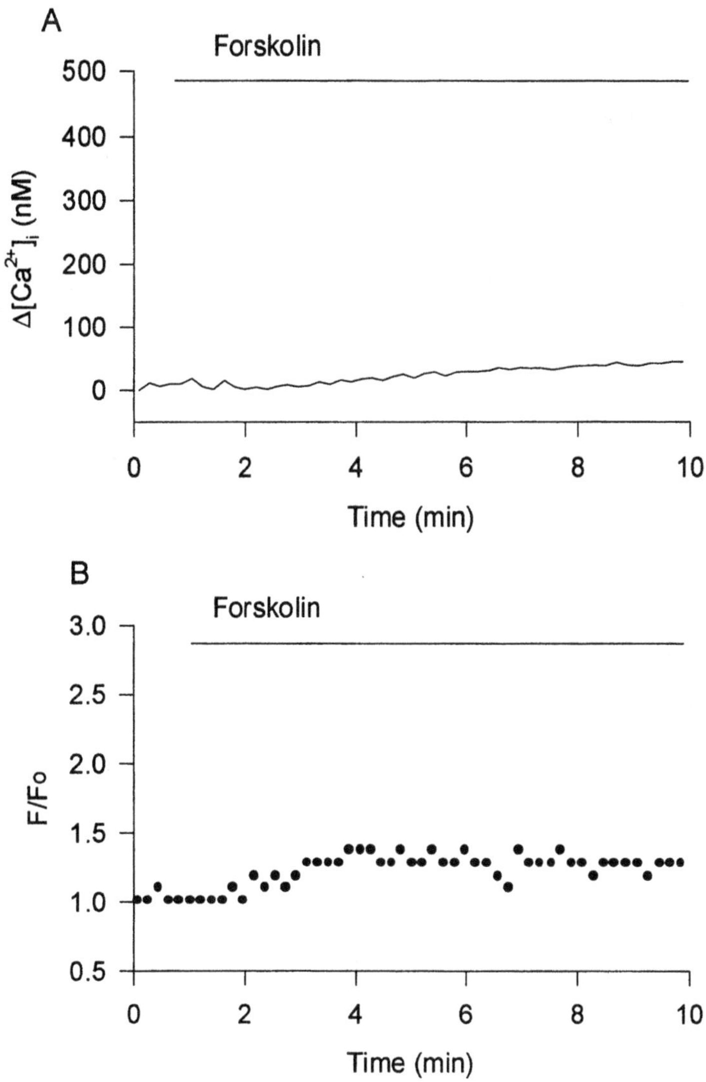


FIG. 4. *The effect of cAMP on CBF. Time course of the effect of 25μM forskolin, a powerful activator of adenylyl cyclase, on* $[Ca^{2+}]_i$ *(A) and CBF (B). Addition forskolin produces a negligible effect on* $[Ca^{2+}]_i$ *and causes moderate enhancement of CBF.*

manifests itself in robust and sustained CBF enhancement. Such a strong CBF enhancement together with the accompanied changes in metachronal wave characteristics [6] may explain the amazing ability of cilia to transport heavy objects at a high speed.

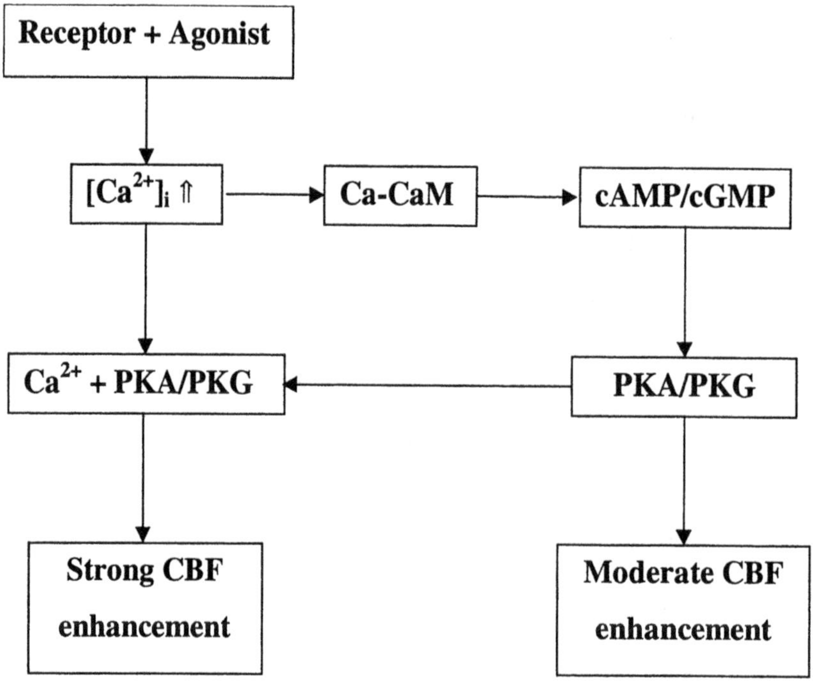


FIG. 5. *Schematic diagram of molecular events in the process of ciliary stimulation.*

**Acknowledgements.** This work was partially supported by The Israeli Science Foundation founded by the Israeli Academy of Sciences and Humanities. A. Braiman gratefully acknowledges the fellowship support of the Kreitman Foundation.

## REFERENCES

[1] AIELLO, E., J. KENNEDY, AND C. HERNANDEZ. Stimulation of frog ciliated cells in culture by acetylcholine and substance P. *Comparative Biochemistry and Physiology* **99**: 497–506, 1991.

[2] ANTONI, F.A. Calcium regulation of adenylate cyclase. Relevance for endocrine control. *Trends in Endocrinology and Metabolism* **8**: 7–14, 1997.

[3] BRAIMAN, A., O. ZAGOORY, AND Z. PRIEL. PKA induces $Ca^{2+}$ release and enhances ciliary beat frequency in a $Ca^{2+}$-dependent and -independent manner. *American Journal of Physiology* **275**: C790-C797, 1998.

[4] DI BENEDETTO, G., C.J. MAGNUS, P.T.A. GRAY, AND A. MEHTA. Calcium regulation of ciliary beat frequency in human respiratory epithelium *in vitro*. *Journal of Physiology* **439**: 103–113, 1991.

[5] GEARY, C.A., C.W. DAVIS, A.M. PARADISO, AND R.C. BOUCHER. Role of CNP in human airways: cGMP-mediated stimulation of ciliary beat frequency. *American Journal of Physiology* **268**: L1021-L1028, 1995.

[6] GHEBER, L. AND Z. PRIEL. Metachronal activity of cultured mucociliary epithelium under normal and stimulated conditions. *Cell Motility and the Cytoskeleton* **28**: 333–345, 1994.

[7] GHEBER, L., Z. PRIEL, C. AFLALO, AND V. SHOHSAN-BARMATZ. Extracellular ATP binding proteins as potential receptors in mucociliary epithelium: characterization using [32P]3'-O-(4-Benzoyl)benzoyl ATP, a photoaffinity label. *Journal of Membrane Biology* **147**: 83–93, 1995.

[8] GRYNKIEWICZ, G., M. POENIE, AND R.Y. TSIEN. A new generation of $Ca^{2+}$ indicators with improved fluorescence properties. *Journal of Biological Chemistry* **260**: 3440–3450, 1985.

[9] HAMASAKI, T., K. BARKALOW, J. RICHMOND, AND P. SATIR. cAMP-stimulated phosphorylation of an axonemal polypeptide that copurifies with the 22S dynein arm regulates microtubule translocation velocity and swimming speed in paramecium. *Proceedings of the National Academy of Sciences of the United States of America* **88**: 7918–7922, 1991.

[10] HAMASAKI, T., K. BARKALOW, AND P. SATIR. Regulation of ciliary beat frequency by a dynein light chain. *Cell Motility and the Cytoskeleton* **32**: 121–124, 1995.

[11] HINRICHSEN, R.D. Calcium and calmodulin in the control of cellular behavior and motility. *Biochimica et Biophysica Acta* **1155**: 277–293, 1993.

[12] KORNGREEN, A. AND Z. PRIEL. Purinergic stimulation of rabbit ciliated airway epithelia: control by multiple calcium sources. *Journal of Physiology* **497**: 53–66, 1996.

[13] KORNGREEN, A. AND Z. PRIEL. Simultaneous measurment of ciliary beating and intracellular calcium. *Biophysical Journal* **67**: 1–4, 1994.

[14] LANSLEY, A. B., AND M.J. SANDERSON. Regulation of airway activity by $Ca^{2+}$: simultaneous measurment of beat frequency and intracellular $Ca^{2+}$. *Biochemical Journal* **77**: 629–638, 1999.

[15] LANSLEY, A.B., M.J. SANDERSON, AND E.R. DIRKSEN. Control of the beat cycle of respiratory tract cilia by $Ca^{2+}$ and cAMP. *American Journal of Physiology* **263**: L232-L242, 1992.

[16] LEVIN, R., A. BRAIMAN, AND Z. PRIEL. Protein kinase C induced calcium influx and sustained enhancement of ciliary beating by extracellular ATP. *Cell Calcium* **21**: 103–113, 1997.

[17] MA, W., A. KORNGREEN, N. UZLANER, Z. PRIEL, AND S. D. SILBERBERG. Extracellular sodium regulates airway ciliary motility by inhibiting $P2X$ receptor. *Nature* **400**: 894–897, 1999.

[18] MAO, H. AND L.B. WONG. Fluorescence and laser photon counting: Measurments of epithelial $[Ca^{2+}]_i$ or $[Na^+]_i$ with ciliary beat frequency. *Annals of Biomedical Engineering* **26**: 666–678, 1998.

[19] MARUYAMA, I. Conflicting effects of noradrenalin on ciliary movement of frog palatine mucosa. *European Journal of Pharmacology* **97**: 239–245, 1984.

[20] NAIRN, A.C. AND M.R. PICCIOTTO. Calcium/calmodulin-dependent protein kinases. *Seminar in Cancer Biology* **5**: 295–303, 1994.

[21] RUNER, T., A. CERVIN, S. LINDBERG, AND R. UDDMAN. Nitric oxide is a regulator of mucociliary activity in the upper respiratory tract. *Otolaryngology - Head-and-Neck-Surgery* **119**: 278–287, 1998.

[22] RUNER, T. AND S. LINDBERG. Effects of nitric oxide on blood flow and mucociliary activity in the human nose. *Annals-of-Otology-Rhinology-and-Laryngology* **107**: 40–46, 1998.

[23] SALATHE, M. AND R.J. BOOKMAN. Coupling of $[Ca^{2+}]_i$ and ciliary beating in cultured tracheal epithelial cells. *Journal of Cell Science* **108**: 431–440, 1995.

[24] SALATHE, M. AND R.J. BOOKMAN. Mode of $Ca^{2+}$ action on ciliary beat frequency in single ovine airway epithelial cells. *Journal of Physiology* **520**: 851–865, 1999.

[25] TAMAOKI, J., A. CHIYOTANI, M. KONDO, H. TAKEMURA, AND K. KONNO. Role of NO generation in b-adrenoceptor-mediated stimulation of rabbit airway ciliary motility. *American Journal of Physiology* **268**: C1342-C1347, 1995.

[26] TAMAOKI, J., M. KONDO, AND T. TAKIZAWA. Effect of cAMP on ciliary function in rabbit tracheal epithelial cells. *Journal of Applied Physiology* **66**: 1035–1039, 1989.

[27] TAMAOKI, J., A. SAKAI, M. KONDO, H. TAKEMURA, AND K. KONNO. Role of nitric oxide in tachykinin-induced increase in potential difference of rabbit tracheal mucosa. *Journal of Physiology* **488**: 115–122, 1995.

[28] TARASIUK, A., M. BAR-SHIMON, L. GHEBER, A. KORNGREEN, Y. GROSSMAN, AND Z. PRIEL. Extracellular ATP induces hyperpolarization and motility stimulation of ciliary cells. *Biophysical Journal* **68**: 1163–1169, 1995.

[29] UZLANER, N. AND Z. PRIEL. Interplay between the NO pathway and elevated $[Ca^{2+}]_i$ enhances ciliary activity in rabbit trachea. *Journal of Physiology* **516**: 179–190, 1999.

[30] VERDUGO, P. $Ca^{2+}$ - dependent hormonal stimulation of ciliary activity. *Nature* **283**: 764–765, 1980.

[31] VERDUGO, P., N.T. JOHNSON, AND P.Y. TAM. b-Adrenergic stimulation of respiratory ciliary activity. *Journal of Applied Physiology* **48**: 868–871, 1980.

[32] VILLALON, M., T.R. HINDS, AND P. VERDUGO. Stimulus-response coupling in mammalian ciliated cells, Demonstration of two mechanisms of control for cytosolic $[Ca^{2+}]$. *Biophysical Journal* **56**: 1255–1258, 1989.

[33] WONG, L.B. AND D.B. YEATES. Luminal purinergic regulatory mechanism of tracheal ciliary beat frequency. *American Journal of Respiratory Cell and Molecular Biology* **7**: 447–454, 1992.

[34] YANG, B., R.J. SCHLOSSER, AND T.V. MCCAFFRY. Dual signal transduction mechanisms modulate ciliary beat frequency in upper airway epithelium. *American Journal of Physiology* **270**: L745-L751, 1996.

[35] ZAGOORY, O., A. BRAIMAN, L. GHEBER, AND Z. PRIEL. The role of calcium and calmodulin in ciliary stimulation induced by acetylcholine. Am. J. Physiol., 2000 (in press).

# A NUMERICAL METHOD FOR SIMULATING FAST-SWIMMING MOTIONS*

RICARDO CORTEZ†

**Abstract.** A numerical method for the simulation of thin, elastic immersed boundaries in a two-dimensional fluid is introduced. The method is Lagrangian and combines the use of vortices and impulse elements (vortex dipoles). Consequently, it is best suited for applications where the Reynolds number is high. The example presented here is the motion of an undulating filament, simulating the swimming of an organism in a slightly viscous fluid.

**Key words.** Particle methods, immersed boundaries.

**AMS(MOS) subject classifications.** 65C20, 76D05.

**1. Introduction.** Swimming motions commonly involve the interaction of very thin flexible bodies with a viscous fluid. Biological examples include organisms propelled by cilia or flagella, as well as larger animals like eels and leeches. Certain fish tails are also thin in comparison with the span direction and therefore fall within this category. There is great interest in the development of numerical techniques for the simulation of such motions, which can be used for the study of evolutionary features of animals, the effects of defects on swimming efficiency (survival), and other important questions. In addition, one expects the same numerical techniques to be applicable to other instances of interactions between fluids and membranes such as blood flow in the heart chambers or arteries, flow around parachutes or fibers in a manufacturing process and others.

In terms of the fluid dynamics involved in these applications, there is a tremendous difference between the flow surrounding micro-organisms and that around larger fast-swimming animals. Namely, the length scales involved in micro-organism motion are so small and the velocities low enough that the Reynolds numbers of the flow are extremely small. Consequently, a mathematical model can be simplified by neglecting inertial forces. On the other hand, in the case of fast-swimming creatures of macroscopic size, the Reynolds numbers are large ($\sim 10^2$ or higher) and the inertial forces must be part of the model. The viscous effects cannot be neglected either, even for small values of the viscosity, since they are significant near the swimming body and are mathematically necessary for the fluid to satisfy the boundary conditions at the surface of the flexible membrane.

The rest of this paper introduces a numerical method that aims to compute membrane/fluid interactions for high Reynolds number flows and incompressible fluids. The approach is to define two components of the

*This work was supported in part by NSF grant DMS-9816951.

†Department of Mathematics, Tulane University, 6823 St. Charles Ave., #424, New Orleans, LA 70118; cortez@math.tulane.edu.

method. The first one computes the forces that develop along the animal. This component is biological in the sense that it contains the internal mechanism which defines the forces that induce the motion. The second component is the fluid dynamics simulation based on the computed forces. This component produces the velocity needed to update the position of the membrane. Various levels of sophistication can be used in each component. Here, admittedly, the force-generating component is primitive and the emphasis is on the fluid dynamics. However, more advanced force-generating mechanisms can easily be made part of the overall model. In fact, the inclusion of such biologically-relevant mechanisms is one of the long-term objectives of this work.

**2. Evolution equations.** The method presented here is Lagrangian. This means that an appropriate domain is tiled with fluid particles carrying vorticity. These particles move with the fluid so that their trajectories must be computed. In addition, the amount of vorticity carried by each particle will change during the evolution of the flow due to the viscous diffusion and to the dynamic forces developed along the immersed membrane.

The membrane is treated as a curve embedded in a two-dimensional fluid. This curve is also transported by the fluid and differs from any other material line by the dynamic introduction of forces along it. The forces are represented not by vortices but by an impulse field [1], [2] and [6]. Impulse (denoted by $\mathbf{m}$) can be thought of as linear momentum and is defined as a vector field whose curl is the vorticity of the flow. In addition, the impulse field is not required to have zero divergence. In this way, the difference between $\mathbf{m}$ and the fluid velocity $\mathbf{u}$ is the gradient of some scalar function

$$\mathbf{m} = \mathbf{u} + \nabla\phi. \tag{2.1}$$

The reason for having two computational variables, vorticity and impulse, is convenience. The equation of motion of impulse introduces body forces naturally. However, if impulse is to remain attached to the membrane, the diffusion should be done in terms of some other variable. For this, vorticity is a good choice.

The Navier-Stokes equations for a viscous incompressible fluid in $\mathbb{R}^2$ are

$$\mathbf{u}_t + \mathbf{u}\cdot\nabla\mathbf{u} = -\nabla p + \nu\Delta\mathbf{u} + \mathbf{F}, \qquad \nabla\cdot\mathbf{u} = 0, \tag{2.2}$$

where $\mathbf{u}$ is the velocity, $p$ the pressure and $\nu$ the kinematic viscosity. We assume here that the fluid density is $\rho = 1$.

Since the body force $\mathbf{F}$ will be accounted for through the impulse, the equation for vorticity is developed here without the force term. By taking the curl of the Navier-Stokes equations and using incompressibility we obtain the equation for the (scalar) vorticity $\omega$:

$$\frac{D\omega}{Dt} = \nu\Delta\omega, \qquad \nabla\cdot\mathbf{u} = 0 \tag{2.3}$$

written here in terms of the material derivative $D/Dt = \partial/\partial_t + \mathbf{u}\cdot\nabla$. The equations indicate that the vorticity is diffused along trajectories of fluid particles. At any instant, given the vorticity, the velocity is recovered by the Biot-Savart law [7]

$$\mathbf{u} = (\partial/\partial_y, -\partial/\partial_x)\ (G * \omega),$$

where $G$ is the Green's function satisfying $\Delta G = -\delta$.

The equation of impulse is derived here excluding the viscous term because the diffusion will be modeled in terms of the vorticity. The main purpose of the impulse is to account for the forces. Substituting Eq. (2.1) into Eq. (2.2) and identifying the pressure with $p = \phi_t + \mathbf{u}\cdot\nabla\phi + \frac{1}{2}|\mathbf{u}|^2$ we arrive at the fluid equations in terms of impulse

$$\frac{D\mathbf{m}}{Dt} = -(\nabla\mathbf{u})^T\mathbf{m} + \mathbf{F}, \qquad \mathbf{u} = \mathbb{P}\mathbf{m}, \tag{2.4}$$

where $\nabla\mathbf{u}$ is a matrix with $ij$-entry given by $\partial u_i/\partial x_j$. The last equation indicates that $\mathbf{u}$ is the projection of $\mathbf{m}$ onto the space of zero-divergence fields. Together with Eq. (2.1) it gives the formula for recovering the velocity induced by a given impulse field.

**3. The numerical method.** The membrane is represented by points placed equally-spaced along it. These points are also where the forces are computed and where the impulse field is defined. A region surrounding the membrane is discretized with vortices whose strengths are initially zero but change in time due to the diffusion (see Figure 1).

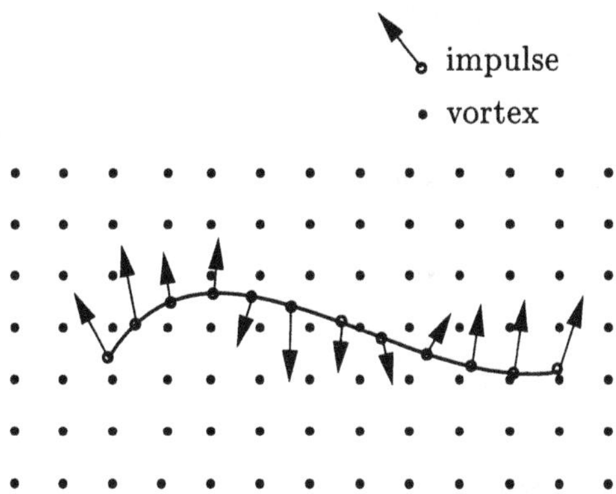


FIG. 1. *Schematic diagram of the immersed membrane, vortices and impulse field.*

If the membrane is defined by the points $\mathbf{x}^k$, the impulse can be represented by the sum

$$\mathbf{m}(\mathbf{x}) = \Delta\ell \sum_{k=1}^{N_i} \mathbf{m}^k f_\delta(\mathbf{x} - \mathbf{x}^k), \tag{3.1}$$

where $\Delta\ell$ is the point separation, $\mathbf{m}^k$ is the strength of the impulse vector at $\mathbf{x}^k$ and $N_i$ is the number of impulse elements. The cutoff function $f_\delta$ is a smooth approximation to the delta function (e.g. a Gaussian), so that $\iint f_\delta(\mathbf{x})d\mathbf{x} = 1$. The parameter $\delta$ refers to the *width* of the function $f_\delta$ and is in practice $O(\Delta\ell)$.

The evolution of the impulse strengths $\mathbf{m}^k$ along particle trajectories is based on Eq. (2.4):

$$\frac{d\mathbf{m}^k}{dt} = -(\nabla\mathbf{u})^T\mathbf{m}^k + \mathbf{f}^k, \tag{3.2}$$

where $\mathbf{f}^k$ is the force at $\mathbf{x}^k$. The velocity contribution from the impulse field is found analytically by extracting the zero-divergence part of the impulse in Eq. (3.1):

$$\text{velocity due to impulse } = \Delta\ell \sum_{k=1}^{N_i} \left[\mathbf{m}^k f_\delta(r) + \nabla(\mathbf{m}^k \cdot \nabla G_\delta(r))\right] \tag{3.3}$$

where $\Delta G_\delta(r) = -f_\delta(r)$ and $r = |\mathbf{x}-\mathbf{x}^k|$. The regularized Green's function $G_\delta = (G * f_\delta)$ is smooth everywhere and is found explicitly for a given choice of cutoff function.

The vorticity carried by the vortices can be approximated similarly by

$$\omega(\mathbf{x}) = h^2 \sum_{i=1}^{N_v} \omega^i f_\delta(\mathbf{x} - \mathbf{x}^i), \tag{3.4}$$

where $h^2$ is the area element occupied by each vortex, $N_v$ is the total number of vortices and $\omega^i$ is the strength of the $i$th vortex. The evolution of these strengths along particle trajectories is simply the diffusion equation (2.3). However, one must be careful to include in the total vorticity the contributions from the vortices and from the impulse. Since vorticity is the curl of impulse, the total vorticity $\Omega$ is given by

$$\begin{aligned} \Omega(\mathbf{x}) &= \omega(\mathbf{x}) + \nabla \times \mathbf{m}(\mathbf{x}) \\ &= h^2 \sum_{i=1}^{N_v} \omega^i f_\delta(\mathbf{x} - \mathbf{x}^i) + \Delta\ell \sum_{k=1}^{N_i} \nabla f_\delta(\mathbf{x} - \mathbf{x}^k) \times \mathbf{m}^k. \end{aligned} \tag{3.5}$$

The rate of change of the vortex strengths $\omega^i$ is found by computing the Laplacian of this expression

$$\frac{d\omega^i}{dt} = \nu\Delta\Omega(\mathbf{x}^i). \tag{3.6}$$

The velocity contribution from the vortices is given by the Biot-Savart law

$$\text{velocity due to vortices } = h^2 \sum_{i=1}^{N_v} \omega^i(\partial/\partial_y, -\partial/\partial_x)G_\delta(\mathbf{x} - \mathbf{x}^i). \tag{3.7}$$

The method consists of the simultaneous solution of the equations $d\mathbf{x}^j/dt = \mathbf{u}(\mathbf{x}^j)$ with Eq. (3.2) and Eq. (3.6), where the velocity is the sum of Eq. (3.3) and Eq. (3.7).

**4. Example.** The example below shows the application of the numerical method to the motion of a swimming organism. The forces applied along the creature have two purposes: keeping its length constant and making the animal's motion track approximately the curve

$$y(s,t) = \epsilon\Big(\frac{L-s}{L}\Big)\sin(\kappa s - \Omega t)$$

where $s$ is the arclength parameter, $0 \leq s \leq L$ with $s = 0$ corresponding to the tail of the creature and $s = L$ to the head. In order to accomplish this, contiguous particles along the organism are connected with springs of a fixed resting length and large stiffness. In addition, the curvature at every point is compared with the curvature of $y(s,t)$ and forces are introduced to keep the two values as close to each other as possible. This strategy is exactly the one employed in [4] and [3]. The parameters used are $L = 0.2$ ft, $\epsilon = 0.02$ ft, $\kappa = 2\pi/L$ and $\Omega = 8\pi$ sec$^{-1}$. The stiffnesses used were as in [4]. The viscosity was set to $\nu = 2 \times 10^{-4}$ ft$^2$/sec (slightly larger than water) which gives a Reynolds number of $R = \frac{\Omega}{\nu\kappa^2} = 127$.

The organism was discretized with 128 particles so that $\Delta s = 1.575 \times 10^{-3}$ ft. The vortices were placed on a uniform rectangular grid of size $h = 4.6875 \times 10^{-3}$ ft covering an area of 0.0545 ft$^2$, requiring 2480 vortices. The cutoff parameter $\delta = 0.014$ ft seemed to work well for both the vorticity and impulse. The cutoff function used was the radial function $f_\delta(r) = \frac{1}{2\pi\delta^2}(4e^{-r^2/\delta^2} - e^{-r^2/2\delta^2})$.

The animal itself is a very strong vortex sheet due to the nature of the forces. Figure 2 shows contours of vorticity at four instants during the motion. Light contours are of positive vorticity and dark contours are of negative vorticity. A feature of this vorticity is that regions of positive and negative vorticity alternate along the organism according to its curvature. These regions seem to slide down the body of the creature until they reach the tail and are shed behind it. This shedding of vorticity is not observed at low Reynolds numbers. Flow circulation regions matching these vortex patterns have been observed experimentally ([5, p. 57]).

**5. Final remarks.** The force model employed in this experiment results in a forward motion at a rate of about 6.5% of the animal's body length per period of the imposed wave. This rate is somewhat low due to the arbitrary nature of the wave. A biologically accurate force model is needed to conduct realistic simulations. One strength of the numerical method presented here is that it can accommodate any force model. The fluid model gives very satisfactory results for large Reynolds number flows. The Lagrangian nature of the method introduces very little numerical dissipation and is relatively insensitive to increases in Reynolds numbers. These properties are not easily attained by grid methods.

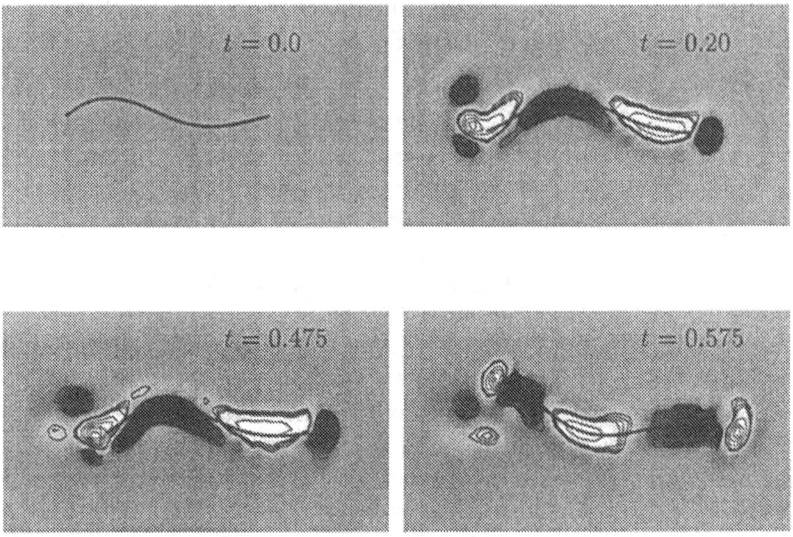


FIG. 2. *Total vorticity in the flow and vorticity contour lines at* $t = 0$, *0.2, 0.475 and 0.575 sec. Positive vorticity is light and negative vorticity is dark.*

## REFERENCES

[1] R. CORTEZ, *An impulse-based approximation of fluid motion due to boundary forces*, J. Comput. Phys. **123** (1996), 341–353.

[2] R. CORTEZ AND D. A. VARELA, *The dynamics of an elastic membrane using the impulse method*, J. Comput. Phys. **138** (1997), 224–247.

[3] L. J. FAUCI AND A. L. FOGELSON, *Truncated Newton methods and the modeling of complex immersed elastic structures*, Comm. Pure Appl. Math. **46** (1993), 787–818.

[4] L. J. FAUCI AND C. PESKIN, *A computational model of aquatic animal locomotion*, J. Comput. Phys. **77** (1988), 85–108.

[5] SIR M. J. LIGHTHILL, *Mathematical biofluiddynamics*, SIAM, Philadelphia, 1975.

[6] M. C. RECCHIONI AND G. RUSSO, *Hamilton-based numerical methods for a fluid-membrane interaction in two and three dimensions*, SIAM J. Sci. Comput. **19** (1998), 861–892.

[7] A. J. CHORIN AND J. E. MARSDEN, *A Mathematical Introduction to Fluid Mechanics*, Springer-Verlag, New York, 1979.

# A FLUID–STRUCTURE INTERACTION MODEL OF CILIARY BEATING

ROBERT H. DILLON* AND LISA J. FAUCI†

**1. Introduction.** The coupled system of a viscous, incompressible fluid and a single, force-generating organism is difficult to analyze, even when the kinetics of an organism's waveform is taken as given. In the past decades, the efforts to describe quantitatively the fluid dynamics of spermatozoa and ciliary propulsion have been very successful. Since the Reynolds number is quite small and inertial effects can be neglected, the linear Stokes flow assumption has been used to investigate the hydrodynamic consequences of flagellar undulations [3]. These investigations have been both analytical and computational. Resistive-force theory, initially developed by Gray and Hancock [12], makes use of the linear Stokes flow assumption, and constructs the flow field by means of distributions of fundamental singularities. Lighthill later improved this theory by incorporating slender body approximations [19], since the diameter of a flagellum is much smaller than its length. More detailed hydrodynamic analysis, such as refined slender body theory and boundary element methods, have produced excellent simulations of both two- and three-dimensional flagellar propulsion in an infinite fluid domain or in a domain with a fixed wall [17, 18, 16, 22]. In addition, ciliary motion, where the effect of a single plane surface at the base of the cilia is included, is studied by Gueron and Liron [14, 15], wherein they also present a thorough survey of flagellar hydrodynamics.

More difficult is the study of the fluid dynamic interactions of organisms with each other and surrounding geometry. Examples of these interactions include sperm motility in the reproductive tract, and metachronal waves exhibited by rows of beating cilia. For recent advances in the mathematical modeling of these complex systems, see [10] and [13].

In both cilia and flagella, the typical "9 + 2" axoneme consists of a central pair of singlet microtubules surrounded by nine outer doublet microtubules and encased by the cell membrane (cf. [20, 23] for review). The microtubules are composed of $\alpha$- and $\beta$- heterodimers that form protofilaments running the length of the microtubules. The A tubule of the outer doublet is made up of 13 protofilaments, the B tubule of 10 or 11 protofilaments. The nine outer doublets are connected by radial spokes to a sheath

*Department of Pure and Applied Mathematics, Washington State University, Pullman, WA 99163; dillon@math.wsu.edu. The work of the first author was supported in part by NSF grant DMS-9805501.

†Department of Mathematics, Tulane University, New Orleans, LA 70118; ljf@math.tulane.edu. The work of the second author was supported in part by NSF grant DMS 9805492 and the NSF Group Infrastructure Grant DMS 9709754.

surrounding the central pair. In addition, the outer doublets are connected by nexin links between adjacent pairs of doublets. Two rows of dynein arms extend from the A tubule of an outer doublet toward the B tubule of an adjacent doublet at regularly spaced intervals. The bending of the axoneme is caused by sliding between pairs of outer doublets. Active sliding is due to the unidirectional ATP-induced force generation of the dynein power stroke. Backward, passive sliding is due to the active sliding of other pairs of doublets within the axoneme. The precise nature of the spatial and temporal control mechanisms regulating the various waveforms is still unknown.

The object of our work is to develop a model for the individual cilium or flagellum which incorporates discrete representations of the dynein arms, the passive elastic structure of the axoneme including the microtubules and nexin links, as well as the surrounding fluid. This model couples the internal force generation of the molecular motors through the passive elastic structure with the external fluid mechanics. The ciliary beat or flagellar waveform is not preset, but is an emergent property of the interacting components of the coupled fluid-axoneme system. The forces due to activation of dynein motors (that cause the microtubules to slide), the passive forces due to the stiffness of the microtubules, and the passive forces due to inter-microtubule linking proteins (radial spokes and nexin), must balance with the viscous forces due to the surrounding incompressible fluid.

A two-dimensional model of the coupled fluid-axoneme system based upon the immersed boundary method was presented in [7]. Here we present an overview of an improved version of the model. A detailed presentation may be found in [5]. The immersed boundary method was originally developed by Peskin [21] to study blood flow in the heart, but has been used to study many other problems in biological fluid dynamics (cf. [1, 11, 8, 2, 6]).

**2. Formulation.** The structure of the two-dimensional model axoneme consists of two model microtubules (see Figure 1). Each microtubule consists of two parallel filaments. Each filament is described by a discrete set of nodes (immersed boundary points) connected by linear elastic spring elements. These elastic springs form vertical connections between neighboring nodes of a filament, as well as diagonal connections across the two filaments of the microtubule. Adjacent pairs of microtubules are interconnected by linear elastic springs representing the nexin and/or radial links of the axoneme. In addition, active dynein motor forces are represented as diagonal forces between adjacent pairs of microtubules. The axoneme is tethered to fixed points in space via strong elastic springs at the base. In the configuration shown in the schematic, the dyneins are attached to fixed nodes on the left microtubule. Contraction of the dynein generates sliding between the two microtubules with the right hand microtubule moving upwards relative to the left. Since the pair of microtubules is tethered at the base, this sliding is converted to bending with the tip moving toward the right. In the second configuration (not shown), the dyneins are attached to fixed nodes on the right hand microtubule and extend downward toward an attachment point on the left hand microtubule. As the dyneins contract, the left hand microtubule moves upward relative to the right hand microtubule and the tip moves toward the left. In either configuration, one end of a dynein can attach, detach, and reattach to attachment sites on the opposite microtubule. The entire structure is embedded in a viscous incompressible fluid.

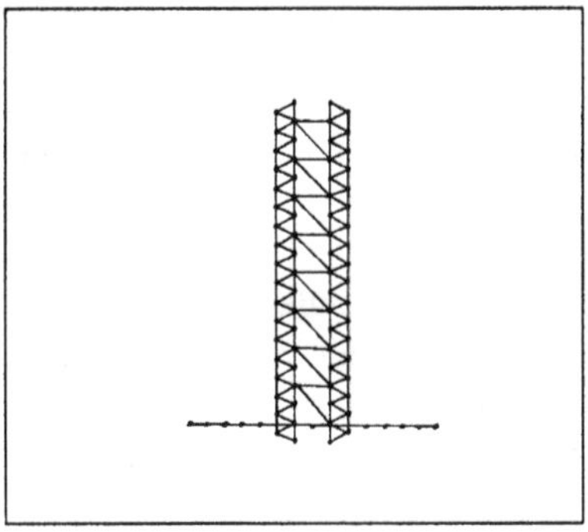

FIG. 1. *Schematic of model cilia in LR mode.*

The fluid is regarded as viscous and incompressible, and the microtubules that comprise the axoneme as elastic boundaries immersed in this fluid. The other protein structures within the axoneme will be represented as forces which act upon the microtubules, as described above. We assume that the flow is governed by the Navier-Stokes equations:

$$\rho \left[ \frac{\partial \mathbf{u}}{\partial t} + \mathbf{u} \cdot \nabla \mathbf{u} \right] = -\nabla p + \mu \Delta \mathbf{u} + \mathbf{F}(\mathbf{x}, \mathbf{t}) \tag{2.1}$$

$$\mathbf{F} = \sum_k \mathbf{F}^{\mathbf{k}}$$

$$\nabla \cdot \mathbf{u} = 0 \,. \tag{2.2}$$

Here $\rho$ is the fluid density and $\mu$ is the viscosity, $\mathbf{u} = (u, v)$ is the fluid velocity, $p$ denotes the pressure, and $\mathbf{F}$ is the force per unit area which is exerted on the fluid by axoneme. This force is split up into the contributions from each of the filaments comprising the microtubules. These forces $\mathbf{F}^{\mathbf{k}}$ due to the $k$th filament include the elastic forces due to the individual microtubule structures, the passive elastic forces due to intermicrotubule links, and the active forces due to dynein arms. $\mathbf{F}$ is a $\delta$-function layer of force supported only by the region of fluid which coincides with material points of the microtubules; away from these points the force is zero.

The state of the fluid at time $t$ is given by its velocity field; the state of each microtubule is given by the configuration of its one-dimensional filaments. Let $\mathbf{X^k(s,t)}$ denote the $k$th filament in our set of microtubules, and let $\mathbf{f^k(s,t)}$ denote the boundary force per unit length along the $k$th filament. (The form of this boundary force will be discussed below.) The mass of the neutrally-buoyant elastic boundary is attributed to the mass of the fluid in which it sits, and thus the forces are transmitted directly to the fluid. The force field $\mathbf{F^k}$ from the filament $\mathbf{X}^k(s,t)$ is therefore:

$$\mathbf{F^k}(\mathbf{x},t) = \int \mathbf{f^k}(s,t)\delta(\mathbf{x} - \mathbf{X^k}(s,t))ds. \tag{2.3}$$

Here the integration is over the $k$th one-dimensional filament comprising an immersed microtubule, and $\delta$ is the two-dimensional Dirac delta-function. The total force $\mathbf{F}(\mathbf{x},\mathbf{t})$ is calculated by adding the forces from each of the filaments.

Each filament of the microtubule is approximated by a discrete collection of points $\mathbf{X}_1, \mathbf{X}_2, ..., \mathbf{X}_n$. This boundary exerts elastic forces on the fluid near each of these points as described above. We imagine that between each pair of successive points $\mathbf{X}_j$, $\mathbf{X}_{j+1}$ on a microtuble filament, there is an elastic spring or 'link' which generates forces to push the link's length toward a specified resting length. The force density vector arising from this spring at the point $\mathbf{X}_j$ has the form:

$$-S(||\mathbf{X}_{j+1} - \mathbf{X}_j|| - \Delta s)\frac{\mathbf{X}_j - \mathbf{X}_{j+1}}{||\mathbf{X}_j - \mathbf{X}_{j+1}||}\ . \tag{2.4}$$

This force is, therefore, a stiffness constant ($S$) times the deviation from rest length ($\Delta s$) that is applied tangent to the spring. In addition to the forces due to springs along the parallel filaments, forces due to horizontal and vertical crosslinks also contribute to the force density. These springs maintain the integrity of an individual microtubule. Additional forces are applied at the base of the microtubule filaments so that they remain tethered.

Our model also accounts for the passive intermicrotubule connections such as nexin or radial spokes, by applying spring forces due to connections between nodes of adjacent microtubules. The spring constants of these connections can be varied to study their effect upon bending behavior.

Most importantly, we need to achieve the relative sliding between adjacent microtubules so that a bend can be initiated. This is achieved by representing the dynein motors as diagonal springs that connect adjacent microtubules. Unlike the microtubule springs or nexin springs, these springs are not passive, but dynamically decrease their resting lengths as they are activated. In addition, their connectivity may be reassessed during the course of a simulation. For instance, as the microtubules slide, the endpoint of a dynein spring may 'ratchet' from one node of the microtubule to another.

The forces due to all of the springs contribute to the force density function at each of the nodes. A discrete version of Equation (2.3) is used to spread this force density defined along microtubules to a force field on the fluid domain.

Our coupled fluid-axoneme system is closed by requiring that the velocity of a material point of a microtubule must be equal to the fluid velocity evaluated at that point:

$$\frac{\partial \mathbf{X}^{\mathbf{k}}(s,t)}{\partial t} = \mathbf{u}(\mathbf{X}^{\mathbf{k}}(s,t),t) = \int \mathbf{u}(\mathbf{x},t)\delta(\mathbf{x} - \mathbf{X}^{\mathbf{k}}(s,t))d\mathbf{x} \, . \tag{2.5}$$

Here the integration is over the entire fluid domain.

We can summarize the immersed boundary algorithm as follows: Suppose that at the end of timestep $n$ we have the fluid velocity field $\mathbf{u}^n$ on a grid, and the configuration of the immersed boundary points comprising the filaments of the microtubules $(\mathbf{X}^{\mathbf{k}})^{\mathbf{n}}$. Then to advance the system by one timestep we:

1. Reassess the connectivity structure of dynein motors and intermicrotubule links.
2. Decide upon the activation of dyneins based upon the configuration of the filaments of the microtubules $(\mathbf{X}^{\mathbf{k}})^{\mathbf{n}}$.
3. Calculate the force densities $\mathbf{f}^{\mathbf{k}}$ from the boundary configuration and the activation states;
4. Spread the force densities to the grid to determine the forces $\mathbf{F}^{\mathbf{k}}$ on the fluid;
5. Solve the Navier-Stokes equations for $\mathbf{u}^{n+1}$;
6. Interpolate the fluid velocity field to each immersed boundary point $(\mathbf{X}^{\mathbf{k}})^{\mathbf{n}}$ and move the point at this local fluid velocity.

It is in step (2) that we may implement different activation strategies. The kinetics of individual dynein activation may depend upon the geometry of the structure, such as local curvature, shear and intermicrotuble spacing. In addition, the activation kinetics may be stochastic. For a comprehensive discussion of the modeling of dynein activation, see [20] and also Charles Lindemann's article in this volume.

The crucial feature of this fluid-structure interaction model is that the axoneme is not the computational boundary in the Navier-Stokes solver. This immersed boundary is the source of a dynamic force field which influences the fluid motion through the force term in the fluid equations. Since the computational domain is a fixed rectangle, we can use an efficient fluid solver designed for a regular grid with simple boundary conditions. For step (5) we use the projection method of Chorin [4] with periodic boundary conditions. Steps (4) and (6) involve the use of a discrete delta function which communicates information between the grid and the immersed boundary points (which do not necessarily coincide with grid points). We refer the reader to [9, 21] for details.

**3. Numerical simulations.** In this section we present numerical results from two illustrative simulations. We use a 128 x 128 finite difference grid for the fluid, 120 immersed boundary points on each of the four filaments. The cilium is approximately 14 $\mu m$ in length. In each simulation, the cilium is initially vertical. During the course of a simulation, the configuration of the dyneins is set in one of two ways:

- **LR mode.** The active dyneins are oriented diagonally downwards from the left microtubule to the right microtubule. As these dyneins contract, the right microtubule slides distally relative to the left microtubule and the structure bends toward the right.
- **RL mode.** The active dyneins are oriented diagonally downwards from the right microtubule to the left microtubule. As these dyneins contract, the left microtubule slides distally relative to the right microtubule and the structure bends toward the left.

The pattern of activation of dyneins is different during the power stroke and the recovery stroke. During the power stroke, each dynein is in an activated state. The contraction of the dynein crossbridges causes bending. The attachment point of each dynein opposite its fixed point on one microtubule is free to ratchet along the neighboring microtubule. Recall that the microtubules are tethered to space at the base, so that maximal sliding occurs at the tip. During the recovery stroke, the dyneins are activated sequentially from base to tip. Once an individual dynein is activated during the recovery stroke, it remains activated for the duration of the recovery stroke. In the simulations presented here, the speed of this activation wave is imposed.

In both simulations, the dyneins are uniformly activated in the power stroke in the LR mode. The base of each dynein is attached to a fixed node on the left microtubule. The head of the dynein is attached to the nearest node on the right microtubule relative to a fixed offset downwards. As the dyneins contract and create sliding between the two microtubules, the tip of the dynein detaches and reattaches to the nearest proximal attachment site on the right. As in [13], a geometric switch is employed in which the dynein configuration (LR mode) switches to the opposite configuration (RL mode) once the tip of the right microtubule has slid distally a prescribed distance relative to the tip of the left microtubule. This switch is set to occur when the most distal dynein becomes attached to a prescribed proximal attachment site. The particular site chosen is arbitrary and determines how far the microtubule will bend before the stroke direction is reversed.

In the first simulation, the dyneins are uniformly activated in each mode LR and RL. In essense, this corresponds to a symmetric beat consisting of power strokes in each direction. The simulation was run for several beats. We show a sequence from the first one and one half beats in Figure 2. The cilium does not return to its initial vertical position and has a slight bend to its profile as the tip moves across the midline in each direction.

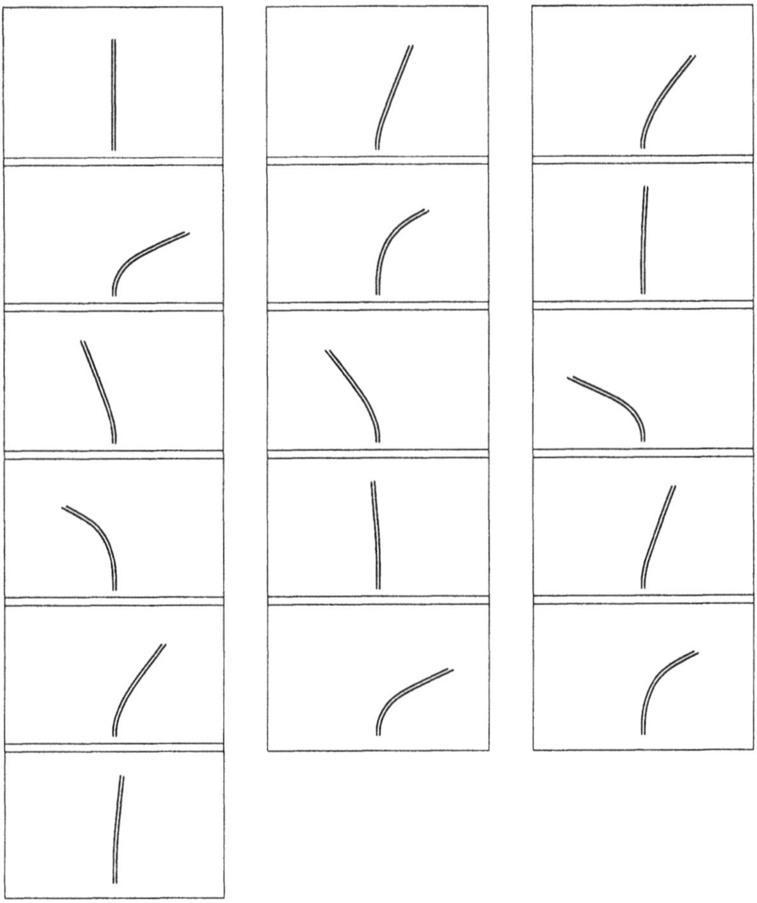

FIG. 2. *Snapshots from cilia beat with uniform dynein activation in each direction.*

In the second simulation, we include a recovery stroke in the RL mode. Figure 3 shows a sequence from a numerical simulation representing one ciliary beat. The duration of the recovery stroke depends upon the speed of the traveling wave of dynein activation. In the results shown here, we do not include the effect of a boundary or wall to which the cilium is attached, but this may be included easily within our framework by including a wall of immersed boundary points [10].

**4. Discussion.** The numerical results demonstrate the feasibility of this fluid-structure interaction model to simulate the ciliary beat. Within the framework of the immersed boundary method, it is possible to study the interaction of many immersed structures simply by including additional

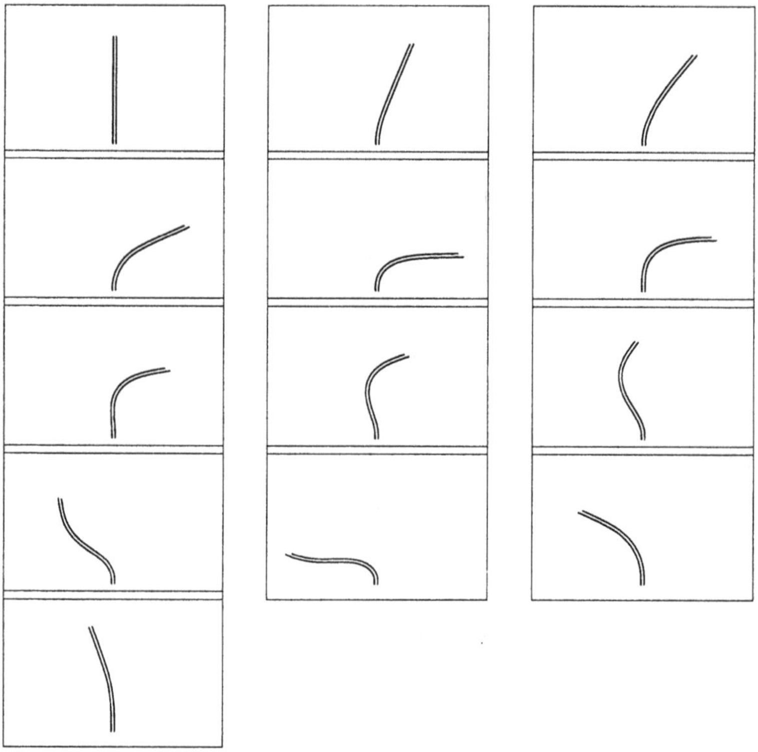

FIG. 3. *Snapshots from cilia beat with uniform dynein activation in the LR mode and traveling wave activation in the RL mode.*

contributions to the force term in the Navier-Stokes equations. The cilia model is robust in the sense that the axonemal structure shows no signs of degradation over several ciliary beats. Thus it will be feasible to study of interaction of several cilia over a number of beats. In future models, the activation of individual dyneins will be dependent upon the configuration of the axoneme at the dynein site. Information concerning the local curvature and spacing between the microtubules is available at each dynein site and can be included in models based on local dynein activation kinetics. In addition, since this model represents a discrete collection of dynein motors, stochastic models of dynein activation can be implemented. At present we are developing a method of controlling the base to tip activation wave during the recovery stroke based on the local curvature so that the wave speed will not be imposed. A similar model will also be used to study flagellar wave forms. A parallel effort is the construction of a three dimensional axoneme based upon the familiar "9 + 2" ultrastructure.

**Acknowledgements.** The authors wish to thank Charlotte Omoto for helpful discussions.

## REFERENCES

[1] K.M. Arthurs, L.C. Moore, C.S. Peskin, E.B. Pitman, and H.E. Layton. Modeling arteriolar flow and mass transport using the immersed boundary method. *J. Comp. Phys.*, **147**:402–440, 1998.

[2] D.C. Bottino. Modeling viscoelastic networks and cell deformation in the context of the immersed boundary method. *J. Comp. Phys*, **147**:86–113, 1998.

[3] C. Brennen and H. Winet. Fluid mechanics of propulsion by cilia and flagella. *Ann. Rev. Fluid Mech.*, **9**:339–398, 1977.

[4] A.J. Chorin. Numerical solution of the Navier-Stokes equations. *Math. Comput.*, **22**:745–762, 1968.

[5] R. Dillon and L.J. Fauci. An integrative model of internal axoneme mechanics and external fluid dynamics in ciliary beating. *in preparation*, 2000.

[6] R. Dillon, L.J. Fauci, A. Fogelson, and D.P. Gaver. Modeling biofilm processes using the immersed boundary method. *J. Comp. Phys*, **129**:57–73, 1996.

[7] R.H. Dillon and L.J. Fauci. Coupling the internal axoneme mechanics of cilia and flagella with external fluid dynamics. In *Proceedings of the International Congress on Cilia, Mucus, and Mucociliary Interactions*, New York, 1997. Marcel Dekker.

[8] C.D. Eggleton and A.S. Popel. Large deformation of red blood cell ghosts in a simple shear flow. *Phys. Fluids*, **10**:1834–1845, 1998.

[9] L.J. Fauci and A.L. Fogelson. Truncated Newton methods and the modeling of complex immersed elastic structures. *Comm. on Pure and Appl. Math.*, **46**:787–818, 1993.

[10] L.J. Fauci and A. McDonald. Sperm motility in the presence of boundaries. *Bull. Math. Biol.*, **57**(5):679–699, 1995.

[11] A.L. Fogelson. Continuum models of platelet aggregation: Formulation and mechanical properties. *SIAM J. Appl. Math*, **52**:1089–1110, 1992.

[12] J. Gray and G. Hancock. The propulsion of sea-urchin spermatozoa. *J. Exp. Biol.*, **32**:802–814, 1955.

[13] S. Gueron, K. Levi-Gurevich, N. Liron, and J.J. Blum. Cilia internal mechanism and metachronal coordination as the result of hydrodynamic coupling. *Proc. Natl. Acad. Sci. USA*, **94**:6001–6006, 1997.

[14] S. Gueron and N. Liron. Ciliary motion modeling, and dynamic multicilia interactions. *Biophys. J.*, **63**:1045–1058, 1992.

[15] S. Gueron and N. Liron. Simulations of three-dimensional ciliary beats and cilia interactions. *Biophys. J.*, **65**:499–507, 1993.

[16] J.J.L. Higdon. The generation of feeding currents by flagellar motion. *J. Fluid Mech.*, **94**:305–330, 1979.

[17] J.J.L. Higdon. A hydrodynamic analysis of flagellar propulsion. *J. Fluid Mech.*, **90**:685–711, 1979.

[18] J.J.L. Higdon. The hydrodynamics analysis of flagellar propulsion: helical waves. *J. Fluid Mech.*, **94**:331–351, 1979.

[19] J.L. Lighthill. Flagellar hydrodynamics. *SIAM Review*, **18**:161–230, 1976.

[20] M. Murase. *The Dynamics of Cellular Motility*. John Wiley, Chichester, 1992.

[21] C.S. Peskin. Numerical analysis of blood flow in the heart. *J. Comp. Phys.*, **25**:220–252, 1977.

[22] N. Phan-thien, T. Tran-cong, and M. Ramia. A boundary-element analysis of flagellar propulsion. *J. Fluid Mech.*, **184**:533–549, 1987.

[23] G.B. Witman. Introduction to cilia and flagella. In R.A. Bloodgood, editor, *Ciliary and Flagellar Membranes*, pp. 1–30, New York, 1990. Plenum.

# ENERGETIC CONSIDERATIONS OF CILIARY BEATING

SHAY GUERON* AND KONSTANTIN LEVIT-GUREVICH†

**Abstract.** The internal mechanism of cilia which is responsible for their beating is among the most ancient biological motors on an evolutionary scale. Ciliary beat cycles consist of two phases: the effective stroke, where the cilium moves approximately as a straight rod, and the recovery stroke, where it bends and rolls close to the surface in a mostly tangential motion.

It is commonly agreed that for efficient functioning, the effective stroke is designed to encounter strong viscous resistance and to generate thrust, whereas the recovery stroke is designed to return to starting position while avoiding viscous resistance as much as possible. Metachronal coordination between cilia, which occurs when many of them beat close to each other, is believed to be an autonomous result of the hydrodynamical interactions in the multiciliary system. Qualitatively, metachronism is understood as a way for minimizing the energy expenditure required for beating.

This paper presents a quantitative investigation of the energetic advantages of metachronism. Using a new method for computing the work by a model cilium beating in a viscous fluid we demonstrate that the energy expenditure during the effective stroke for a single cilium is approximately five times the amount of work done during the recovery stroke. Investigation of multicilia configurations shows that having neighboring cilia beat metachronally is energetically advantageous and perhaps crucial for multiciliary functioning. Finally, the model is used to approximate the number of dynein arm attachments that are likely to occur during the effective and recovery strokes of a beat cycle.

**1. Introduction.** On an evolutionary scale, cilia are among the oldest and most frequently occurring locomotion organelles. Powered by a sliding filaments internal mechanism, they beat in periodical two-phase patterns. During the effective stroke they move approximately as a straight rod and during the recovery stroke they bend, roll close to the cell surface in a mostly tangential direction,and return to a ready position to initiate the next beat cycle. The primary function of cilia is to generate fluid flow in a preferred direction. Such flow can be utilized for propelling a cell through a fluid medium, for moving food particles in a desired direction, or to move mucus layers (see Sleigh, 1962, for a detailed description).

Typical velocities and dimensions indicate that cilia live in a low Reynolds number (Stokes flow realm) inertialess environment. The evolution of asymmetrical beat patterns can be explained by the reversal property of Stokes flows, which implies that symmetrical beats create a zero average flow flux. Qualitative understanding of the two phases of the ciliary beats suggests (e.g., Sleigh, 1962) that for optimal functioning they are designed to:

*Department of Mathematics, University of Haifa, Haifa, 31905, ISRAEL; E-mail: shay@math.haifa.ac.il.

†Department of Mathematics, Israel Institute of Technology, Haifa, 32000, ISRAEL; E-mail: konstant@math.technion.ac.il.

(a) Take advantage of the viscous resistance during the effective stroke to generate maximal thrust.

(b) Minimize viscous resistance during the return to a "ready" position by using a relatively slow and mainly tangential recovery stroke.

In the present study we introduce a method for computing the work done (hereafter referred to as energy expenditure) against viscous forces by a model cilium, and incorporate this technique into our recently developed model of ciliary motility (Gueron et al., 1997; Gueron and Levit-Gurevich, 1998). Metachronal coordination between cilia is a situation where cilia beat with a constant phase difference between adjacent rows in such a way that their tips form a moving wave pattern. The reason why and how arrays of cilia beat in a metachronal pattern is not fully understood, (Machemer, 1972) but there is evidence that metachronism may occur in part as the result of hydrodynamic coupling (Gueron et al., 1997; Gueron and Levit-Gurevich, 1998). Since the mathematical model accounts for the interactions between cilia in multiciliary configurations, we can assess the energetic efficiency of such interactions. Using the new energy equation with the original model we now answer the following questions:

1. How much work is done by the cilia against the viscous forces during a beat cycle? In particular, what portion of this energy is spent during the effective stroke ($E_{eff}$) and the recovery ($E_{rec}$) stroke?
2. How does energy expenditure depend on hydrodynamic interactions between neighboring cilia? Does it pay energetically for one cilium to synchronize with an adjacent cilium? Does metachronal coordination save energy in a multiciliary configuration, and, if so, to what extent?

These questions have been discussed in the literature only qualitatively since an appropriate modeling framework became available only recently. One indication that synchrony is energetically efficient comes from the early work of Taylor (1951) who showed that when a pair of two-dimensional sheets move in a sinusoidal pattern in a viscous fluid, minimal energy dissipation occurs when they are in synchrony. One should note, however, that this is not a dynamical model since the wave patterns were assumed to be fixed a priori and did not change as a result of hydrodynamical coupling.

Satir et al. (1997) have proposed a stochastic model relating the present understanding of the internal structure of cilia to the formation and propagation of bends along a cilium. They concluded from their model that only a small proportion (about 1%) of the dynein arms need to be active at any one moment in order to generate ciliary motion. From an evolutionary perspective it would be surprising if such an early developed structure would have so much potentially wasteful engines. Their model, however, does not take into account the hydrodynamic interactions of the cilium with the surrounding fluid or with neighboring cilia. Using the equations described in this paper we are able to obtain estimates of the percentage of dynein arms that must be active during any phase of the beat cycle, and conclude that activity of most of the dynein arms is required to generate the observed motions. Moreover, according to the model, ciliary interactions and metachronal coordination markedly increase the energetic efficiency, and in this sense appear to play a very important role in the functioning of large arrays of cilia.

**2. Computation of energy expenditure during a beat cycle.** The procedure used to compute the work done against the viscous resistance of the fluid in which the cilia are beating will be described briefly. Using the equations and notation introduced by Gueron and Liron (1992) we denote the drag force per unit length exerted by the cilium on the surrounding fluid by $\boldsymbol{\phi}(s,t)$, the fluid viscosity by $\mu$, and the cilium's velocity by the vector $\boldsymbol{V}(s,t)$. The variable $s$ denotes the centerline arc length and $t$ denotes time. The power exerted by the cilium at any moment in time is given by

$$H(t) = \int_0^L (\boldsymbol{\phi}(s,t) \cdot \boldsymbol{V}(s,t))\, ds. \tag{1}$$

The scalar product in equation (1) can be expressed by

$$\boldsymbol{\phi} \cdot \boldsymbol{V} \;=\; -C_T V_T^2 - C_N V_N^2 + C_T G_T V_T + C_N G_N V_N.$$

where the subscripts $T$ and $N$ denote the tangential and normal components of the vectors, respectively. The quantity $\boldsymbol{G}$ denotes the velocity of the external fluid flow and is a function of the drag forces per unit length induced on an element of the cilium by the external flow. The coefficients $C_T$ and $C_N$ are the tangential and normal GL drag coefficients (see Appendix for details). Finally, the energy spent (mechanical work done) by the cilium during the time interval $[t_1, t_2]$ can be written as

$$\begin{aligned} E(t_1,t_2) &= \int_{t_1}^{t_2} H(t)dt = \int_{t_1}^{t_2} \left\{ \int_0^L (\boldsymbol{\phi}(s,t) \cdot \boldsymbol{V}(s,t))ds \right\} dt \\ &= \int_{t_1}^{t_2} \left\{ \int_0^L \Big( -C_T V_T^2(s,t) - C_N V_N^2(s,t) + C_T G_T(s,t) V_T(s,t) \right. \\ &\qquad\qquad \left. + C_N G_N(s,t) V_N(s,t) \Big) ds \right\} dt, \end{aligned} \tag{2}$$

It is important to note that the expression for $E(t_1, t_2)$ depends only on the velocity of the cilium, though in a complicated manner, as $\boldsymbol{G}$ is an implicit function of the cilium's velocity. Since the positions of the cilium at different stages of the beat cycle can be used for approximating its velocity, it follows that $E(t_1, t_2)$ can be computed using only observed ciliary beat patterns. Such computation is independent of any model for the internal ciliary mechanism (engine hereafter) which generates the observed motion.

## 3. Results.

**3.1. Energy expenditure calculated from observed beats.** We used the positions of the cilium of *Paramecium* during a beat cycle (Sleigh, 1968) to approximate the momentary velocity at each location along the cilium. The reconstructed beat pattern is displayed in Figure 1 (panel c). These positions were used for computing the drag forces induced on the cilium. The computational technique is described in detail by Gueron & Levit-Gurevich (1998). Finally, the work done by the cilium against the viscous resistance during parts of the beat cycle, was computed by using Eq. (1). We computed that the energy expenditure is $\approx 9 \cdot 10^{-16} Joule$ during the effective stroke and $\approx 2 \cdot 10^{-16} Joule$ during the recovery stroke. This compares reasonably well with Sleigh's (1962) estimate for the energy expenditure of the cilia of *Paramecium* during he effective stroke. Sleigh estimated that if the cilium exerts a bending couple of $5 \times 10^{-17} N \cdot m$ in order to overcome the viscous resistance during the effective stroke, and sweeps an angle of $\approx 90°$ within $\approx 0.008 sec$ during this phase, then work is done at a rate of $\approx 1.25 \times 10^{-14} Joules/cilium/sec$. The overall work (per cilium) during the effective stroke is therefore approximated by $\approx 10^{-16} Joule$. This simplistic approximation is appropriate only for an *isolated* cilium during the effective stroke, and *does not* take into account the effects of the flat surface from which the cilia emerge. We also point out that the discussion of an isolated cilium is merely a theoretical one, since what we see in nature is already the overall result of the interaction of this cilium with many neighboring cilia.

Our model enables us to compute the energy expenditure at any desired time interval, whenever the ciliary positions are given. This can therefore provide additional information which, to the best of our knowledge, is not available from experiments. We computed the ratio of the work done during the effective and the recovery strokes, and obtained the ratio

$$\frac{\text{work during the recovery stroke}}{\text{work during the effective stroke}} \approx 0.22.$$

This approximately five-fold difference indicates the gain in energy expenditure that is achieved by having the recovery stroke occur more slowly than the effective stroke and close to the surface of the cell.

**3.2. Energy expenditure obtained with a model internal engine.** The computations performed in section 3.1 used data obtained from observations on a cilium during its beat cycle, and are independent of any model for the internal engine of the cilium. We now repeat these computations, while this time using the model equations presented in Gueron et al. (1997) and Gueron and Levit-Gurevich (1998). These model the effective and recovery strokes and the geometrically-dependent switches that initiate the transition from one phase to the other. Figure 2 (top line) shows

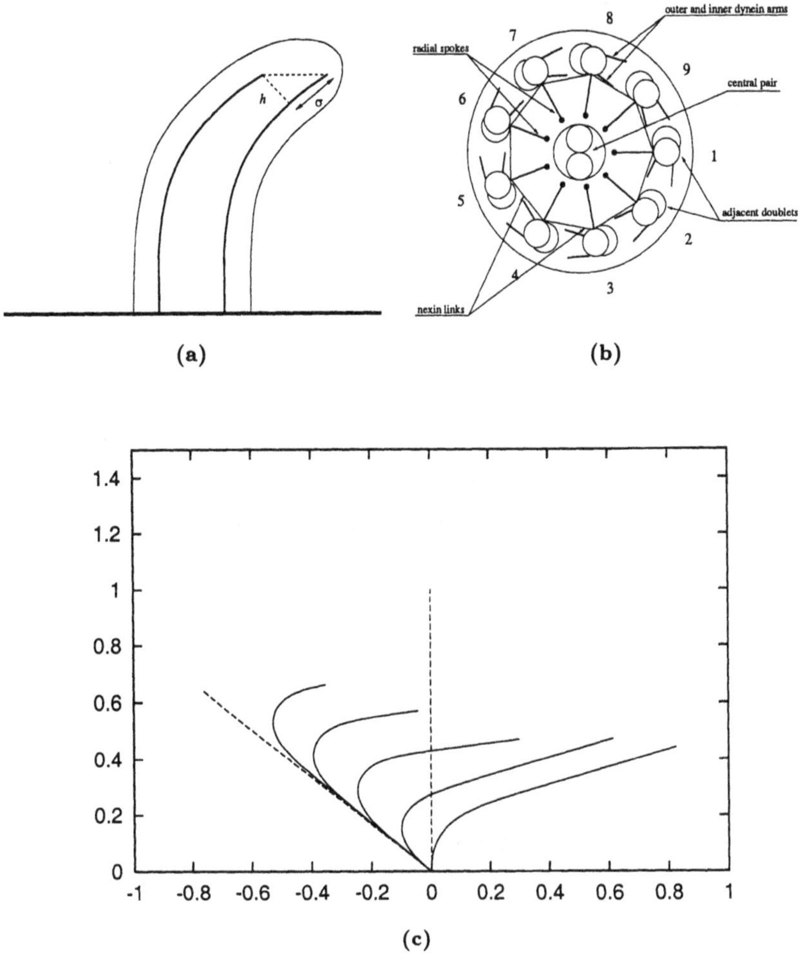


FIG. 1. *a). A schematic representation of how cilia bend by the sliding of its filaments. b). A schematic representation of a cross section of a cilium, illustrating the 9+2 structure. Inner and outer dynein arms attach to adjacent doublets. The* 9 *doublets are labeled. During the effective stroke, filaments 1–4 are actively sliding and filaments 5–9 are passive. During the recovery stroke, filaments 6–9 are active and filaments 1–5 are passive. c). A reconstructed diagram of the beat cycle of Paramecium. Data were digitized, smoothed and reconstructed from the diagrams of Sleigh (*1962*). Five positions during the recovery stroke are displayed. The axes are in units of ciliary length.*

the shape of the beat produced by this model engine at three different viscosities. The computation of energy expenditure by this model cilium in water agreed within a few percent with the energy expenditure computed by using the observed ciliary velocities. Thus this model engine is suitable for investigating the energy expenditure in multiciliary configurations and at different fluid viscosities.

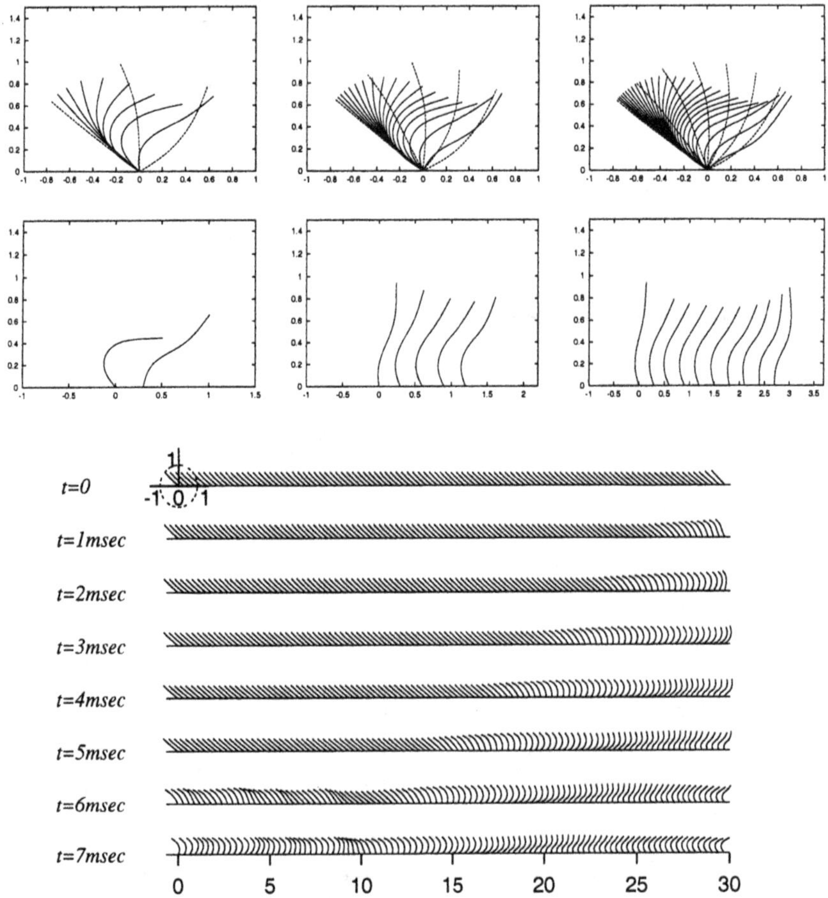


FIG. 2. *Ciliary beats in different cilia configurations. The top line displays three isolated cilia beating in different viscosities ($\mu_{water}$, $2\mu_{water}$ and $3\mu_{water}$, correspondingly, from left to right). The solid and dashed lines represent the recovery and effective stroke, respectively. The time interval between snapshots is $3ms$. The middle line displays one snapshot of the two, five, and ten cilia configurations. The bottom line displays eight snapshots of the hundred cilia configuration. Note that coordination similar to metachronism occurs in the multicilia configurations. The axes are in units of ciliary length. These figures are reproduced from Gueron and Levit-Gurevich (1998).*

**3.2.1. Effects of increased viscosity on energy expenditure.** Using our model engine, we found that the overall energy, as well as the ratio $E_{rec}/E_{eff}$, remain practically fixed in the viscosity range $\mu_{water} \leq \mu \leq 5\mu_{water}$ (see Figure 2 (panels a–c) for the beat patterns in different viscosities). This robust behavior in different environmental conditions suggests a possible important feature of the ciliary structure. It occurs here due to the particular changes in the beat pattern and frequency that result from the changed external conditions.

It is important to point out that this robust behavior is actually the result of the modeling paradigm that underlies our model engine. The forces that generate the motion are quite accurately reconstructed by our hydrodynamic equations. However, there are various ways to model their control, which affects the response of the engine to external load. Our modeling assumption is that the internal mechanism does not change its properties due to a changed load. Therefore, the load dependence is reflected only implicitly, through the geometric switches between the effective and the recovery strokes.

A different modeling approach would be to replace our configuration-dependent engine with a velocity-dependent one, such as Brokaw's (1985) model engine, which includes the property that the dyneins generate more force when the viscous drag is increased. Such an approach is likely to produce a different response to changes in the viscosity than the present configuration-dependent engine.

**3.2.2. The effects of cilia interactions.** The beat patterns of multicilia configurations having two, five, ten and one hundred cilia in a row are shown in Figure 2. The intercilia spacing is 0.3 ciliary length, i.e. $3.6\mu m$. Computation of the energy expenditure per cilium shows that as the number of cilia in the row increases there is an approximately three-fold decrease in the energetic cost of a beat cycle, with the maximum increase in efficiency being attained for a row of ten cilia (Figure 3).

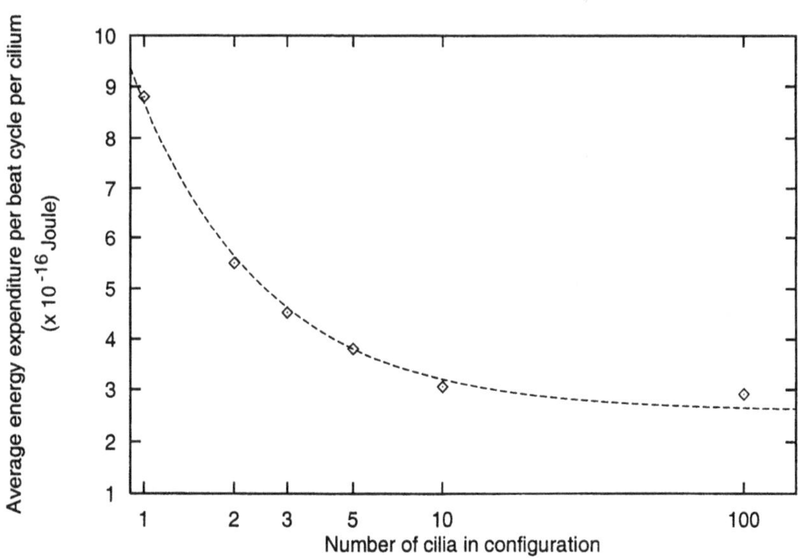


FIG. 3. *The average per-cilium energy expenditure during the beat cycle, as a function of the number of cilia in multicilia configurations. The interciliary spacing in all cases is* 0.3 *ciliary length. The horizontal axis is a logarithmic scale. The computations are done for the viscosity of water.*

In addition to ascertaining the overall increase in efficiency with increasing number of cilia in a row, it was of interest to determine the effect of row length on the energy expenditure during the effective stroke as compared to the recovery stroke. As shown in Figure 4, the energy used during the recovery stroke decreases by approximately a factor of two as the number of cilia in a row increases, and the position of the cilium within the row has a small or negligible effect on the energy expenditure during the recovery stroke. The energy expenditure during the effective stroke, however, decreases by a factor of four with increasing row length, and the cilia at the leading edge of the row (i.e., if the effective stroke is towards the left, then cilium number 1 is at the left end of the row) expend more energy during their effective stroke than do the cilia further inside the row. The ratio $E_{rec}/E_{eff}$ varies between 0.18 and 0.3, depending on the number of cilia in the row and their spacing. These computations indicate that the alignment of cilia along a row markedly increases the efficiency of beating.

As noted earlier (Gueron et al., 1997; Gueron and Levit-Gurevich, 1998) multicilia configurations reach their steady-state beat patterns within a few cycles from their initial resting state. When two closely spaced cilia start beating, they synchronize completely within two cycles, and when many cilia are aligned in a row a beat pattern resembling a metachronal wave rapidly evolves, as shown in Figure 2. Our results indicate that the energy expended per cilium decreases very rapidly during this transition from a resting state to an actively beating state, and, as shown in the next section, that this decrease in energy expenditure per cilium may be crucial for enabling motion.

**3.2.3. Does interciliary coupling increase the overall efficiency of the ciliary system?** It is interesting to investigate the energetic effect of the hydrodynamic coupling between the cilia, and in particular, to assess efficacy of metachronal coordination which the cilia interactions develop autonomously. It was already shown (GLG98) that the beats that emerge in multicilia configurations change in pattern and frequency as a function of the number of cilia in the row (e.g., from $29.5Hz$ for a single cilium in water to $42Hz$ for a ten cilia configuration). Further, our current computations show that the energy required per cilium per cycle decreases when the number of cilia is increased. However, this result alone does not reflect directly on the efficiency of the system, and a target quantity must first be defined. When the swimming of *Paramecium* is considered, for instance, the energy required for swimming a unit distance would be a most appropriate measure of efficiency. However, such computation is currently beyond the capability of the present model, and we settle for a more modest attempt. Recalling that the primary function of cilia is to generate fluid flow in a preferred direction, we compute the net fluid mass moving in the direction of the effective stroke through a test area near the beating ciliary array, per unit of spent energy. If this quantity increases with the number

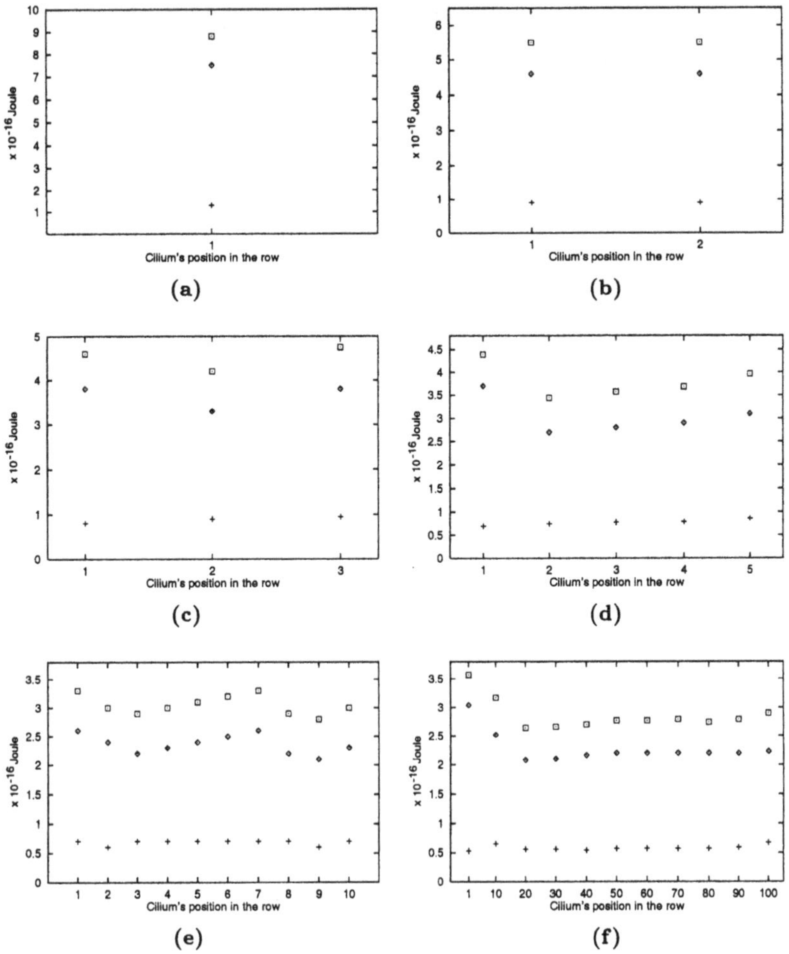


FIG. 4. *The average energy expenditure in multicilia configurations: one, two, three, five, ten, and hundred (panels a-f, respectively). The* $\diamond$ *symbols represent the effective stroke. The* $+$ *symbols represents the recovery stroke. The* $\square$ *symbols represents to whole beat cycle. The horizontal axis represents the position of the cilia in the row configuration. The vertical axis is the energy expenditure* ($\times 10^{-16}$*Joule*). *The interciliary spacing in all cases is* 0.3 *ciliary length. The computations are done for the viscosity of water.*

of cilia in the row configuration, it would give some indication that the interactions between the cilia are beneficial. To that end, we take advantage of the fact that our model allows us to compute the velocity of the fluid at any point in space.

We now define the test area that we use for this presentation. Suppose that the cilia are located along the line $y = 0$ in the $x-y$ plane, the effective stroke is in the positive direction of the $x$ axis, the anchor of the leftmost cilium is at the origin, and that of the rightmost cilium is at $(d \times (n_{cilia} -$

1), 0, 0), where $n_{cilia}$ is the number of cilia in the configuration and $d$ is the interciliary spacing. For the present study, we define our test area as the rectangle $R$ of dimensions $2d \times 1.5$ (in units of ciliary length), whose vertices have the coordinates $(D, 0, -d)$, $(D, 0, d)$, $(D, 1.5, d)$, $(D, 1.5, -d)$ where $D = d \times (n_{cilia} - 1) + 1$.

Figure 5 displays the result of the calculation of the fluid mass per energy spent for different ciliary row configurations. It demonstrates that the ciliary interactions that evolve contribute to an appreciable increase in the fluid propulsion per unit of energy spent, for rows of cilia with more than about 10 cilia per row.

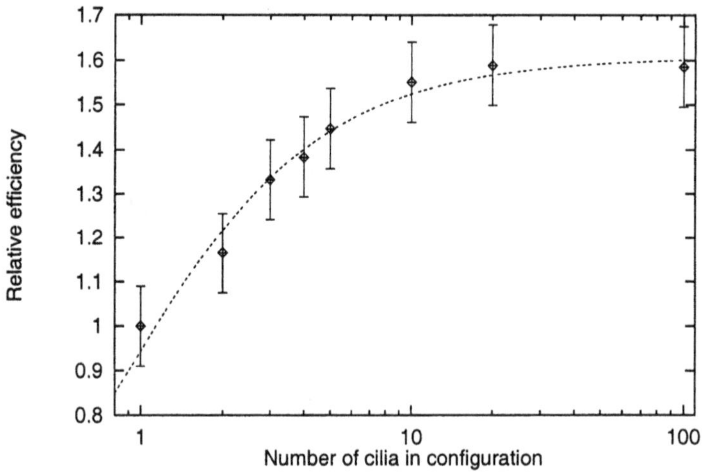


FIG. 5. *The relative efficiency of fluid propulsion, as a function of the number of the cilia in the row. Efficiency is defined here as the net fluid mass moved in the direction of the effective stroke through a test area near the ciliary array, per unit of spent energy (see explanation in text). The results are scaled with respect to those obtained for a single cilium. The horizontal axis displayed is a logarithmic scale.*

**4. Estimating the rate of dynein arm attachments.** The hydrolysis of ATP by the dynein arms located along the length of the cilium provides the energy for ciliary motility. The sequence of attachments along individual doublets generates filament sliding and hence bend formation. Although the precise details of bend formation and propagation are not fully understood, one can nevertheless use the present model to estimate the number of dynein arms required for motion generation.

Let $E_{ATP} = 6.0192 \cdot 10^4$ *Joule* be the amount of energy produced by the hydrolysis of one mole of ATP (Brokaw, 1989). The number of molecules in one mole is given by Avogadro's constant $N_{av} = 6.022 \cdot 10^{23}$. Thus, the hydrolysis of one ATP molecule releases $E_{ATP}/N_{av} \approx 10^{-19}$ *Joule/molecule*. We now consider the cilium of *Paramecium* whose typical length is $L = 12\mu m$. The distance between two adjacent dyneins is approximately $d = 24nm$ (Satir, 1994), and therefore the number of dynein pairs along one doublet is $L/d = 500$ (or $2L/d = 1000$ dyneins if we count inner and outer arms). During the effective stroke, filaments 1–4 are actively sliding and filaments 5–9 are passive (See Figure 1, panel b). During the recovery stroke, filaments 6–9 are active and filaments 1–5 are passive (Sleigh & Barlow, 1982; Satir, 1994). Therefore, only 2000 dynein pairs (i.e., 4000 dynein arms) can be reactivated during each phase. Satir (1994) estimated that the duty phase of the dyenin mechanochemical cycle is about 1–2 $msec$, whereas the whole cycle is about 33–34 $msec$. Since the beat duration of the cilia of *Paramecium* is $\approx 35msec$ (beat frequency is $\approx 28\ Hz$), it follows that each dynein arm can be reactivated at most once during the beat cycle.

The attachment of one dynein arm "costs" one ATP molecule (2 molecules for pair). If we denote the energy spent during some time interval by $A$, and the number of dynein arms attachments by $N_{att}$, we have the following relation:

$$(3) \qquad A = \frac{E}{E_{ATP}}\ moles, \qquad N_{att} = A \cdot N_{av} = \frac{E \cdot N_{av}}{E_{ATP}}$$

We can now use relation (3) together with the results concerning ciliary energy expenditure. Figure 6 displays the average amount of energy spent (panel b), the average amount of hydrolyzed moles of ATP (panel c), and the average number of dynein arms attachments (panel d), during the effective and the recovery strokes and for several multicilia configurations. One can observe that $\approx 85\%$ of the overall number of dynein attachments occur during the effective stroke.

For the isolated cilium, our computations show that roughly 7500 attachments are required during the effective stroke, which implies that each arm (of doublets $1 - -4$) attaches twice during the effective stroke that lasts $\approx 8ms$. The conclusion is that this theoretical isolated cilium cannot beat - at least not at the observed pattern and frequency - unless it has neighboring cilia. On the other hand, note that in the one hundred cilia configuration, which is energy-wise, more efficient, only $\approx$ 2000 arm attachments are required during the effective stroke. This number is feasible as it requires only one attachment per arm of doublets $1 - -4$ during this phase. Note also that in order to achieve $\approx$ 2000 arm attachments, all of the available dynein arms must participate.

We point out that our computations assume perfect mechanical efficiency, whereas the actual efficiency must be smaller. Indeed, Rikmenspoel (1964) speculated that the efficiency of the energy conversion is roughly 20%. Taking into account imperfect efficiency, this would require that there are either more dynein arm attachments per beat (which, as argued above, is physically impossible) or that cilia interactions would occur in closely packed two dimensional arrays decrease the required energy.

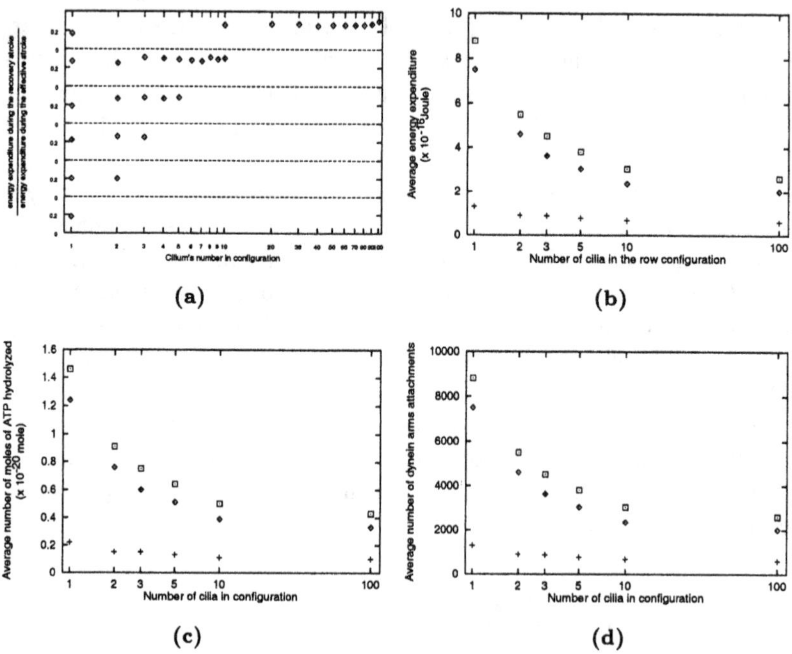


FIG. 6. *Results for the following multicilia configurations are considered: one, two, three, five, ten, and a hundred, where the interciliary spacing in all cases is* 0.3 *ciliary length. The horizontal axis represents the position of the cilia in the row configuration, and it is displayed on a logarithmic scale. Panel a: the ratio of the average energy expenditure during the recovery and the effective strokes, as function of cilium's position in the row configuration. Panels b–d: the average value of the energy that is spent (panel b), the average amount of hydrolyzed moles of ATP (panel c), and the average number of dynein arms attachments (panel d). The* ◇ *symbols represent the effective stroke. The* + *symbols represents the recovery stroke. The* □ *symbols represents to whole beat cycle. The horizontal axis represents the position of the cilia in the row The vertical axes are measured in* $\times 10^{-16}$*Joule (panel b),* $\times 10^{-20}$*mole (panel c), and nondimensional units (panel d). The number of cilia in each row is* 1, 2, 2, 5, 10 *and* 100 *going from the bottom of the panel to the top. The computations are done for the viscosity of water.*

**5. Discussion.** The results of our study can be summarized by:

1. We obtained the estimation $E_{rec}/E_{eff} \approx 0.2$ from observations of the beat pattern alone, independently of a model for the internal ciliary engine. This supports the qualitative understanding of the opposite roles of the effective and the recovery strokes.
2. The geometric switch ciliary engine model turned out to produce robust behavior under changed environmental conditions such as increased viscosity. For cilia, this may be considered as an important biological feature.
3. The results for the effects of cilia interactions indicate that:
   (a) The average energy expenditure per beat cycle is smaller in multicilia configurations and in densely packed configurations.

Thus, metachronism can be understood as a way for minimizing energy expenditure.

(b) The energy expenditure of each cilium in the row configuration depends on its position in the row. Those located at the edges spend more energy than the others. However, the ratio $E_{rec}/E_{eff}$ is roughly the same for all cilia.

(c) It is advantageous or even crucial for a cilium to have neighboring cilia beating metachronally. Cilia are not likely to be able to beat if they are isolated.
In this respect it is important to point out that the calculations in Section 3.1 were based on the model for internal engine presented in Gueron et al. (1997), and in Gueron and Levit-Gurevich (1998). This model was based on reconstructing the forces that are responsible for the ciliary motion, and the data was extracted from Sleigh's diagrams of a real cilium as it was observed in an experiment. This cilium, however, was not isolated: it was picked up from an unspecified multi-cilia configuration. Therefore, the observed beat was already the result of the interactions with the neighboring cilia. Thus, what we can actually conclude is that an isolated cilium cannot move at the beat frequency and pattern that we observe when this cilium has nearby beating neighboring cilia. Of course, this does not necessarily imply that an isolated cilium cannot move at all. In fact, individual flagella do, and some researchers are now able to study the motion of a very small number of cilia in experimental setups.

4. The estimate by Satir et al. (1997) that $\approx 1\%$ of the dynein arms are involved at any moment in the generation of ciliary motility presents an evolutionary question, since it would be surprising to evolve a structure with so much redundancy. The results of our hydrodynamic and energy expenditure computations instead suggest that it is more likely that most if not all of the dynein arms are active during each beat.

**Acknowledgments.** This research was supported by the U.S.-Israel Binational Science Foundation Grant number 97-400, and by a grant from the Israeli Ministry of Science.
K. Levit-Gurevich acknowledges the support of the Technion Graduate School, and the Selim and Rachel Benin Fund, New York UJA Federation.

## APPENDIX

The dynamic ciliary model with the internal engine that was actually used here has been introduced in Gueron et al. (1997). An updated and much more detailed description of this model's equations and the related numerical procedures can be found in Gueron and Levit-Gurevich (1998). We settle here with a very brief description.

**A. The hydrodynamic description.** The GL relation between the drag force and the velocity is

$$\text{(A.1)} \quad \phi_T(s,t) = -C_T V_T(s,t) + g_T(s,t), \quad \phi_N(s,t) = -C_N V_N(s,t) + g_N(s,t).$$

where $g_T = C_T G_T$, $g_N = C_N G_N$ and $\boldsymbol{G}$ denotes the velocity field induced at $s$ by "far segments" of the cilium, neighboring cilia or external flow. These are expressed in terms of appropriate singular solutions of the Stokes flow equations. The tangential ($C_T$) and normal ($C_N$) resistance coefficients are $C_T = 0.0025\frac{kg}{m\cdot sec}$ and $C_N = 0.0035\frac{kg}{m\cdot sec}$.

**B. Dynamic equations and the model for the internal engine.** The normal component of the forces induced within the cilium, $\boldsymbol{F}$, is modeled (in nondimensional form) by

$$\text{(B.1)} \qquad F_N = \frac{E_b}{S_0 L^2}\alpha_{ss} + S,$$

whereas the tangential component, $F_T$, is obtained from inextensibility condition

$$\text{(B.2)} \qquad \begin{aligned} F_{T_{ss}} &= (1 + C_{TN})F_{N_s}\alpha_s + C_{TN}\, F_T(\alpha_s)^2 \\ &\quad + F_N\alpha_{ss} - C_{TN}\, g_N\alpha_s + g_{T_s}. \end{aligned}$$

Here $\alpha(s,t)$ denotes the angle between the tangent to the cilium's centerline and a fixed horizontal axis. $S = S(s,t)$ is the active shear force due to the internal sliding filaments mechanism and the radial spokes system. $S_0$ is a typical fixed magnitude of the internal shear force (used for scaling), $L$ is the length of the cilium, and $E_b$ is its elastic bending resistance, $C_{TN} = C_T/C_N$, and $C_{NT} = C_N/C_T$, and the subscript $s$ denotes space derivative. Propagation in time is implemented by

$$\text{(B.3)} \qquad \begin{aligned} & F_{N_{ss}} + (1 + C_{NT})F_{T_s}\alpha_s + F_T\alpha_{ss} \\ &\quad = -(C_N\omega L^2/S_0)\alpha_t + C_{NT}\, F_N(\alpha_s)^2 + C_{NT}\, g_T\alpha_s + g_{N_s}, \end{aligned}$$

The active shear force (engine) generated within the cilium, namely $S(s,t)$, is modeled by:

$$\text{(B.4)} \qquad \begin{aligned} S_{eff/rec}(s,t) &= \pm 1 \left\{ \bar{C}_N\omega_{eff/rec}\frac{(s^2-1)}{2} \right. \\ &\quad \left. \times \left[1 + (A_1 + A_2(\alpha(0,t) - \tfrac{\pi}{2})^2)\right] + B_{eff/rec}\cdot\kappa(s,t) \right\}. \end{aligned}$$

Here, $\kappa$ is the curvature of a cilium ($\kappa(s,t) = \alpha_s(s,t)$), $\omega_{eff/rec}$ is a typical velocity during the effective and the recovery strokes, respectively, $\bar{C}_N = C_N L^2/S_0$ $A_1$, $A_2$ and $B_{eff/rec}$ are parameters used for scaling and have different values during the effective and the recovery strokes.

**C. Boundary and initial conditions.** The boundary conditions used in the model correspond to a cilium that is pinned at its basal end and free at its distal end:

$$\begin{aligned} \alpha_s(0,t) &= 0, \quad \alpha_{sss}(0,t) = -\frac{S_0 L^2}{E_b} S_s(0,t), \\ \alpha_s(1,t) &= 0, \quad \alpha_{ss}(1,t) = -\frac{S_0 L^2}{E_b} S(1,t), \\ F_{T_s}(0,t) = F_{N_s}(0,t) &= 0, \quad F_N(1,t) = F_T(1,t) = 0, \end{aligned} \tag{C.1}$$

The initial conditions correspond to an erect cilium:

$$\alpha(s,0) = \frac{\pi}{2}. \tag{C.2}$$

**D. The parameters' values.** The values of the parameters used in our model are: $S_0 = 10^{-12} Newtons$, $L = 12\mu m$, $E_b = 25 \cdot 10^{-24} Newtons\ m^2$, $C_N = 0.0035 \frac{kg}{m \cdot sec}$, $C_T = 0.0025 \frac{kg}{m \cdot sec}$.
The parameters of model engine are: $\omega_{eff} = 11000 \frac{\circ}{sec}$, $\omega_{rec} = 2290 \frac{\circ}{sec}$, $\alpha_{left} = 130°$ and $\alpha_{right} = 20°$. During the effective stroke $A_1 = 0.26$, $A_2 = -0.17$, and during the recovery stroke $A_1 = 2$ for $0 \leq s \leq 0.1L$ and $A_1 = 1$ for $0.1L \leq s \leq L$, $A_2 = -2$, and $B_{eff} = 0$, $B_{rec} = 2$.

## REFERENCES


[1] BAUM, G., PRIEL, Z., ROTH, Y., LIRON, N., AND OSTFED, E.G. (editors) (1997). *Cilia, Mucus, and Mucociliary Interactions.* (Marcel Dekker, New-York).

[2] BROKAW, C.J. (1985). Computer simulation of flagellar movement. VI. simple curvature-controlled models are incompletely specified. *Biophys. J.* **48**:633–642.

[3] BROKAW, C.J. AND JOHNSON, K.A. (1989). Dynein-induced microtubule sliding and force generation. *Cell Movement* **1**:191–198.

[4] GUERON, S. AND LEVIT-GUREVICH, K. (1998). Computation of the internal forces in cilia: Application to ciliary motion, the effects of viscosity, and cilia interactions. *Biophys. J.* **74**:1658–1676.

[5] GUERON, S., LEVIT-GUREVICH, K., LIRON, N., AND BLUM, J.J. (1997). Cilia internal mechanism and metachronal coordination as the result of hydrodynamical coupling. *Proc. Natl. Acad. Sci. USA.* **94**:6001–6006.

[6] GUERON, S. AND LIRON, N. (1992). Ciliary motion modeling, and dynamic multicilia interactions. *Biophys. J.* **63**:1045–1058.

[7] HAMASAKI, T., HOLWILL, M.E.J., BARKALOW, K. AND SATIR., P. (1995). Mechanochemical aspects of axonemal dynein activity studies by in vitro microtubule translocation. *Biophys. J.* **69**:2569–2579.

[8] HOLWILL, M.E.J., FOSTER, G.F., HAMASAKI, T., AND SATIR, P. (1995). Biophysical aspects and modelling of ciliary motility. *Cell Motil. Cytoskel.* **32**:114–120.

[9] MACHEMER, H. (1972). Ciliary activity and the origin of metachronism in Paramecium: effects of increased viscosity. *J. Exp. Biology.* **57**:239–259.

[10] RIKMENSPOEL, R. (1964). Measurement of motility and energy metabolism of bull spermatozoa. *Trans. NY Acad. Sci.* **26**:1072–1086.

[11] SATIR, P. (1994). Biochemical events in the production of ciliary movement. in *Biomechanics of active movement and division of cells*, ed. Akkas, N. (Springer, Heidelberg), pp. 465–470.

[12] SATIR, P., HAMASAKI, T. AND HOLWILL, M.E.J. (1997). Modeling outer dynein arm activity and its relation to the ciliary beat cycle. in *Cilia, Mucus, and Mucociliary Interaction*, eds. Baum, G., Priel, Z., Roth, Y., Liron, N., and Ostfed, E.G. (Marcel Dekker, New-York), pp. 13–19.

[13] SLEIGH, M.A. (1962). *The Biology of Cilia and Flagella.* (Pergamon Press, Oxford).

[14] SLEIGH, M.A. (1968). Patterns of ciliary beating. *Aspects of Cell Motility (22nd Symposium of the Society for Experimental Biology)*, ed. Miller, P.L., pp. 131–150.

[15] SLEIGH, M.A. AND BARLOW, D.I. (1982). How are different ciliary beat patterns produced? *Symposia of the Society for Experimental Biology.* **35**:139–157.

[16] TAYLOR, G.I. (1951). Analysis of the swimming of microscopic organisms. *Proc. R. Soc. Lond. A.* **209**:447–461.

# FLUID DYNAMICS OF ANIMAL APPENDAGES THAT CAPTURE MOLECULES: ARTHROPOD OLFACTORY ANTENNAE

M.A.R. KOEHL*

**1. Introduction.** Appendages bearing arrays of hair-like structures serve important biological functions in many animals from a variety of phyla. For example, feathery gills take up oxygen, hairy olfactory antennae capture odorant molecules, filamentous suspension-feeding structures catch single-celled algae, setulose appendages create ventilatory or feeding currents, and bristly legs or wings propel little animals through the surrounding water or air. To perform any of these functions, an array of hairs must interact with the fluid around it. Therefore, in order to elucidate basic rules governing how hairy appendages work, we have been studying the fluid dynamics of arrays of cylinders. The purpose of this paper is to provide a brief overview of what mathematical and physical models have taught us about fluid motion around and through arrays of hairs, and of how those insights have helped us unravel ways in which the function of hairy appendages depends on their structure and behavior. I will focus on examples of appendages that capture molecules: the olfactory antennae of various arthropods.

**2. Fluid flow near hairs in an array.**

*Overview.* The Reynolds numbers of the hairs on the types of appendages listed above are generally of order $10^{-5}$ to 10 (reviewed in Rubenstein and Koehl, 1977; LaBarbera, 1984; Cheer and Koehl, 1987a; Shimeta and Jumars, 1991; Loudon, et al., 1994). Reynolds number (Re= $UL/\nu$, where $U$ is velocity, $L$ is a linear dimension, in this case hair diameter, and $\nu$ is the kinematic viscosity of the fluid) represents the ratio of inertial to viscous forces for a particular flow situation. At the low and intermediate Re's at which these hairs operate, viscosity is very important in determining flow patterns, although inertia cannot be ignored for hairs operating at the upper end of this Re range. When a hair moves through a viscous fluid, the layer of fluid in contact with its surface does not slip relative to the surface and a velocity gradient develops in the fluid around the hair. The lower the Re, the thicker this layer of sheared fluid is relative to the dimensions of the hair. If the layers of fluid moving along with the hairs in an array are thick relative to the gaps between the hairs, then little fluid may leak through the array.

We have defined the "leakiness" of a gap between neighboring hairs in an array as the ratio of the volume of fluid that flows through the gap in a

*Department of Integrative Biology, University of California, Berkeley, CA 94720-3140, U.S.A.

unit of time to the volume of fluid that would have flowed (at freestream velocity) through a space of that width if there were no hairs present (Cheer and Koehl, 1987a). A hair-bearing appendage with a high leakiness functions like a sieve. In contrast, fluid moves around rather than through an array of hairs with low leakiness, hence an appendage bearing such an array is functionally a paddle even though it is full of holes.

If we consider appendages like olfactory antennae or gills that capture molecules on hair-like structures, the volume of fluid that can be filtered per time depends on the leakiness of the array of hairs on the appendage. Furthermore, the steepness of the velocity gradients near individual hairs affects the flux of molecules to the surfaces of the hairs (e.g. Adam and Delbruck, 1968; Berg and Purcell, 1977; Murray, 1977; Schmidt and Ache, 1979; Futrelle, 1984; Atema, 1985; Moore, et al., 1991; Koehl, 1996).

*Approaches to Studying Flow Through Hairy Appendages.* Fluid movement around and through arrays of hairs can be studied using various empirical and theoretical tools. Although technically challenging, we can measure the kinematics of hair-bearing appendages on real organisms as well as the fluid velocities around them (e.g. Koehl and Strickler, 1981; Koehl, 1993; 1995; Mead, et al., 1999). However, the details of velocity profiles around individual microscopic hairs on rapidly-moving appendages are difficult to resolve. Another limitation of working with living organisms is that the range of morphologies and behaviors available in nature limits our ability to quantify the consequences of particular aspects of structure or motion; when using real organisms, it is usually not possible to systematically vary only one parameter at a time while holding all others constant to quantify the effects of each.

In contrast, mathematical and physical models permit us to study combinations of morphology and behavior not found in nature, thereby enabling us to tease out the roles played by specific aspects of structure or motion and to quantify the effects on performance of changes in defined parameters. Mathematical and physical models also enable us to quantify aspects of the behavior of a system that are difficult to measure on living organisms, such as velocity gradients around individual microscopic hairs on a flapping appendage.

Repeatable velocity and force measurements with fine spatial and temporal resolution that cannot be made on uncooperative, delicate microscopic organisms are technically possible with large physical models if they are scaled properly. If a model and a real appendage are geometrically similar and operate at the same Re, they are dynamically similar (i.e. the ratios of velocities and forces at analogous points in the fluid around the model and the appendage are the same). Therefore, a large model moved at a conveniently slow speed in a Newtonian viscous syrup can be used to study the fluid dynamics of a microscopic appendage operating at the same Re in water or air.

### 3. Models of fluid motion between cylinders.

*Mathematical Models of Filters and Porous Screens.* Various methods have been used to calculate fluid motion through filters composed of arrays of fibers (reviewed in e.g. Fuchs, 1964; Davies, 1973; Laws and Livesey, 1978). For example, Tamada and Fujikawa (1957) modeled the two-dimensional flow at low Re through an infinite row of cylinders. All the fluid must flow through the gaps between hairs in an infinite array, therefore such models are not appropriate for predicting the flow between hairs on appendages of finite width because fluid can flow around as well as through such appendages. Mesh-like structures of finite width have been modeled as porous plates, and a variety of methods for calculating the flow through and around them have been used (e.g. Taylor and Batchelor, 1949; Spielman and Goren, 1968; and Koo and James, 1973). One approach that has been used to estimate the flow through biological filters is a modification of Tamada and Fujikawa's model (Silvester, 1983) .The predictions of Silvester's model agree reasonably well with water flow measured through the silk feeding nets spun by aquatic caddisfly larvae (Loudon, 1990), but underestimate flow through feathery moth antennae and through bristly filter-feeding fans of aquatic black fly larvae (Cheer and Koehl, 1987b). Flow through biological filters has also been estimated using Darcy's law (Cheer and Koehl, 1987b), which yields leakiness values for various insect filters that are orders of magnitude too low when compared with empirical data (Cheer and Koehl, 1987b). These comparisons of model results with biological data suggest that porous plate models are only appropriate for arrays of hairs that form screens, such as caddis fly nets or appendages with hairs bearing bristles that completely span the gaps between adjacent hairs. In contrast, the modeling approach of Cheer and Koehl (1987a; 1987b) yields much better matches with empirical measurements of flow through hairy appendages.

*Mathematical Model of Flow Between a Pair of Cylinders.* Cheer and Koehl (1987a) calculated fluid movement between and around a pair of circular cylinders at low Re using a two-dimensional model. Fluid velocities near the cylinders were calculated in bipolar coordinates using Stoke's low-Reynolds-number approximation of the Navier-Stokes equation, and fluid velocities far from the cylinders were calculated in polar coordinates using Oseen's low-Re approximation, which includes inertial effects. These flow fields were combined using a matched asymptotic expansion technique, and the velocity profiles predicted by this model were used to determine the leakiness of pairs of hairs operating at a variety of biologically-relevant Re's and spacings (described by the ratio of the width of the gap between neighboring hairs to the diameter of a hair ($G/D$)).

The calculations of Cheer and Koehl (1987a) reveal that a transition in the function of an array of hairs, between operating like a solid paddle and functioning like a leaky sieve, occurs as Re is changed between $10^{-2}$ and 1. The leakiness of a pair of hairs depends not only on Re, but also on the spacing ($G/D$) of the hairs. At very low Re (Re $\leq 10^{-3}$), little fluid moves through the gap between adjacent hairs (even when $G/D$ is high), thus appendages whose hairs operate at such low Re's should behave like paddles. In contrast, at Re near 1, little fluid is dragged along with cylinders as they move, and hairy appendages should be sieve-like unless $G/D$ is very low (Fig. 1). The more-closely-spaced the hairs (i.e. the lower the $G/D$), the higher the Re at which the transition to being leaky occurs. The model of Cheer and Koehl predicts that, at Re's of order $10^{-1}$ to 1, the leakiness of appendages with closely-spaced hairs is very sensitive to changes in velocity or changes in $G/D$ (Fig. 1).

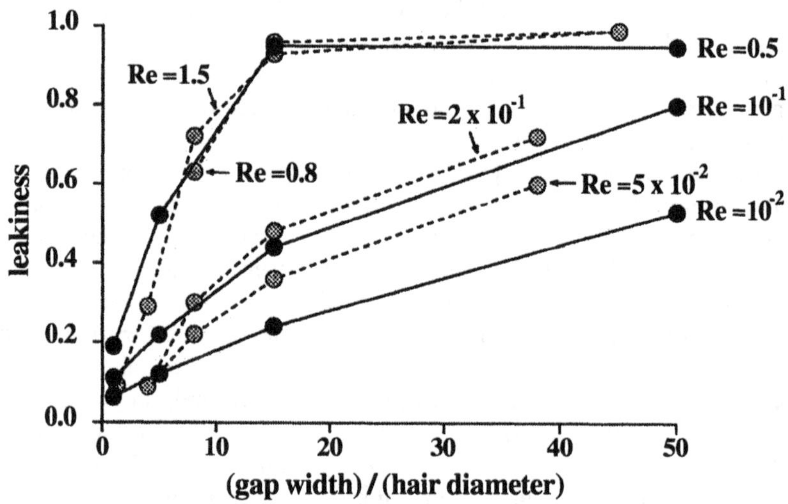


FIG. 1. *Leakiness (the ratio of the volume of fluid that flows in a unit of time through the gap between neighboring hairs to the volume of fluid that would have flowed at freestream velocity through a space of that width if there were no hairs present) plotted as a function of G/D, the ratio of the width of the gap between adjacent hairs to the diameter of a hair. Black circles connected by solid lines indicate leakiness values calculated using the model of Cheer and Koehl (1987a). Grey circles connected by dashed lines represent leakiness measured during towing experiments with comb-like physical models (Hansen and Tiselius, 1982). Reynolds number (Re) for each line was calculated using hair diameter for the linear dimension.*

The model of Cheer and Koehl can be used to estimate the flow through feathery appendages composed of hairs that bear bristles (Cheer and Koehl, 1987b). First the velocity profile between neighboring hairs is calculated; this velocity profile is then used to determine the freestream velocity encountered at defined positions along the length of a bristle, and the velocity profile between adjacent bristles is then calculated. Leakiness calculated using this approach provides a good match with measured values for a variety of feathery appendages of insects and crustaceans (Cheer and Koehl, 1987b; Koehl, 1993; 1995).

*Physical and Numerical Models of Flow Through Rows of Hairs.* The model described in Cheer and Koehl (1987a) only deals with two cylinders at Re $< 1$, whereas most hairy appendages bear more than two hairs and some animals operate their hairs at Re $> 1$. Hansen and Tiselius (1992) measured flow through physical models bearing a comb-like row of cylinders that were towed through mixtures of glycerol and water to yield a range of Re's. Their leakiness results are similar to those calculated by Cheer and Koehl (1987a) (Fig. 1). Abdullah and Cheer (unpublished; described in Koehl, 1993) used a numerical model to compute the steady, two-dimensional flow near cylinders at Re's between 0.5 and 4 to investigate the consequences of adding more cylinders to a row. Their model predicts that the addition of more hairs to a row reduces leakiness if Re $<$ 1, but increases leakiness if Re $> 1$; measurements of flow through physical models have shown the same phenomenon (reviewed in Koehl, 1995).

Leonard (1992) and Hansen and Tiselius (1992) measured flow through comb-like physical models at Re's up to 40, higher than those at which the mathematical model of Cheer and Koehl (1987a) is applicable. These physical models revealed that leakiness is very sensitive to hair spacing only when hairs are very close together at Re's of order 10, and that the critical $G/D$ below which gap width affects leakiness becomes lower as Re is increased (Koehl, 1995).

Hair-bearing appendages generally operate near solid boundaries (such as the body surface of an animal, the wall of a burrow, or the substratum in an animal's habitat). Loudon, et al. (1994) measured the leakiness of pairs of cylinders towed along or towards a wall. We found that, at Re's of $10^{-2}$ and lower, leakiness was increased by movement near a wall. In contrast, we found that leakiness of models towed near a wall at Re's of $10^{-1}$ and 1 matched the theoretical values for cylinders in an unbounded fluid. This suggests that a behavior that is effective at altering leakiness at very low Re (i.e. moving close to the body surface or another wall) does not change leakiness at intermediate Re.

The mathematical model of Cheer and Koehl (1987a; 1987b) is two-dimensional, thus it does not address the consequences of fluid flowing around the tips of hairs of finite length. Our experiments with physical models indicated that the leakiness of the gap between cylinders operating at Re's $\leq 10^{-2}$ is lower for cylinders of finite length than for those infinitely long because fluid is free to flow around the tips of the appendage as well as around the sides (Koehl, 1995). The physical models of Hansen and Tiselius (1992) showed that increasing hair length-to-diameter ratio can decrease comb leakiness at Re $\geq 10^{-2}$.

*Summary.* Both mathematical and physical models of rows of cylinders show that a transition in the leakiness of an array of hairs occurs at low or intermediate Re. Furthermore, mathematical and physical models of hair-bearing appendages of particular species of animals also show a transition between non-leaky paddle-like behavior and leaky sieve-like function if speed is increased in this Re range (Koehl, 1995; 1998). These various models predict that the finer the mesh (i.e. the lower the $G/D$) of the appendage, the higher the Re at which this transition between paddle-like and sieve-like function occurs. The models also predict that, at Re's of order 1, leakiness is very sensitive to hair spacing and to flow velocity if the $G/D$ of the hairs is small. Armed with the insights that models of flow between cylinders have provided about how morphology and behavior affect the leakiness of hair-bearing appendages, let us now consider the molecule-capturing performance of hairy olfactory antennae of different designs.

## 4. Olfactory antennae.

*Background.* Many animals use chemical cues in the water or air around them while performing ecologically-important activities such as locating food, selecting habitats, detecting predators or competitors, and communicating with conspecifics (e.g. reviewed in Cardé, 1984; Atema, 1985; 1995; Zimmer-Faust, 1989; Murlis, et al., 1992; Koehl, 1996). The arrival of odorant molecules from the environment to the surface of a chemosensory structure is a critical step in the process of olfaction. From a fluid dynamic standpoint, this process can be broken into two regimes based on scale: 1) large-scale turbulent flow of water or air transports the odorant from the source to the immediate vicinity of the sensor; and 2) small-scale viscous flow and molecular diffusion transport the odorant to the surface of the sensor (e.g. DeSimone 1981; Koehl 1996). The first step in filtering the spatial and temporal information contained in an odor plume (e.g. Murlis, 1986; Weissburg and Zimmer-Faust, 1994; Willis, et al., 1994; Atema, 1995) is therefore due to the physical manner in which an olfactory structure affects fluid flow across its surfaces (e.g. Moore, et al., 1989; Atema, 1995; Koehl, 1996).

The olfactory antennae of many organisms are appendages bearing arrays of microscopic hairs containing chemosensory neurons (e.g. Zacharuk, 1985; Halberg, et al., 1992); such antennae pick up chemical signals from the surrounding fluid, water or air. Water or air moves relative to an olfactory antenna as an animal locomotes, experiences ambient currents or wind, creates flow past itself, or flicks the antenna. The Re's of the olfactory hairs on a variety of arthropods range from $\leq 10^{-4}$ to 10 (Loudon, et al., 1994). In this Re range, the viscous flow near sensory hairs is laminar. Therefore, molecular diffusion is the mechanism that moves odorants across streamlines towards or away from the surface of a chemosensory hair.

Some researchers have argued that the diffusion of molecules to olfactory receptors determines the temporal pattern of the onset of neural responses to odorants (e.g. Getchell & Getchell 1977; DeSimone 1981; Nachbar & Morton 1981; Moore et al. 1989), whereas others have suggested that molecular diffusion does not limit the access of odor molecules to receptors (Futrelle 1984; Mankin and Mayer 1984; Boeckh, et al. 1965).

Calculations of odorant interception by a chemosensory hair modeled as an isolated, infinitely-long cylinder indicated that increasing the flow velocity encountered by the hair has a relatively small effect on the rate of chemical signal interception by the hair (Adam and Delbruck, 1968; Murray, 1977). In contrast, increasing the fluid velocity encountered by a chemosensory antenna made up of an array of cylinders (rather than a single cylinder) may have a large effect on chemical signal interception by a sensory hair in that array *if* the velocity increase occurs in the Re range in which the leakiness of that array of hairs is sensitive to velocity (Koehl, 1996).

In spite of the importance of small-scale fluid flow to molecule capture, information has been lacking about velocity profiles near chemosensory hairs on real olfactory appendages.

## 5. Leakiness of arrays of hairs on olfactory appendages.

*Olfactory Antennules of Crustaceans.* Many malacostracan crustaceans, such as lobsters, stomatopods ("mantis shrimp"), and crabs, capture odorant molecules with appendages called antennules (Fig. 2A, C, E). The lateral branch of such an antennule bears chemosensory hairs called "aesthetascs" (Heimann, 1984; Laverack, 1988) that serve an olfactory function (evidence reviewed by e.g. Atema, 1977; 1995; Gleeson, 1982; Grunert and Ache, 1988; Hallberg, et al. 1992; Atema and Voigt, 1995). Lateral branches of crustacean antennules are often used as model systems for studying the electrophysiology of olfactory neurons (reviewed by e.g. Ache, 1982; 1988; Atema, 1985; Marscall and Ache, 1989; Michel, et al., 1991; Atema and Voigt, 1995). Aesthetasc arrangement varies between species. For example, the aesthetascs on the antennules of a spiny lobster are arranged in complex, zig-zag rows surrounded by a cage of larger non-sensory guard hairs (Fig. 2B), the aesthetascs on the antennules of a stomatopod are arranged in simple, sparse rows (Fig. 2D), and the aesthetascs on the antennules of a crab are arranged in a densely-packed tuft resembling a toothbrush (Fig. 2F). We have used a combination of experiments and mathematical models to investigate the functional consequences of these different antennule designs.

A variety of crustaceans, such as lobsters, stomatopods, and crabs, flick their olfactory antennules through the surrounding water. It has been suggested that antennule flicking is a mechanism of increasing water penetration into an array of aesthetascs, thereby enhancing the access of odors to receptor cells (e.g. Snow, 1973; Schmidt and Ache, 1979; Atema, 1985; Gleeson, et al., 1993). Several lines of evidence are consistent with this idea. Pulses of flowing water (to mimic flicking) onto the aesthetascs of antennule preparations increased penetration of tracer molecules into the water between aesthetascs (Moore, et al., 1989; 1991) and enhanced the response of lobster olfactory receptor neurons to changes in odor concentration (Schmidt and Ache, 1979).

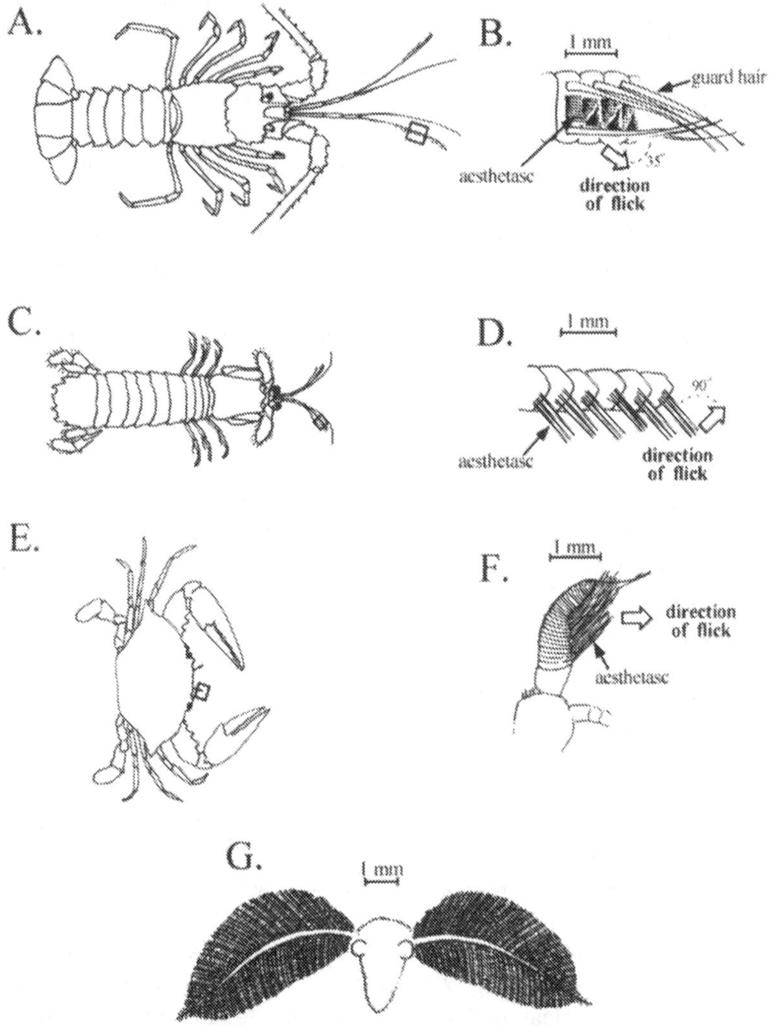


FIG. 2. *Examples of arthropods with hair-bearing olfactory appendages. The boxes around antennules in A, C, and E indicate the region of the antennule diagrammed in B, D, and F, respectively. A. Spiny lobster, Panulirus argus. B. Magnified view of a section of the lateral filament of a P. argus antennule. The lateral filament flicks downward, with the aesthetascs at an angle of ~35° to the direction of motion (Gleeson, et al., 1993). C. Stomatopod ("mantis shrimp" Squilla empusa. D. Magnified view of a section of the aesthetasc-bearing filament of a stomatopod antennule, Gonodactylus mutatus. The antennule flicks laterally, with the aesthetascs perpendicular to the direction of movement (Mead, et al., 1999). E. Blue crab, Callinectes sapidus. F. Magnified view of the tip of the antennule of a C. sapidus. The antennule can flick in many directions, but the aesthetascs point in the direction of motion during a flick (Martinez, Lee, and Koehl, unpublished data). G. Head of a male silkworm moth, Bombyx mori, showing the olfactory antennae. When the male fans his wings, air moves from front to back across the antennae (Loudon and Koehl, in press).*

**Lobster antennules.** We conducted a high-speed video analysis of the kinematics of flicking by the antennules of spiny lobsters, *Panulirus argus*, and a morphometric analysis of the hairs on the lateral branches of these antennules (Goldman and Koehl, submitted) (Fig. 2A, B). We found that during the rapid flick downstroke, the Re of the aesthetascs was 2, whereas during the slower upstroke, the Re was only 0.5. *P. argus* aesthetascs are closely packed ($G/D < 1$), hence we might expect a transition between leaky behavior during the downstroke and non-leaky behavior during the upstroke.

We tested this prediction by measuring the flow around a dynamically-scaled physical model of the lateral branch of a *P. argus* antennule (Koehl and Hu, unpublished data). The model was towed through a tank of viscous sugar syrup (technical details described in Loudon, et al., 1994) at the same orientation and Re used by *P. argus* antennules during the flick downstrokes and upstrokes. A sheet of laser light illuminated one plane normal to the long axis of the antennule model, and video records of the motions of neutrally-buoyant marker particles in the fluid in that plane were digitized. Examples of tracings of particle paths around the model during the rapid flick downstroke and the slower recovery stroke are shown in Figure 3. Fluid clearly moves through the array of aesthetascs during the leaky downstroke, but flows around the array during the non-leaky upstroke. This transition in leakiness of the aesthetasc array results in a flushing of old water out of the aesthetasc array during the downstroke, and a retention of a new water sample within the array during the upstroke. Therefore, because *P. argus* antennules operate at this critical range of aesthetasc Re's in which leakiness is sensitive to speed, they are able to take discrete water samples to smell with each flick, akin to sniffing.

We found that the Reynolds number (Re), and thus the flow regime, of *P. argus* aesthetascs during flicking was maintained as body size changed (Goldman and Koehl, unpublished data). Although we only examined animals representing a 33% change in body length, our results are consistent with those of Best (1995), who found only a slight change in aesthetasc Re over a 20-fold range of body lengths for the clawed lobster, *Homarus americanus.* Such maintenance during lobster growth of aesthetasc Re in the critical range in which leakiness is sensitive to flicking speed, suggests that the ability to take discrete water samples (i.e. to sniff) is maintained during the ontogeny of these animals. It also suggests that taking discrete samples in space and time may be important to processing information about the distribution of odorants in the environment.

**Stomatopod antennules.** Stomatopods ("mantis shrimp") have antennules bearing simple rows of aesthetascs (Fig. 2C, D). We conducted a high-speed video analysis of antennule flicking by stomatopods, *Gonodactylus metatus*, and a morphometric analysis of their aesthetascs; we examined animals representing a ten-fold range of body lengths, from small juveniles to adults (Mead, et al. 1999). We found that *G. mutatus* flick their anten-

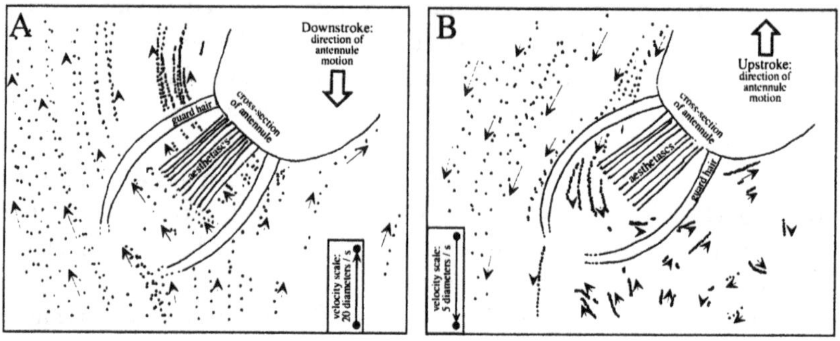


FIG. 3. *Diagrams of the fluid flow near a dynamically-scaled physical model of the lateral filament of the antennule of a spiny lobster, Panulirus argus, during the rapid flick downstroke (A) and the slower upstroke (B). The antennule filament is seen in cross-section, and the direction of antennule motion is indicated by the large arrow. One plane in the fluid was illuminated by a sheet of laser light. A video camera moving with the model recorded the motions relative to the model of neutrally-buoyant marker particles in the fluid (small arrows indicate the directions of fluid motion relative to the model). The dots on the diagrams indicate the positions of marker particles after successive time intervals; the closer the dots in a row within one diagram, the slower the flow at that position in the flow field. The velocity scale (shown in a box in each diagram) indicates the distance between particle positions in one time interval if the particle were moving 20 mm/s (A) or 5 mm/s (B). (Koehl and Hu, unpublished data).*

nules laterally, moving more rapidly during the outstroke than during the return stroke. Unlike lobsters, however, these stomatopods do not maintain the Re of their aesthetascs as they grow: the aesthetascs of small juveniles operate at Re's of order $10^{-1}$, whereas those of adults operate at Re's of order 1. However, the $G/D$ for aesthetascs in adjacent rows is much greater for juveniles than for adults.

We measured the flow around dynamically-scaled physical models of adult and juvenile *G. mutatus* antennules (Mead and Koehl, in press), using the techniques described above for lobster models. Our measurements of the leakiness of the aesthetasc arrays on these models when operating at a variety of different aesthetasc Re's showed that the adults and juveniles each operate at the Re's that maximize the ratio of leakiness during the outstroke to leakiness during the return stroke. The Re range that maximizes the ability of the antennule to take discrete water samples is lower for juveniles than for adults because juveniles have larger $G/D$'s than do adults. Thus, like lobsters, stomatopods appear to maintain the ability to sniff during their ontogeny.

**Crab Antennules.** Brachyuran crabs, such as the blue crab *Callinectes sapidus*, flick small antennules bearing dense arrays of hairs (Fig. 2E, F). Our morphometric analysis of *C. sapidus* antennules and our high-speed video study of their flicking revealed that the aesthetascs operate at Re's of order 1 (Martinez, Lee, and Koehl, unpublished data). Videomicrog-

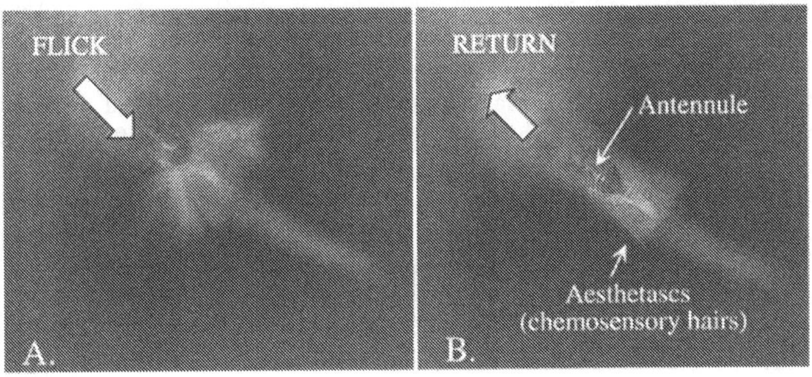


FIG. 4. *Frames of a video of an end-on view of the antennule of a blue crab, Callinectes sapidus, when exposed to flow that simulates the water motion relative to the antennule during the flick (A) and during the return stroke (B). The aesthetascs are passively splayed apart by the flow relative to the antennule during the flick (A), and are pushed together during the return stroke (B) (Koehl and Martinez, unpublished data).*

raphy of antennules exposed to water currents simulating the flow they experience during flicking revealed that the aesthetascs are flexible and are splayed apart during the flick downstroke, but are pushed together during the return stroke (Fig. 4). Because the aesthetascs operate at a Re range in which changes in the gap width between closely-spaced hairs can produce very large changes in leakiness (Cheer and Koehl, 1987; Koehl, 1995; 1996) (see Fig. 1), such aesthetasc splaying could lead to a substantial increase in flow between the hairs during the flick downstroke. We are currently using dynamically-scaled physical models to test this prediction.

*Moth Antennae.* Moths have feathery olfactory antennae (Fig. 2G) composed of chemosensory hairs (sensilla) borne on rows on larger branches. Previous models of molecule interception by a moth sensillum have assumed a single infinitely-long hair (Adam and Delbruck, 1968; Murray, 1977), although Murray (1977) pointed out that the flow near a sensillum surrounded by others would be slower than around an isolated cylinder. Indeed, measurements (Vogel, 1983) and mathematical modeling (Cheer and Koehl, 1987b) of the airflow through the feathery antennae of male *Acitas luna* moths indicate that the leakiness of these olfactory structures is only ~0.08 at air velocities they would encounter on a flying moth.

We have used a combination of experiments and mathematical modeling to study the airflow through the olfactory antennae of male silkworm moths, *Bombyx mori* (Loudon and Koehl, unpublished data). Most of the sensory hairs on the antennae of *B. mori* adult males are sensitive to a specific molecule, the sex pheromone emitted by conspecific females (e.g.

Gnatzy, et al. 1984; Steinbrecht 1992; Pophof, 1997). Silkworm moths do not fly, but males fan their wings when exposed to the female pheromone. We have used hot-film anemometry to measure the air velocities encountered by the antennae on fanning male moths, and have used these data to estimate (employing the approach of Cheer and Koehl, 1987b) the velocity profiles between the sensilla. We found that the difference between air penetration through an antenna on a moth fanning his wings versus on a moth walking without fanning is impressive: wing fanning produces air flow that is 15 times faster than that generated by walking at top speed, but the air velocities this produces through the gaps between sensilla is predicted to be ~560 times faster. This enormous increase in flow between sensilla occurs because the velocity at which moths fan is fast enough to produce a transition to leaky behavior for the array of closely-spaced hairs on their antennae.

*Summary.* The diverse examples of olfactory antennules and antennae described above appear to operate in a range of hair spacings ($G/D$) and Re's at which changes in velocity or in $G/D$ can have a big effect on leakiness. Therefore, by altering the fluid velocity encountered by an olfactory appendage (via flicking in lobsters, stomatopods, and crabs; by wing fanning in moths) or by changing the hair spacing (by passive splaying during the flick in crabs), the animals can alter the penetration of ambient fluid between their chemosensory hairs.

In order to assess how such changes in leakiness affect the molecule-capturing performance of olfactory antennae of various designs, we are using mathematical modeling to estimate the rates at which odorant molecules diffuse to the surfaces of chemosensory hairs under different flow conditions.

## 6. Molecule capture by chemosensory hairs on olfactory Antennae.

*Molecule Capture by Stomatopod Antennules.* In order to model the diffusion of odorant molecules to the surfaces of olfactory hairs, we must first determine the spatial distribution of odorant concentrations in the fluid arriving at the appendage bearing the sensory hairs. Although early analyses of animal navigation in odor plumes treated the plumes as evenly-diffusing clouds (e.g. Bossert and Wilson, 1968), more recent studies have revealed the heterogeneous distribution of odor concentrations in plumes in both water and air, and have suggested that animals might use this information in locating odor sources (e.g. Murlis and Jones, 1981; Kramer, 1986; 1992; Murlis, 1986; Murlis, et al., 1990; 1992; Mafra-Neto and Cardé, 1994; Weissburg and Zimmer-Faust, 1994; Willis, et al., 1994; Atema, 1995; Vickers and Baker, 1997; Baker, et al., 1998). Therefore, before calculating the transport of odorant molecules to stomatopod aesthetascs, we estimated realistic spatial distributions of odorant concentrations in the fluid encountered by stomatopod antennules. We used acoustic doppler velocimetry to measure water velocity profiles and turbulence in stomatopod microhabitats in the field (Koehl and Cooper, unpublished data). Similar velocity profiles and fluctuations were then produced in a large laboratory flume (at the Environmental Fluid Mechanics Laboratory, Stanford) and planar laser induced fluorescence was used to study the spatial and temporal distribution of concentrations in dye plumes in the flume (Koehl, Koseff, Crimaldi, Cooper, McCay, Wiley, and Moore, unpublished data.) Our measurements revealed fine filaments of high concentration in the water arriving at animal antennules (Fig. 5).

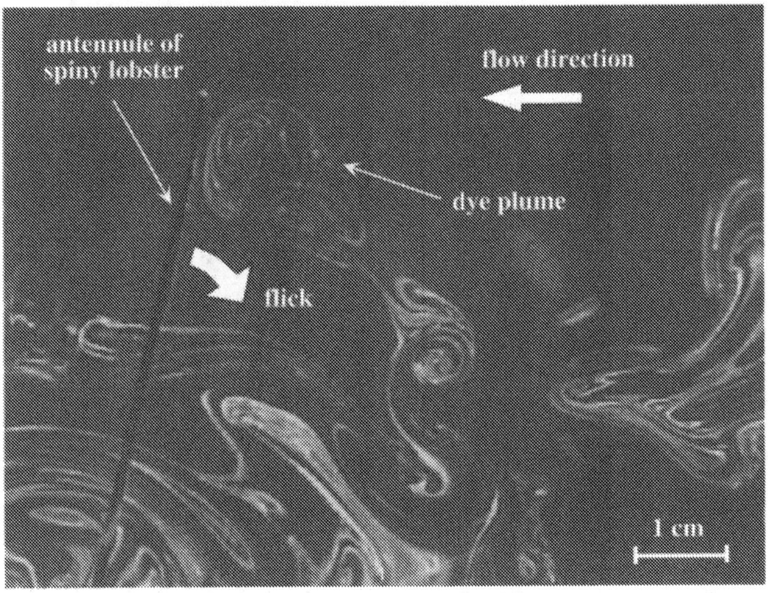


FIG. 5. *Frame of a video of a rhodamine dye plume in a flume (described in the text). A sheet of laser light illuminated the vertical plane in the flume in which the antennule of a spiny lobster was flicking (flick downstroke direction indicated by large arrow); that plane was parallel to the direction of water flow in the flume (indicated by large arrow). The antennule was ~1.5 m downstream from the "odor" source, a porous plate flush with the floor of the tank through which a solution of rhodamine "odor" was oozing. The lighter the water in the image, the higher the concentration of rhodamine dissolved in it. Note the fine filaments of high rhodamine concentration interspersed between thin layers of clean water. (Koehl, Koseff, Crimaldi, Cooper, McCay, Wiley, and Moore, unpublished data.)*

A two-dimensional advection-diffusion model was used to calculate the fate of odor filaments in water through which stomatopod antennules flick (Stacey, Mead, and Koehl, submitted). The velocity fields measured around dynamically-scaled physical models of stomatopod antennules (described above) were used to drive the advective component of the advection-diffusion equation. Odor filaments of various widths were carried in this flow past the arrays of aesthetascs on an antennule; the concentration distribution in a filament at the start of each calculation was based on concentration measurements obtained by digitization of plume images similar to that shown in Fig. 5. Odorant molecules in a filament underwent molecular diffusion (diffusion coefficient of $10^{-5}$ cm$^2$/s) as the filament was carried in the flow through the aesthetasc array, and molecules encountering the surface of an aesthetasc were removed from the water (Fig. 6).

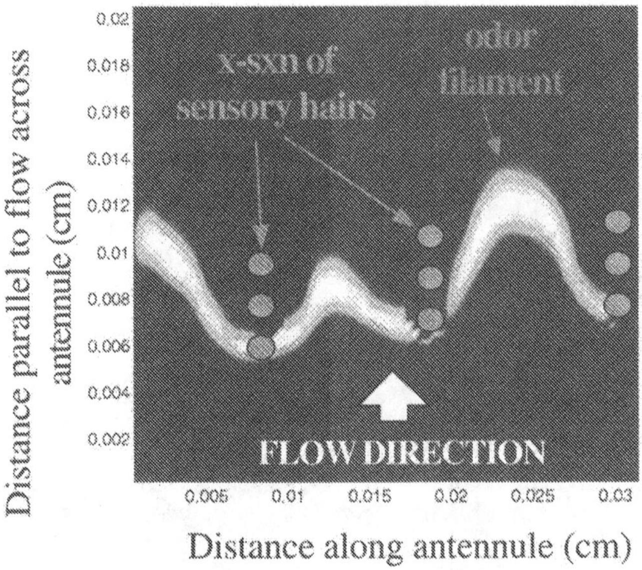


FIG. 6. *Diagram of one time step in the calculation of the molecular diffusion of an odor filament carried in water moving through the array of aesthetascs on the antennule of an adult stomatopod, Gonodactylus mutatus (Stacey, Mead, and Koehl, submitted). The long axis of the antennule is horizontal in the diagram, and the aesthetascs (which stick out of the page towards the reader) are shown in cross-section. The antennule is flicking downwards in the diagram, hence the flow direction relative to the aesthetascs is upwards in the diagram. The water velocity field around the aesthetascs was measured around a dynamically-scaled physical model (Mead and Koehl, in press). Odor filaments of various dimensions (digitized from images such as Fig. 5) were used in the calculations; a very thin filament is shown here. The lighter the water in the image, the higher the concentration of "odorant" dissolved in it. Any odorant molecule hitting the surface of an aesthetasc is removed from the water.*

Numerical simulations of such odor filament advection and diffusion enabled us to calculate the flux (molecules per time per area) of odorant molecules arriving at the surface of an aesthetasc as a function of time. One important aspect of the time course of odorant arrival at an aesthetasc is the peak flux attained. The importance of peak flux is indicated by experiments showing that chemosensory neurons in lobster antennules produce more spikes when the peak concentration of odorant pulses delivered to them is higher (Gomez and Atema, 1996). The onset slope of a plot of odorant flux to the aesthetasc as a function of time is also likely to be important; Moore's (1994) mathematical model of adaptation and deadaptation in olfactory receptor neurons suggests that the rate at which odorant concentration increases can determine which chemosensory neurons fire. We used our advection-diffusion model to calculate peak odorant flux and onset slope for the aesthetascs on the antennules of adult stomatopods and on juveniles. We compared these parameters for an antennule encountering an odor filament during the rapid outward flick with those for an antennule encountering such a filament during the slower return stroke. For adults, the peak flux during an outstroke is 12 times greater than the peak flux during a return stroke, and the onset slope is 26 times greater during the outstroke than during the return. For juveniles, whose aesthetasc rows are spaced with a greater $G/D$ than are those of adults, the differences between outstroke and return are even more striking: the peak flux is 26 times greater and the onset slope is 82 times greater during the rapid flick outstroke than during the slower return stroke.

Our advection-diffusion model of molecule transport to the surfaces of the aesthetascs of stomatopod antennules indicates that differences in water flow through aesthetasc arrays on these olfactory structures do indeed translate into differences in molecule capture by these antennules.

*Molecule Capture by Moth Antennae.* We have also used mathematical modeling to explore the effect on molecule capture by silkworm moth antennae of the increased antennal leakiness produced by wing fanning (Loudon and Koehl, in press). Because moths can sense pheromone molecules in very low concentrations in the air around them, we calculated the probabilities of encountering the chemosensory hairs (sensilla) for individual molecules undergoing random thermal motion in the air flowing through an antenna.

To calculate the probability of a pheromone molecule reaching the surface of a sensillum by diffusion once it has entered the gap between adjacent sensilla, we considered the one-dimensional random walk of odorant molecules perpendicular to the direction of air flow between the sensilla. The step length and frequency for the simulated random walks were determined, as explained in Berg (1993), from the diffusion coefficient ($2.5 \times 10^{-6}$ $m^2/s$ in air at $20^oC$) and molecular weight (238 g/mole) of bombykol, the primary molecular component of the sex pheromone of *B. mori* (Adam and Delbruck, 1968). We divided the gap between neighboring sensilla into 30 lanes and used our calculated velocity profiles between sensilla (described above) to determine the mean air velocity in each lane. Computer simulations were run of random walks by bombykol molecules entering the gap in each lane; molecules "walked" between lanes, so the velocity of the air carrying the molecule through the gap at each step depended upon the lane in which the molecule was located at the start of that step. The proportion of simulated walks (500 walks were run per lane) that resulted in contact with either sensillum bounding a gap was used to estimate the probability of capture for molecules that entered the gap in each lane.

Our calculations showed that wing fanning can increase the rate of pheromone capture by the sensilla on *B. mori* antennae by two orders of magnitude over that of antennae on moths standing or walking in still air without fanning. Although a smaller proportion of the pheromone molecules in a parcel of air traveling through a gap between sensilla have time to diffuse to either of the sensilla when the air velocity is high than when it is low, faster air carries more molecules per time through the gap. Thus, as we found for stomatopod antennules, differences in air flow through arrays of sensory hairs on moth antennae do produce differences in molecule capture rates by these olfactory structures.

**7. Conclusions.** Appendages bearing arrays of hairs are important not only because they are found in so many different taxa, but also because they serve a variety of essential biological functions, such as feeding, locomotion, and molecule capture. Mathematical and physical modeling, done in conjunction with kinematic and morphometric studies of hairy legs on real animals, have revealed a number of patterns in how the flow through arrays of hairs depends on their structure and behavior. Rows of hairs operating at Re $> 1$ function like leaky sieves (unless the hairs are very close together), whereas those operating at Re $\leq 10^{-2}$ function like paddles, with very little fluid moving between adjacent hairs. The Re range in which the transition in leakiness occurs depends on the geometry of the appendage: the wider the gap ($G$) between neighboring hairs relative to hair diameter ($D$), the lower the Re at which this transition between paddle-like and sieve-like function occurs. At Re's of order 1, the leakiness of arrays of closely-spaced hairs (i.e. arrays with low $G/D$) is very sensitive to hair spacing and to flow velocity. The chemosensory hairs on a variety of olfactory antennae operate in a range of Re's and hair spacings in which changes in velocity and changes in hair spacing can have profound effects on the flow near the hairs, and thus on their performance in capturing molecules from the surrounding fluid.

**Acknowledgements.** This research was supported by grants from the Office of Naval Research, U.S.A. (#N00014-96-1-0594 and #N00014-98-1-0775, to M. Koehl) and the Hasselblad Foundation (to M. Koehl and C. Loudon), by an Augmentation Award for Science and Engineering Research Training, Department of Defense, U.S.A. (#N00014-97-1-0726, to M. Koehl), and by a fellowship from the John D. and Catherine T. MacArthur Foundation (to M. Koehl). I am grateful to D. Abdullah, B. Best, A. Cheer, T. Cooper, J. Crimaldi, J. Goldman, W. Hu, J. Koseff, C. Loudon, M. Martinez, K. Mead, P. Moore, M. Stacey, and M. Wiley for collaborating on various aspects of the work cited here; and to L. Castillo, L. Gorsky, and U. Lee for help with digitizing.

## REFERENCES

ACHE, B.W. (1982). Chemoreception and thermoreception. pp. 369–393 In H.L. Atwood and D.C. Standeman, eds. *The Biology of the Crustacea*, Vol. **3**., New York, Academic Press.

ACHE, B.W. (1988). Integration of chemosensory information in aquatic invertebrates. In J. Atema and et. al., eds. *Sensory Biology of Aquatic Animals.* New York, Springer-Verlag.

ADAM, G. AND M. DELBRUK (1968). Reduction in dimensionality in biological diffusion processes. pp. 198–215 In *Structural Chemistry and Molecular Biology.* A. Rich and N. Davidson, eds., W.H. Freeman and Co., San Francisco.

ATEMA, J. (1977). Functional separation of smell and taste in fish and crustacea. pp. 165–174 In J. LeMagnen and L. MacLeod, eds. *Olfaction and Taste IV.* London, Information Retrieval.

ATEMA, J. (1985). Chemoreception in the sea: Adaptations of chemoreceptors and behavior to aquatic stimulus conditions. *Soc. Exp. Biol. Symp.* **39**:3887–3423.

ATEMA, J. (1995). Chemical signals in the marine environment: Dispersal, detection and temporal analysis. pp. 147–159 In *Chemical Ecology: The Chemistry of Biotic Interaction.* Academy Press, Washington, D.C.

ATEMA, J. AND R. VOIGT (1995). Behavior and Sensory Biology. pp. 313–348 In J.R. Factor, eds. *Biology of the Lobster Homarus americanus.* San Diego, CA, Academic Press.

BAKER, T.C., H.Y. FADAMIRO, AND A.A. COSSE (1998). Moth uses fine tuning for odour resolution. *Nature* **393**:530.

BERG, H.C. AND E.M. PURCELL (1977). Physics of Chemoreception. *Biophys. J.* **20**:193–217.

BOECKH, J., K.E. KAISSLING, AND D. SCHNEIDER (1965). Insect olfactory receptors. *Cold Spring Harbor Symp. Quant. Biol.* **30**:263–280.

BOSSERT, W.H. AND E.O. WILSON (1963). The analysis of olfactory communication among animals. *J. Theor. Biol.* **5**:443–469.

CARD, R.T. (1984). Chemo-orientation in flying insects. pp. 109–134. In W.J. Bell and R.T. Cardé, eds. *Chemical Ecology of Insects.* Elsevier Press, Amsterdam.

CHEER, A.Y.L. AND M.A.R. KOEHL (1987a) Paddles and rakes: Fluid flow through bristled appendages of small organisms. *J. Theor. Biol.* **129**:185–199.

CHEER, A.Y.L. AND M.A.R. KOEHL (1987b). Fluid flow through filtering appendages of insects. *IMA J. Math. Appl. Med. Biol.* **4**:185–199.

DAVIES, C.N. (1973). *Air Filtration.* Academic Press, New York.

DESIMONE, J.A. (1981). Physicochemical principles in taste and olfaction, pp. 213–229. In *Taste and Olfaction.* Academic Press, New York.

FUCHS, N.A. (1964). *The Mechanics of Aerosols.* Oxford University Press, Oxford.

FUTRELLE, R.P. (1984). How molecules get to their detectors: The physics of diffusion of insect pheromones. *Trans. Neorosci.* April, 116–120.

GETCHELL, T.V. AND M.L. GETCHELL (1977). Early events in vertebrate olfaction. *Chem. Senses* **2**:313–326.

GLEESON, R.A. (1982). Morphological and behavioral identification of the sensory structures mediating pheromone reception in the blue crab, *Callinectes sapidus. Biol. Bull.* **163**:162–171.

GLEESON, R.A., W.E.S. CARR AND H.G. TRAPIDO-ROSENTHAL (1993). Morphological characteristics facilitating stimulus access and removal in the olfactory organ of the spiny lobster, *Panulirus argus*: Insight from the design. *Chemical Senses.* **18**:67–75.

GNATZY, W., W. MOHREN, AND R.A. STEINBRECHT (1984). Pheromone receptors in Bombyx mori and *Antheraea pernyi.* II. Morphometric analysis. *Cell Tiss. Res.* **235**:35–42.

GOLDMAN, J.A. AND M.A.R. KOEHL. Fluid dynamic design of lobster olfactory organs: High-speed kinematic analysis of antennule flicking in *Panulirus argus: Chemical Senses* (submitted).

GOMEZ, G. AND J. ATEMA (1996). Temporal resolution in olfaction: Stimulus integration time of lobster chemoreceptor cells. *J. Exp. Biol.* **199**:1771–1779.

GRUNERT, U. AND B.W. ACHE (1988). Ultrastructure of the aesthetasc (olfactory) sensilla of the spiny lobster *Panulirus argus*. *Cell Tissue Res.* **251**:95–103.

HALLBERG, E., K.U.I. JOHANSSON AND R. ELOFSSON (1992). The aesthetasc concept: Structural variations of putative olfactory receptor cell complexes in crustaceans. *Microsc. Res. Techn.* **22**:336–350.

HANSEN, B. AND P. TISELIUS (1992). Flow through the feeding structures of suspension feeding zooplankton: A physical model approach. *J. Plankton Res.* **14**:821–834.

HEIMANN, P. (1984). Fine structure and molting of the aesthetasc sense organs on the antennules of the isopod, *Asellus aquaticus* (Crustacea). Cell Tissue Res. **235**:117–128.

KOEHL, M.A.R. (1993). Hairy little legs: Feeding, smelling, and swimming at low Reynolds numbers. *Contemp. Math.* **141**:33–64.

KOEHL, M.A.R. (1995). Fluid flow through hair-bearing appendages: Feeding, smelling, and swimming at low and intermediate Reynolds number. In C.P. Ellington, T.J. Pedley, eds., *Biological Fluid Dynamics, Soc. Exp. Biol. Symp.* **49**:157–182.

KOEHL, M.A.R. (1996). Small-Scale fluid dynamics of olfactory antennae. *Mar. Fresh. Behav. Physiol.* **27**:127–141.

KOEHL, M.A.R. (1998). Small-Scale hydrodynamics of feeding appendages of marine animals. *Oceanogr.* 11:10–12.

KOEHL, M.A.R. AND J.R. STRICKLER (1981). Copepod feeding currents: Food capture at low Reynolds number. *Limnol. Oceanogr.* **26**:1062–1073.

KOO, J.-K. AND D.F. JAMES (1973). Fluid flow around and through a screen. *J. Fluid Mech.* **60**:513–538.

KRAMER, E. (1986). Turbulent diffusion and pheromone-triggered anemotaxis, pp. 59–67. In T.L. Payne, M.C. Birch and C.E.J. Kennedy, eds. *Mechanisms in Insect Olfaction.* Oxford: Clarendon Press.

KRAMER, E. (1992). Attractivity of pheromone surpassed by time-patterned application of two nonpheromone compounds. *J. Insect Behav.* **5**:83–97.

LABARBERA, M. (1984). Feeding currents and particle capture mechanisms in suspension feeding animals. *Am. Zool.* **24**:71–84.

LAVERACK, M. S. (1988). The diversity of chemoreceptors, pp. 287–317 In Atema, J. [ed.], *Sensory Biology of Aquatic Animals.* Springer-Verlag, New York.

LAWS, E.M. AND J.L. LIVESEY (1978). Flow through screens. *Ann. Rev. Fluid Mech.* **10**:247–266.

LEONARD, A.B.P. (1992). *The Biomechanics, Autecology and Behavior of Suspension-Feeding in Crinoid Echinoderms.* Ph.D. Dissertation, University of California, San Diego.

LOUDON, C. (1990). Empirical test of filtering theory: Particle capture by rectangular-mesh nets. *Limnol. Oceanogr.* **35**:143–148.

LOUDON, C., B.A. BEST, AND M.A.R. KOEHL (1994). When does motion relative to neighboring surfaces alter the flow through an array of hairs? *J. Exp. Biol.* **193**:233–254.

LOUDON, C. AND M.A.R. KOEHL. Sniffing by a silkworm moth: Wing fanning enhances air penetration through and pheromone interception by antennae. *J. Exp. Biol.* (in press).

MAFRA-NETO, A. AND R.T. CARD (1994). Fine-scale structure of pheromone plumes modulates upwind orientation of flying moths. *Nature* **369**:142–144.

MANKIN, R.W. AND M.S. MAYER (1984). The insect antenna is not a molecular sieve. *Experientia* **40**:1251–1252.

MARSCALL, H.-P. AND B.W. ACHE (1989). Response dynamics of lobster olfactory neurons during simulated natural sampling. *Chem. Senses* **14**:725.

MEAD, K.S., M.A.R. KOEHL, AND M.J. O'DONNELL. (1999). Stomatopod sniffing: The scaling of chemosensory sensillae and flicking behavior with body size. *J. Exp. Mar. Biol. Ecol.* **241**:235–261.

MEAD, K.S. AND M.A.R. KOEHL. Stomatopod antennule design: The asymmetry, sampling efficiency, and ontogeny of olfactory flicking. *J. Exp. Biol.* (in press).

MICHEL, W.C., T.S. MCKLINTOCK, AND B.W. ACHE (1991). Inhibition of lobster olfactory receptor cells by an odor-activated potassium conductance. *J. Neurophysiology* **65**:446–453.

MOORE, P.A. (1994). A model of the role of adaptation and disadaptation in olfactory receptor neurons: implications for the coding of temporal and intensity patterns in odor signals. *Chemical Senses.* **19**:71–86.

MOORE, P.A., J. ATEMA AND G.A. GERHARDT (1991). Fluid dynamics and microscale chemical movement in the chemosensory appendages of the lobster, *Homarus americanus. Chemical Senses* **16**:663–674.

MOORE, P.A., G.A. GERHARDT, AND J. ATEMA (1989). High resolution spatio-temporal analysis of aquatic chemical signals using microchemical electrodes. *Chemical Senses* **14**:829–840.

MURLIS, J. (1986). The structure of odour plumes, pp. 27–38. In T.L. Payne, ed. *Mechanisms of Insect Olfaction*, Clarendon Press, NJ.

MURLIS, J., J.S. ELKINTON, AND R.T. CARD (1992). Odor plumes and how insects use them. *Ann. Rev. Entomol.* **37**:505–532.

MURLIS, J. AND C.D. JONES (1981). Fine-scale structure of odour plumes in relation to insect orientation to distant pheromone and other attractant sources. *Physiol. Entomol.* **6**:71–86.

MURLIS, J., M.A. WILLIS AND R.T. CARDÉ (1990). Odour signals: patterns in time and space, pp. 6–17 In K. Doving, ed. *Proceedings of the X International Symposium on Olfaction and Taste, Oslo.*

MURRAY, J.D. (1977). Reduction of dimensionability in diffusion processes: Antenna receptors of moths, pp. 83–127 In *Lectures on Nonlinear-Differential-Equation Models in Biology*, Oxford University Press, Oxford.

NACHBAR, R.B. AND T.H. MORTON (1981). A gas chromatograph (GPLC) model for the sense of smell: Variations of olfactory sensitivity with conditions of stimulation. *J. Theor. Biol.* **84**:387–407.

POPHOF, B. (1997). Olfactory responses recorded from sensilla coeloconica of the silkworm *Bombyx mori. Physiol. Entomol.* **22**:239–248.

RUBENSTEIN, D.I. AND M.A.R. KOEHL (1977). The mechanisms of filter feeding: Some theoretical considerations. *Amer. Natur.* **26**:981–994.

SCHMIDT, B.C. AND B.W. ACHE (1979). Olfaction: Responses of a decapod crustacean are enhanced by flicking. *Science* **205**:204–206.

SHIMETA, J. AND P.A. JUMARS, (1991). Physical mechanisms and rates of particle capture by suspension-feeders. *Oceanogr. Mar. Biol. Annu. Rev.* **29**:191–257.

SILVESTER, N.R. (1983). Some hydrodynamic aspects of filter feeding with rectangular-mesh nets. *J. Theor. Biol.* **103**:265–286.

SNOW, P.J. (1973). The antennular activities of the hermit crab, *Pagarus alaskensis* (Benedict). *J. Exp. Biol.* **58**:745–765.

SPEILMAN, L.A. AND S.L. GOREN (1968). Model for predicting pressure drop and filtration efficiency in fibrous media. *Environ. Sci. Technol.* **2**:279–287.

STACEY, M., K.S. MEAD, AND M.A.R. KOEHL. Molecule capture by olfactory antennules: Mantis shrimp. *J. Math. Biol.* (submitted).

STEINBRECHT, R.A. (1992). Experimental morphology of insect olfaction: tracer studies, x-ray microanalysis, autoradiography, and immunochemistry with silkmoth antennae. *Microscopy Res. and Tech.* **22**:336–350.

TAMADA, K. AND H. FUJIKAWA (1957). The steady two-dimensional flow of viscous fluid at low Reynolds numbers passing through an infinite row of equal parallel circular cylinders. *Quart. J. Mech. Appl. Math.* **10**:425–432.

TAYLOR, G.I. AND G.K. BATCHELOR (1949). The effects of wire gauze on small disturbances in uniform stream. *Quart. J. Mech. Appl. Math.* **2**:1–29.

VICKERS, N.J. AND T.C. BAKER (1997). Flight of *Heliothis virescens* males in the field in response to sex pheromone. *Physiol. Entomol.* **22**:277–285.

VOGEL, S. (1983). How much air passes through a silkmoth's antenna? *J. Insect Physiol.* **29**:597–602.

WEISSBURG, M.J. AND R.K. ZIMMER-FAUST (1994). Odor plumes and how blue crabs use them in finding prey. *J. Exp. Biol.* **197**:349–375.

WILLIS, M.A., C.T. DAVID, J. MURLIS AND R.T. CARDÉ (1994). Effects of pheromone plume structure and visual stimuli on the pheromone-modulated upwind flight of male gypsy moths (Lumantria dispar) in a forest (Lepidoptera, Lymantriidae). *J. Insect Behav.* **7**:385–409.

ZACHARUK, R.Y. (1985). Antennae and sensilla, pp. 1–69. In G.A. Kerkut and L.I. Gilbert, eds. *Comprehensive Insect Physiology, Biochemistry and Pharmacology,* New York, Pergamon Press.

ZIMMER-FAUST, R.K. (1989). The relationship between chemoreception and foraging behavior in crustaceans. *Limnol. Oceanogr.* **34**:1367–1374.

# CARTESIAN GRID METHODS FOR FLUID FLOW IN COMPLEX GEOMETRIES*

RANDALL J. LEVEQUE† AND DONNA CALHOUN‡

**Abstract.** Biological fluid dynamics typically involves geometrically complicated structures which are often deforming in time. We give a brief overview of some approaches based on using fixed Cartesian grids instead of attempting to use a grid which conforms to the boundary. Both finite-difference and finite-volume methods are discussed, as well as a combined approach which has recently been used for computing incompressible flow using the streamfunction-vorticity formulation of the incompressible Navier-Stokes equations.

**Key words.** Cartesian grids, fluid dynamics, advection-diffusion, immersed boundary method, immersed interface method.

**1. Introduction.** Biological fluid dynamics typically involves geometrically complicated structures which are often deforming in time. Solving fluid dynamics equations in such geometries can be extremely challenging, and methods based on body-fitted grids which conform to the geometry can be difficult to implement. We give a brief overview of some approaches based on using fixed Cartesian grids instead of forcing the grid to conform to the boundary. Peskin's *immersed boundary method* outlined in Section 4 is the best-known example of this class in the field of biological fluid dynamics and has been successfully used by many researchers. The *immersed interface method* discussed in Section 5 is a related approach which can often achieve better accuracy near the boundary. These are both finite-difference methods.

For advection-diffusion equations, finite-volume methods may be better suited, as described in Section 6. These allow exact conservation on grid cells to be maintained and flux boundary conditions to be easily imposed. High-resolution "shock-capturing" methods developed for conservation laws are also quite useful for problems with discontinuities or steep gradients in the solution.

In Section 7 we describe a hybrid approach for computing incompressible flow using the streamfunction-vorticity formulation of the Navier-Stokes equations. Vorticity is advected and diffused on a finite-volume grid, with flux at solid wall boundaries as required to satisfy the no-slip boundary conditions. The streamfunction is computed from the vorticity

*The work of the both authors was supported in part by DOE grant DE-FG03-96ER25292 and NSF grants DMS-9505021 and DMS-9626645.

†Department of Applied Mathematics, University of Washington, Box 352420, Seattle, WA 98195-2420.

‡Courant Institute of Mathematical Sciences, New York University, 251 Mercer St., New York, NY 10012.

on a finite-difference grid and then differenced to obtain velocities needed in the finite-volume method.

**2. Finite-difference and finite-volume methods.** We will discuss both finite-difference and finite-volume approaches, and first review the essential ideas of each approach.

A finite-difference discretization of a differential equation is based on replacing derivatives by differences of values defined on a discrete set of *grid points*. The approximate solution generated consists of values $U_{ij}^n$ (in two dimensions) representing pointwise approximations to the solution at grid points $(x_i, y_j)$ at time $t_n$,

$$U_{ij}^n \approx u(x_i, y_j, t_n). \tag{2.1}$$

A finite-volume method, on the other hand, is based on approximating cell averages of the solution over grid cells,

$$U_{ij}^n \approx \frac{1}{\Delta x \Delta y} \int_{y_{j-1/2}}^{y_{j+1/2}} \int_{x_{i-1/2}}^{x_{i+1/2}} u(x, y, t_n)\, dx\, dy. \tag{2.2}$$

For differential equations which arise from conservation laws and fluxes (including advection-diffusion equations and most equations in fluid dynamics), this is a very natural approach. The differential equation often arises from a more fundamental integral form stating that the integral of $u$ over any finite volume changes with time only due to the fluxes across the boundary of the volume. If the volume is a single grid cell, then this leads to an updating formula for the cell average based on numerical flux formulas. This has some advantages over more familiar finite-difference formulations. For problems where some quantity should be conserved it is often easy to guarantee exact conservation numerically, This is particularly important in nonlinear wave-propagation problems with shock waves, where the conservation form must be respected in order to obtain correct solutions (see [43], for example). It can also be important in other problems, such as when we wish to exactly conserve the total mass of some chemical advecting in a flow. Another advantage of finite-volume methods is that physical boundary conditions are often given in terms of fluxes at the boundary. These are often more naturally discretized in the context of finite-volume methods than with finite differences.

In some contexts, however, finite-difference methods are more natural and may be easier to use. We have used a mixture of the two approaches in developing Cartesian grid methods, using each where it is most appropriate. (Another standard approach for partial differential equations is the finite element method, based on variational principles. We don't discuss this class of methods here.)

**3. Cartesian grids.** Figure 1 shows a comparison of body-fitted grids with Cartesian grids, in both the finite-difference and finite-volume frameworks. The top row shows body-fitted grids near a curved wall. A finite-difference method might update the values at one grid point based on values at the 4 nearest neighbors, as indicated by the 5-point stencil shown. A finite-volume method, on the other hand, would update the cell average of $u(x, y, t)$ over the shaded grid cell based on fluxes through the four edges indicated. If the fluxes are based on values in this cell and the four neighboring cells, then again a 5-point stencil is obtained.

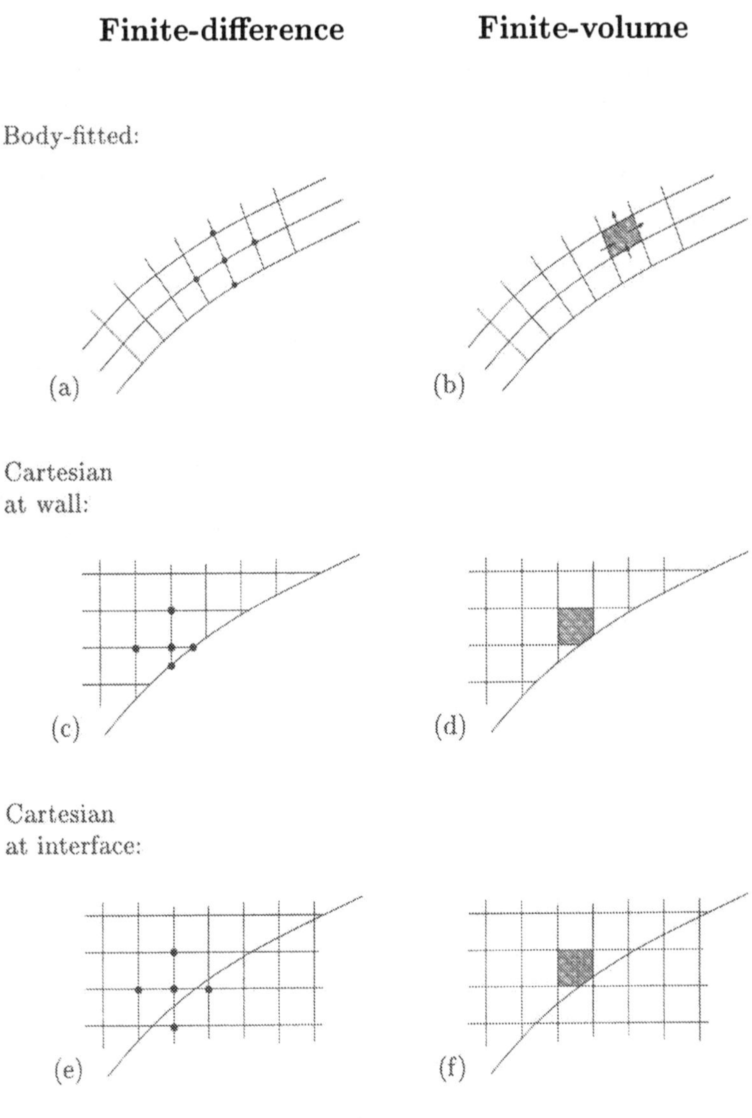


FIG. 1. *Comparison of different types of grids for boundaries or interfaces.*

The middle row of Figure 1 shows a Cartesian grid at the same boundary. With a finite-difference method one might develop a modified 5-point method based on the unequal distances from the grid point shown to the points which lie on the wall. With the finite-volume approach one instead has an irregular-shaped grid cell which requires computing a flux at the edge corresponding to the physical boundary as well as at the sides adjacent to other cells. Note that the finite-volume method requires a *flux* at the wall (Neumann boundary conditions) whereas the finite-difference method requires a pointwise value at the wall (Dirichlet boundary condition), although with additional work either boundary condition can be imposed on either grid.

The bottom row of Figure 1 shows a different situation in which there is an interface immersed in the domain rather than a wall. This interface might be a membrane immersed in the fluid or an interface between two different fluids, for example. In this case we typically have partial differential equations which must hold on each side of the interface, though perhaps with discontinuous coefficients across the interface, or with an additional source term appearing only at the interface. This generally leads to nonsmoothness in the solution at the interface. With a finite-difference method we might attempt to use a 5-point stencil with equally-spaced points on both sides of the interface as indicated in Figure 1(e). This can be successful as long as the nonsmoothness is properly accounted for. With a finite-volume method one might view grid cells as being confined to one side or the other of the interface, as indicated in Figure 1(f), with special formulas for the flux across the interface, or one could view the uniform Cartesian cells as spanning the interface, with appropriate source terms incorporated in addition to the fluxes.

**4. Embedded bodies and the immersed boundary method.** For problems with boundaries which cut through the grid, rather than using grids of the type shown in Figure 1(c) or (d) an alternative is to embed the physical domain into a larger computational domain and view the physical boundary as an interface of the type shown in Figure 1(e) or (f). Grid points or cells are introduced which lie outside the physical domain. If the equations can be extended in a suitable way, it may be simpler to solve the extended equations on a purely Cartesian grid. A variety of methods of this type have been introduced, such as fictitious domain methods (e.g., [34, 35]) and Mayo's method for elliptic equations [51, 52].

The *immersed boundary method* is a numerical method of this type developed primarily by Peskin and his co-workers for solving the incompressible Navier-Stokes equations in complicated time-varying geometries. The original application, which is still being studied with this method, was to model a beating heart in order to study the motion of both natural and artificial heart valves [54–58]. Since then it has been used for many other applications, including aquatic locomotion [24], blood platelet aggregation

[27–29], and wave motion in the cochlea [6, 7]. Similar approaches have also been used outside of biophysics, e.g., to sedimentation [65] and bubble dynamics [69, 70].

For a problem such as heart modeling, the geometry is not only extremely complex but is also time-varying as the heart goes through its pumping cycle and the valves open and close. Attempting to use a grid that conforms to the physical boundary would be extremely difficult. Peskin's idea was to instead view the entire heart as being immersed in a Cartesian box of fluid (a square in the original 2D model or a cube in the current 3D model). The incompressible Navier-Stokes equations are solved on a uniform Cartesian grid in this box and the outer heart wall (as well as interior walls separating chambers, valve leaflets, etc.) are viewed as being immersed within the fluid. These structures are modeled by a separate Lagrangian set of grid points that move with the local fluid velocity in order to impose the requirement that no fluid can flow through the wall.

In each time step, the Navier-Stokes equations are solved on the uniform grid to determine the fluid velocities at each of these grid points. These velocities are then interpolated to the immersed boundary points to determine its motion. Of course the boundary must also have an effect on the fluid, and this is modeled by adding a forcing term to the Navier-Stokes equations to represent the force exerted on the fluid by the immersed boundary. This is a singular (delta function) force with support only along the boundary. The strength of this force depends on the nature and position of the boundary. For example, an elastic material which is stretched exerts a restoring force that can be determined from the configuration of the boundary. The heart wall also exerts force due to muscle contraction that can be determined based on position and phase of the heart cycle.

The incompressible Navier-Stokes equations then take the form

$$\begin{aligned} \vec{u}_t + (\vec{u} \cdot \nabla)\vec{u} + \nabla p &= \nu \Delta \vec{u} + f \\ \nabla \cdot \vec{u} &= 0 \end{aligned} \tag{4.1}$$

where

$$f(\vec{x}, t) = \int_\Gamma F(s,t)\delta(\vec{x} - \vec{X}(s,t))\, ds. \tag{4.2}$$

Here $\vec{x} = (x, y)$ is the spatial variable in 2D, $\Gamma$ is the immersed boundary consisting of points $\vec{X}(s,t)$ parameterized by $s$ at time $t$, $\delta$ is the 2D delta function and $F(s,t)$ is the force density at $\vec{X}(s,t)$. These equations are coupled with the equations for the motion of $\Gamma$,

$$\frac{\partial}{\partial t}\vec{X}(s,t) = u(\vec{X}(s,t), t) \tag{4.3}$$

and equations for determining $f(s,t)$ from the current boundary configuration $\vec{X}(s,t)$. We solve these equations in a box $\Omega$ that contains $\Gamma$, with some convenient boundary conditions on the boundary of $\Omega$, *e.g.*, periodic boundary conditions as used by Peskin [54].

The box is then discretized by a uniform grid with grid points $\vec{x}_{ij} = (x_i, y_j)$ with $x_i = ih$ and $y_j = jh$. The immersed boundary $\Gamma$ is represented by a set of points $\vec{X}_k = (X_k, Y_k)$, $k = 1, 2, \ldots, M$ with spacing which is $O(h)$. The Navier-Stokes equations (4.1) are discretized using any standard technique on the uniform grid. The only difficulty is that the equation for updating $u_{ij}$, the velocity at $\vec{x}_{ij}$, will typically involve $f_{ij}$, the forcing function evaluated at this point. But since the singular force (4.2) has support only on $\Gamma$, this will typically be zero at each $\vec{x}_{ij}$. We need some method to spread the singular force (4.2) out to neighboring grid points on the regular grid. This can be done by replacing the delta function by some approximation, a sharply peaked function with unit integral and support that extends only distance $O(h)$ from the origin. In one space dimension a simple choice is the hat function

$$d_h(x) = \begin{cases} \frac{1}{h^2}(h - |x|) & \text{if } |x| < h \\ 0 & \text{otherwise} \end{cases} \tag{4.4}$$

though other choices are generally used in practice. In two space dimensions, the product $d_h(x)d_h(y)$ can be used. The force $f_{ij}$ used in the discrete equations at $\vec{x}_{ij}$ is then given by

$$f_{ij} = \Delta s \sum_{k=1}^{M} F_k(t) d_h(x_i - X_k) d_h(y_j - Y_k) \tag{4.5}$$

where $\vec{X}_k = (X_k, Y_k)$ and $F_k(t)$ is the force at this point.

Moving the boundary requires interpolating the velocities from the grid to $\Gamma$. One way to do this is to use the same discrete delta function defined above and set

$$U_k = h^2 \sum_{i,j} u_{ij} d_h(x_i - X_k) d_h(y_j - Y_k). \tag{4.6}$$

Even when a highly accurate Navier-Stokes solver is used on the uniform grid, there remains the interesting and difficult issue of the accuracy of the immersed boundary procedure, *i.e.*, the replacement of the boundary conditions by a singular force that is spread to the uniform grid by discrete delta functions. This has been analyzed in some simple cases by Beyer and LeVeque [8], who considered a 1D model problem of the form

$$u_t = u_{xx} + c(t)\delta(x - \alpha(t)) \qquad 0 \le x \le 1 \tag{4.7}$$

with the boundary conditions $u(0,t) = u(1,t) = 0$. Here the point $\alpha(t)$ plays the role of the immersed boundary $\Gamma$ and $c(t)$ is the strength of the singular source at $\alpha$. In the simplest case $\alpha(t) \equiv \alpha$ is fixed and $c(t)$ is a smooth function that is given *a priori*. A discretization in the spirit of the immersed boundary method would be the Crank-Nicolson method

$$\frac{1}{\Delta t}(U_j^{n+1} - U_j^n) = \frac{1}{2}(D_x^2 U_j^n + D_x^2 U_j^{n+1}) + c(t_{n+1/2})d_h(x_j - \alpha) \tag{4.8}$$

where $D_x^2$ is the second difference operator

$$D_x^2 U_j = \frac{1}{h^2}(U_{j+1} - 2U_j + U_{j-1})$$

and $d_h$ is a discrete delta function such as (4.4). If $c(t) \equiv 0$ then the Crank-Nicolson method is well known to be second order accurate in both space and time. Beyer and LeVeque [8] showed that the method (4.8) remains second order accurate for smooth functions $c(t)$ provided that the discrete delta function $d_h$ satisfies certain moment conditions.

When $\alpha(t)$ is moving, it was found that a correction term must be added to the natural extension of the Crank-Nicolson method in order to maintain second order accuracy [8]. This is because the time derivative of the solution at a fixed grid point is not continuous in time as the point $\alpha(t)$ passes this grid point. However, again the jump can be determined directly from the equations and used to correct the formulas. This is an indication that there may be difficulties in achieving second order accuracy with the immersed boundary method in fluid dynamics problems, even with an improved delta function.

In two dimensions it appears to be very difficult to obtain second-order accuracy near the boundary by using a simple discrete delta function. Zhilin Li [47] found this to be so even for the simplest case of a steady state Poisson problem

$$u_{xx} + u_{yy} = f \tag{4.9}$$

where $f$ is again given by a singular source on some curve $\Gamma$,

$$f(x, y) = \int_\Gamma F(s)\delta(x - X(s))\delta(y - Y(s))\, ds. \tag{4.10}$$

The immersed boundary method would correspond to using the discrete equations

$$(D_x^2 + D_y^2)u_{ij} = f_{ij}$$

where $D_x^2 + D_y^2$ is the standard 5-point Laplacian and $f_{ij}$ is given by (4.5). There does not seem to be a simple way to define $d_h$ so that this yields second order accuracy for arbitrary curves $\Gamma$.

It is, however, possible to obtain second order accuracy at all uniform grid points by a different construction of the $f_{ij}$, as discussed in the next section.

**5. The immersed interface method.** In an attempt to obtain better accuracy for problems with delta-function source terms on an interface cutting between grid points, a different approach was developed by Li [47] and LeVeque and Li [45], first in the context of elliptic equations such as the Poisson problem mentioned above. This approach was called the *immersed interface method* since it is similar in spirit to the immersed boundary method but turns out to be applicable to a variety of interface problems beyond immersed boundaries in incompressible flow. The basic idea is most easily explained by considering the 1D model problem (4.7) with $\alpha(t) \equiv \alpha$. Suppose we want to compute $u_t(x_i, t)$ at some grid point $x_i$ at time $t$. If $\alpha$ lies between grid points then $u_t = u_{xx}$ since the source term vanishes at $x_i$. We can usually approximate $u_{xx}$ to second order by the second difference

$$D_x^2 u(x_i, t) = \frac{1}{h^2}(u(x_{i+1}, t) - 2u(x_i, t) + u(x_{i+1}, t)).$$

This is second order accurate provided $\alpha$ does not lie between $x_{i-1}$ and $x_{i+1}$. If it does, then the nonsmoothness of $u(x, t)$ at $\alpha$ destroys the accuracy of this approximation. The solution to (4.7) is continuous and so is $u_{xx}$, but the first derivative $u_x$ has a discontinuity at $x = \alpha$ of magnitude $-c(t)$. We can use this information to obtain a good approximation to $u_{xx}$ at grid points near $\alpha$.

Suppose, for example, that $x_j < \alpha < x_{j+1}$ and we wish to approximate $u_{xx}(x_j, t)$. Expanding in Taylor series about $\alpha$ shows that

$$u(x_j, t) = u(\alpha, t) + (x_j - \alpha)u_x^-(\alpha, t) + \frac{1}{2}(x_j - \alpha)^2 u_{xx}(\alpha, t) + O(h^3)$$
$$u(x_{j+1}, t) = u(\alpha, t) + (x_{j+1} - \alpha)u_x^+(\alpha, t) + \frac{1}{2}(x_{j+1} - \alpha)^2 u_{xx}(\alpha, t) + O(h^3).$$

We use the notation

$$u_x^-(\alpha, t) = \lim_{x \to \alpha-} u_x(x, t), \qquad u_x^+(\alpha, t) = \lim_{x \to \alpha+} u_x(x, t)$$

for the limits as $x$ approaches $\alpha$ from the left and right. (Note that continuity of $u_{xx}$ allows us to simply write $u_{xx}(\alpha, t)$.)

We also have

$$u_x^-(\alpha, t) = u_x(x_j, t) + (\alpha - x_j)u_{xx}(x_j, t) + O(h^2)$$
$$u_x^+(\alpha, t) = u_x(x_j, t) + (\alpha - x_j)u_{xx}(x_j, t) + [u_x] + O(h^2)$$

where $[u_x] = u_x^+(\alpha, t) - u_x^-(\alpha, t)$ is the jump in $u_x$ at $\alpha$. Combining these expressions gives

$$u(x_{j+1}, t) = u(x_j, t) + hu_x(x_j, t) + \frac{1}{2}h^2 u_{xx}(x_j, t) + (x_{j+1} - \alpha)[u_x] + O(h^3).$$

Combining this with the standard Taylor series expansion of $u(x_{j-1},t)$ about $u(x_j,t)$,

$$u(x_{j-1},t) = u(x_j,t) - hu_x(x_j,t) + \frac{1}{2}h^2 u_{xx}(x_j,t) + O(h^3)$$

shows that

$$u_{xx}(x_j,t) = D_x^2 u(x_j,t) - \frac{1}{h^2}(x_{j+1} - \alpha)[u_x] + O(h).$$

Since we know that

$$[u_x] = -c(t),$$

this gives an approximation to $u_{xx}(x_j,t)$:

$$u_{xx}(x_j,t) = \frac{1}{h^2}(u(x_{j+1},t) - 2u(x_j,t) + u(x_{j-1},t)) + \frac{1}{h^2}(x_{j+1} - \alpha)c(t) + O(h).$$

Similarly, at $x_{j+1}$ we find that

$$u_{xx}(x_{j+1},t) = D_x^2 u(x_{j+1},t) + \frac{1}{h^2}(\alpha - x_j)c(t) + O(h).$$

At all other grid points $x_i$ we can use $D_x^2 u(x_i,t)$ alone to obtain a second order accurate approximation.

It is interesting to note that these formulas can be combined in the following general formula valid for all $i$, including $j$ and $j+1$:

$$(5.1)\quad u_{xx}(x_i,t) = D_x^2 u(x_i,t) + c(t)d_h(x_i - \alpha) + \begin{cases} O(h) & \text{if } i = j,\ j+1 \\ O(h^2) & \text{otherwise} \end{cases}$$

where $d_h(x)$ is the hat function (4.4). Note that this is a discrete delta function, so using the approximation (5.1) in a Crank-Nicolson method for (4.7) gives precisely the method (4.8) with the particular choice (4.4) for the discrete delta function. (See [8, 45] for more details.)

It turns out that we only need an $O(h)$ local truncation error at these grid points in order to obtain $O(h^2)$ global errors, since there are only two grid points involved independent of $h$. It is this type of analysis that was used in [8] to show that the immersed boundary method is second order accurate in some cases for the 1D model problem.

In two dimensions consider the Poisson problem (4.9). It can be shown that the solution $u$ will be continuous across $\Gamma$ while the normal derivative of $u$ will have a jump discontinuity of magnitude $F(s)$. To devise a finite difference scheme, we need to approximate $u_{xx}+u_{yy}$ to at least $O(h)$ at each grid point $(x_i, y_j)$ near the interface. The standard 5-point approximation $D_x^2 u_{ij} + D_y^2 u_{ij}$ can be used provided that the interface $\Gamma$ does not come between any of the points in this stencil. If it does, then we must add correction terms arising from jumps in derivatives across $\Gamma$, just as in the 1D expression (5.1). This correction can be derived by expanding $u$ at each grid point in the stencil about some common nearby point on $\Gamma$, e.g., the closest point on $\Gamma$ to $(x_i, y_j)$. The resulting formula for approximating $u_{xx} + u_{yy}$ will have the form

$$u_{xx} + u_{yy} = D_x^2 u_{ij} + D_y^2 u_{ij} + C_{ij} + O(h)$$

where the correction term $C_{ij}$ is zero away from $\Gamma$ and near $\Gamma$ will have magnitude $O(1/h)$ and depend on the source strength $F(s)$. For details see [45]. If this approximation is used to discretize

$$u_{xx} + u_{yy} = 0$$

then we will obtain the system

$$D_x^2 u_{ij} + D_y^2 u_{ij} = -C_{ij}$$

which can be viewed as a discretization of the original equation (4.9). But in this case, unlike the 1D model problem, it is not possible to derive the correct $C_{ij}$ simply by using the formula (4.5) with an appropriate choice of $d_h$. The expression for $C_{ij}$ can be obtained from the more general case considered in [45]. The resulting formula involves not only the jump in normal derivative of $u$ along the interface but also the tangential derivative of this jump and the curvature of the interface. These can all be obtained from the original problem.

By computing the correct values $C_{ij}$, we are able to obtain a second order accurate method that is in the spirit of the immersed boundary method, but in a case where the immersed boundary method itself fails to deliver this accuracy. All that is required for this approach is *a priori* knowledge of the jumps in certain derivatives across $\Gamma$. This can often be determined from the PDE, as in the above examples.

Returning to the incompressible Navier-Stokes equations, we find that this is also true. In this case the force density $F(s,t)$ is a vector. By decomposing this into components normal and tangential to $\Gamma$, it is possible to determine jumps in the pressure $p$ and in derivatives of $p$ and the velocities. It should be possible to use this information to derive appropriate correction terms for a second order Navier-Stokes solver on a uniform grid.

**5.1. Stokes flow.** So far this has been completely carried out only for the Stokes equations,

$$\begin{aligned} p_x &= \mu \Delta u + f^{(1)} \\ p_y &= \mu \Delta v + f^{(2)} \\ u_x + v_y &= 0 \end{aligned} \tag{5.2}$$

where $f = (f^{(1)}, f^{(2)})$ has the form (4.10). This problem can be reduced to a set of Poisson problems ($\Delta p = 0$ and then $\mu\Delta u = p_x$ and $\mu \Delta v = p_y$) with known jumps in the functions and derivatives across $\Gamma$.

Development of methods for this problem with a moving interface are developed in [46]. The method is compared with the immersed boundary method for a test problem studied by Tu and Peskin [68] consisting of a two-dimensional "balloon" relaxing to a circular shape in a highly viscous fluid. It is shown that the new approach yields second order accuracy in both pressure and velocity while an immersed boundary approach with the discrete delta function (4.4) gives only first order accuracy in velocities and $O(1)$ errors in pressure near the discontinuity. Two-dimensional bubbles are also considered in [46], in which the singular force at the interface arises from surface tension rather than from an elastic membrane.

**5.2. Stiffness and implicit methods.** As in the immersed boundary method of Peskin, the moving interface tracked in [46] is modeled by a set of marker points which move with the local fluid velocity as interpolated from the uniform grid on which velocity is computed. An explicit method might be written symbolically as

$$\vec{X}^{n+1} = \vec{X}^n + \Delta t \mathcal{U}(\vec{X}^n), \tag{5.3}$$

where $\vec{X}^n$ represents a vector of marker positions at time $t_n$. The nonlinear function $\mathcal{U}(\vec{X})$ is the mapping from marker locations at some time to the resulting velocities at the same points. Evaluating $\mathcal{U}(\vec{X})$ typically requires evaluating forces at the points $\vec{X}$, computing the resulting source terms $F_{ij}$ on the uniform grid, solving the fluid equations, and interpolating velocities back to the marker points $\vec{X}$. Often this problem is very stiff and the time step $\Delta t$ in the explicit method (5.3) must be taken very small in order to avoid instabilities in which small oscillations in the interface grow exponentially. Often a semi-implicit approach is used with the immersed boundary method to improve stability, see for example [68, 53, 55, 64].

For the Stokes flow problem in [46] a fully implicit method has been used, based on the trapezoidal method

$$\vec{X}^{n+1} = \vec{X}^n + \frac{1}{2}\Delta t\Big(\mathcal{U}(\vec{X}^n) + \mathcal{U}(\vec{X}^{n+1})\Big). \tag{5.4}$$

The potential difficulty with this approach is that $\mathcal{U}(\vec{X})$ is expensive to evaluate and it is impossible to explicitly compute the Jacobian of this mapping as would be required for a Newton-type method.

In [46], a quasi-Newton method has been used with good results. With this approach an approximation to the Jacobian is built up in the process of iterating to solve the nonlinear system (5.4), using rank 1 or rank 2 updates of the previous approximation. In the first time step the procedure takes quite a few iterations to converge, but in future time steps a good initial approximation to the Jacobian is available from the previous time step. For the bulk of the computation only 2 or 3 iterations per time step are required. This approach has also been successfully used for Hele-Shaw flow [36] and for Stefan problems with moving phase boundaries. It could probably be used for many other problems of this type, also in conjunction with the immersed boundary method or other methods for computing the mapping $\mathcal{U}(\vec{X})$.

Another device used in [46] to reduce the expense of the implicit method is to represent the interface by relatively few marker points, with the distance between them much larger than the mesh spacing of the Cartesian grid. Cubic splines are then used to interpolate and compute jump conditions as needed at various points on the interface in implementing the immersed interface method.

**5.3. Discontinuous coefficients.** Besides providing the possibility of a more accurate version of the immersed boundary method, the immersed interface approach has the advantage that it can be applied to a wide class of interface problems where discontinuities in the solution or its derivatives across $\Gamma$ arise from factors in the equations other than a singular source term along $\Gamma$. For example, the Poisson problem

$$\nabla \cdot (\beta \nabla u) = f$$

with variable coefficients $\beta(x, y)$ which are discontinuous across an interface $\Gamma$ has a solution $u$ which is continuous but has a discontinuous normal derivative across $\Gamma$. This arises from the physical jump condition that the flux $\beta \frac{\partial u}{\partial n}$ must be continuous, which gives the jump condition required to implement the immersed interface method. Problems of this form with discontinuous coefficients arise in many applications, e.g., electrostatic or steady state heat distribution problems in nonhomogeneous media, and also in the Stokes flow problem described above if the fluids inside and outside the balloon or bubble have different viscosities.

LeVeque and Li [45] showed how second-order accuracy could be obtained using the same idea of a Taylor series expansion as described in Section 5 for this case, with the 5-point Laplacian replaced by a 6-point stencil whose weights depend on the coefficients $\beta$ and the location and curvature of the interface. Unfortunately, choosing the appropriate six points to use and obtaining a method with uniformly good stability properties and smooth behavior as the interface moves turns out to be difficult for some problems. The problems with this approach have been the subject of more recent research, and several improvements and alternatives have been proposed which give more robust results. See for example [9, 26, 48, 50, 72, 74]. Wiegmann and Bube [73] have also extended this approach to a nonlinear equation with discontinuous coefficients and singular sources.

**6. Advection-diffusion equations.** We now turn to some problems where the finite-volume formulation of Figure 1(d) is quite natural to use. Suppose we wish to model the advection and diffusion of some substance in a given velocity field (e.g., the transport of oxygen or other substances in the blood or tissue, the dispersal of a pheromone). In this section we assume the velocity field is known, and in the next section we will illustrate how this technique can be applied together with immersed interface ideas to compute time-varying velocities in incompressible flow using a streamfunction-vorticity formulation.

The finite-volume framework is natural if we want to insure exact conservation. Also, for the advection portion of the algorithm we can use high-resolution "shock-capturing" finite-volume methods which can deal with steep gradients or even discontinuities in the concentration without introducing spurious numerical oscillations or excessive numerical diffusion. The CLAWPACK software [40, 44], which applies to more general nonlinear hyperbolic systems of equations, is used as the basis for the Cartesian grid method we have developed.

Cartesian grid finite-volume methods for fluid dynamics with embedded boundaries have been the subject of much work recently, see for example [2, 4, 5, 18, 19, 31, 33, 38, 41, 42, 59, 61, 62]. Here we briefly review the method developed in [10]. We solve the advection-diffusion equation

$$\kappa q_t + u q_x + v q_y = (D q_x)_x + (D q_y)_y \tag{6.1}$$

in two space dimensions, where $q(x,y,t)$ is a density or concentration, $(u(x,y,t),\ v(x,y,t))$ is the specified velocity field, $D(x,y)$ is the diffusion coefficient, and $\kappa(x,y)$ is a capacity function. This function could represent, for example, heat capacity if $q$ is the temperature, or porosity in a porous medium if $q$ is the concentration of a transported solute. A fundamental feature of the numerical method we have developed is the use of a discrete version of $\kappa$ which also incorporates information about the fraction of a computational grid cell which lies in the physical domain. This is used to obtain a method which is conservative as well as accurate and stable on the cut cells.

The advection-diffusion equation (6.1) is solved using a fractional step approach in which we alternate between solving the advection equation

$$\kappa q_t + u q_x + v q_y = 0 \tag{6.2}$$

and the diffusion equation

$$\kappa q_t = (D q_x)_x + (D q_y)_y. \tag{6.3}$$

This allows a high-resolution explicit method to be used for the advection equation while an implicit method is used for the diffusion equation.

**6.1. Advection.** We assume the incompressible velocity field is specified via a streamfunction $\psi(x,y)$ with

$$u = \psi_y, \qquad v = -\psi_x.$$

We assume that $\psi$ is known at all grid points in the rectangular computational domain, with $\psi$ identically constant in any "solid body" region where there is no fluid. The points where we need $\psi$ are the corners of the finite-volume grid cells, e.g., $\psi_{i-1/2,j-1/2}$ is the value at the lower left corner of the $(i,j)$ cell. Then differencing $\psi$ between any two corners gives the integral of the normal velocity along the edge in between, so we easily obtain average edge velocities

$$\begin{aligned} U_{i-1/2,j} &= \frac{1}{\Delta y}(\psi_{i-1/2,j+1/2} - \psi_{i-1/2,j-1/2}) \\ V_{i,j-1/2} &= -\frac{1}{\Delta x}(\psi_{i+1/2,j-1/2} - \psi_{i-1/2,j-1/2}). \end{aligned} \tag{6.4}$$

If we have accurate pointwise values of $\psi$ then the average velocities computed by (6.4) will be accurate even if the boundary cuts through an edge of the cell. In this case the velocity does *not* approximate the fluid velocity but rather the average over the edge taking into account the fact that the normal velocity is zero over part of the edge. See [10] for more discussion and the relation to "Darcy velocities" in porous media flow. These velocities are exactly what are needed in computing fluxes across the edge in the high-resolution methods for the advection equation. Moreover, these velocities are divergence free on the discrete grid in the sense that

$$\frac{1}{\Delta x}(U_{i+1/2,j} - U_{i-1/2,j}) + \frac{1}{\Delta y}(V_{i,j+1/2} - V_{i,j-1/2}) = 0, \tag{6.5}$$

as is easily verified.

How $\psi$ is determined depends on the problem. It might be specified analytically for steady-state flow in a simple geometry. Or we might compute it numerically based on some model of the fluid dynamics. In [10] we consider flow in a porous medium with embedded objects, in which case $\psi$ is obtained using Darcy's law by solving a Poisson problem. Accurate pointwise values of $\psi$ on the Cartesian grid can be obtained using a variant of the immersed interface method.

In Section 7 we discuss extensions of this approach to solving the streamfunction-vorticity formulation of the time-dependent incompressible Navier-Stokes equations. In this case the streamfunction must be recomputed in each time step based on the vorticity distribution. Figures 3 through 5 show examples discussed later in which this solution is coupled with an advection equation for a tracer in the fluid. Streamlines of the flow at one instant are shown as contour lines of $\psi$, and the concentration of the advected tracer at this instant are also shown in these figures. The gray tracer is injected at the left boundary starting at time $t = 0$. In these examples there is no diffusion, $D = 0$ in (6.1), and the sharp front in the concentration is well captured.

**6.2. Diffusion.** We also solve the diffusion equation in a finite volume setting, using the capacity function to handle the irregular geometry. The diffusion step of the fractional step scheme is given by

$$\kappa q_t = \nabla \cdot (D(x,y)\,\nabla q) \tag{6.6}$$

subject to Neumann boundary conditions on the boundary of embedded objects. This equation is discretized by using numerical fluxes at cell edges to update the average value of $q$ in each cell. These fluxes are the approximation at cell edges to the quantity $-D\partial q/\partial n$, where $\partial q/\partial n$ is the derivative normal to the cell edge. At the edges of regular cells, these fluxes are computed numerically by simply differencing the values of $q$ in cells adjacent to the edge. At the edges of irregular cells, we use bilinear interpolation to approximate the normal derivatives at partial cell edges. The details of these approximations can be found in [10].

Using these numerical fluxes, we discretize (6.6) using the second order, implicit Crank-Nicolson scheme. Non-zero fluxes at the boundary of the embedded objects are incorporated into the discretization as known source terms. This discretization leads to a linear system of equations which is not symmetric, because of the presence of the irregular cells, and so cannot be solved using standard fast solvers. We use instead a linear iterative method designed for non-symmetric systems. In particular, we have found that the stabilized bi-conjugate gradient method (BiCGSTAB) of Van der Vorst [20] works quite well.

Because we only use bilinear interpolation to compute fluxes at irregular cell edges, our method is formally only first order at these edges. However, numerical results show that the method achieves global second order accuracy. Another method developed by Johansen and Collela [38] is formally second order and could be considered as an alternative to our approach.

**7. Streamfunction-vorticity equations.** In the previous section, we described a numerical method for solving advection-diffusion equations in complex geometries. However, we only considered transport in a steady state velocity field $\vec{u}(x,y) = (u(x,y), v(x,y)) = (\psi_y(x,y), -\psi_x(x,y))$. We now want to consider methods for computing a time dependent velocity field. Also, since fluid viscosity and friction at boundaries plays an important role in many biological fluid dynamics problems, we also want to include the effects of fluid viscosity and impose a no-slip boundary condition on the velocity field at solid boundaries.

We begin by considering the three-dimensional Navier-Stokes equations, given by

$$\begin{aligned} \vec{u}_t + (\vec{u}\cdot\nabla)\vec{u} + \nabla p &= \nu\nabla^2\vec{u} \\ \nabla\cdot\vec{u} &= 0, \end{aligned} \tag{7.1}$$

where $\vec{u}(x,y,z,t) = (u(x,y,z,t), v(x,y,z,t), w(x,y,z,t))$ is the three-dimensional velocity, $p(x,y,z,t)$ is the pressure field, and $\nu$ the kinematic viscosity. We have assumed that the density of the fluid is constant and equal to 1. This is sometimes called the *primitive* variable formulation since desired quantities such as velocity and pressure are solved for directly. This formulation serves as the basis for the projection method, a computational method developed by Chorin [13] and in common use in the biological fluid dynamics community [21, 24].

One potential drawback to solving the primitive variable equations directly is that we are not given any explicit boundary conditions on pressure. To eliminate the pressure from equations (7.1), we can take the curl of the momentum equation and obtain an equation for the *vorticity* $\vec{\omega} = \nabla \times \vec{u}$:

$$\vec{\omega}_t + (\vec{u} \cdot \nabla)\, \vec{\omega} - (\vec{\omega} \cdot \nabla)\, \vec{u} = \nu \nabla^2 \omega \tag{7.2}$$

This equation is most tractable in two space dimensions, where $\vec{\omega} \equiv (0, 0, \omega)$ and the vorticity transport equation (7.2) reduces to

$$\omega_t + (\vec{u} \cdot \nabla)\omega = \nu \nabla^2 \omega \tag{7.3}$$

where now, $\vec{u} = (u, v)$ and $\omega = v_x - u_y$ is the scalar vorticity. Because the velocity field is continuous and divergence free, we can introduce a function $\psi(x, y, t)$ for which

$$u = \psi_y, \qquad v = -\psi_x. \tag{7.4}$$

Using the definition of the vorticity, we have that $\psi$ satisfies the elliptic equation $\nabla^2 \psi = -\omega$. The function $\psi(x, y, t)$ has the property that

$$(\vec{u} \cdot \nabla)\psi = 0 \tag{7.5}$$

so that level curves of $\psi$ trace out instantaneous particle paths or *streamlines* of the flow. For this reason, $\psi$ is called a *streamfunction.*

The streamfunction-vorticity equations for incompressible flow are then given by

$$\begin{aligned} \omega_t + (\vec{u} \cdot \nabla)\omega &= \nu \nabla^2 \omega \\ \nabla^2 \psi &= -\omega \\ u = \psi_y, \qquad v &= -\psi_x \end{aligned} \tag{7.6}$$

subject to boundary conditions which we discuss momentarily. For a more detailed discussion of these equations, one can consult any one of a number of elementary fluid dynamics texts, including [1] and [15].

The streamfunction-vorticity equations have certain computational and modeling advantages over the primitive-variable formulation. First, these equations are naturally decoupled into an advection-diffusion equation for the vorticity and an elliptic equation for the streamfunction. Second, in two space dimensions, the vorticity is a conserved quantity and so we can take advantage of high resolution finite volume advection algorithms to solve the advection-diffusion equation for vorticity. Finally, the numerically computed velocity field obtained by differencing the streamfunction as in (6.4) satisfies the discrete divergence free condition (6.5) exactly. From a modeling point of view, one may actually prefer to compute vorticity and streamlines directly, rather than velocity or pressure. For example, in [71], Wang uses this formulation to compute vortex shedding behind flapping insect wings. In [12, 11], Cheer and Koehl compute the mass flux between filtering appendages, which they represent as solid objects embedded in a flow field. Such a computation can be done quite easily if the streamfunction for the flow is known. Finally, quantities of aerodynamic interest, such as lift and drag forces are easily computed from a given vorticity field.

Boundary conditions appropriate for the above equations are derived from desired boundary conditions on the primitive variables. A no-flow condition $\vec{u} \cdot \vec{n} = 0$, where $\vec{n}$ is the vector normal to a boundary, is imposed by requiring that the tangential derivative of $\psi$, $\partial\psi/\partial\tau$, along the boundary be zero, or that $\psi$ be constant along solid boundaries. Mathematically, this corresponds to imposing a Dirichlet condition on $\psi$ at the solid boundaries. To impose the no-slip condition, we must require that $\vec{u} \cdot \vec{\tau} = 0$, or that $\partial\psi/\partial n = 0$. This corresponds to a Neumann condition on $\psi$. If our domain is multiply-connected, we must also impose a circulation condition on each object to insure the uniqueness of the streamfunction solution. The fact that we must satisfy both $\psi =$ constant and $\partial\psi/\partial n = 0$ on $\psi$ appears to overdetermine the elliptic problem for the streamfunction. However, when we consider the elliptic problem *coupled* with the vorticity transport equation, we see that we have exactly the right number of boundary conditions. This is seen most easily by eliminating $\omega$ from the equations and considering the resulting biharmonic equation for $\psi$.

When considering how to solve the coupled system of equations numerically, however, we are faced with the problem that we do not have any explicit boundary conditions on $\omega$, and seemingly too many for $\psi$. Mathematically, a shear gradient in the boundary layer at a solid boundary is coincident with the presence of non-zero vorticity there. Although physically the mechanism by which vorticity appears near solid boundaries is not so clear, it is nonetheless common to appeal to the notion of "vorticity generation" near solid boundaries and to consider numerical methods which introduce vorticity into the domain at these boundaries in an amount that exactly cancels the slip velocity, thereby satisfying the boundary condition $\psi_n = 0$. Considerable research has gone into deriving schemes for introducing vorticity into the domain or coming up with conditions on the vorticity that insure that the no-slip boundary condition will be satisfied. See for example [66, 37, 14, 60, 3, 23]. Our approach is in the spirit of vorticity generation, in that we derive a flux condition on the vorticity that introduces vorticity into the domain and cancels the slip velocity at the boundary.

For the remainder of this section, we discuss our approach to solving equations (7.6) on a Cartesian grid in the presence of embedded complex geometry. A typical example of the geometry we have considered so far is the multiply-connected exterior region shown in Figure 2. Although our methods are applicable to very general geometries, we focus here on the exterior flow problem for a couple of reasons. First, the problem of flow around isolated objects arises in a number of problems in bio-fluid mechanics, including flow around swimming organisms [22], filtering appendages [12, 11, 39], through tissue past capillaries [67] and flapping wings [71]. Second, we wanted to be able to compare our results to standard benchmark problems, of which the classic problem is flow past a cylinder. Finally, we wish to discuss the additional constraints that one must impose on the streamfunction in a multiply-connected domains.

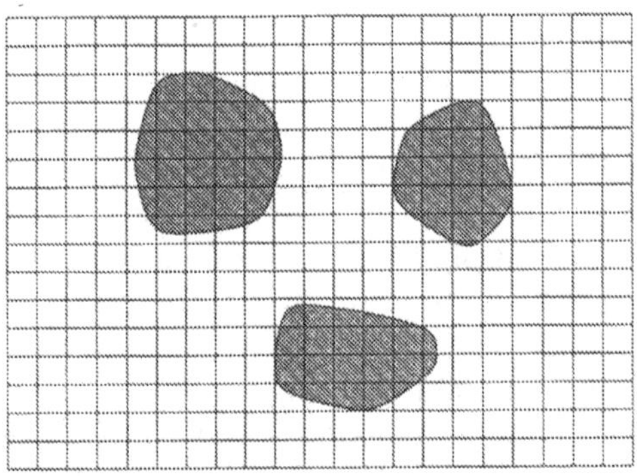

FIG. 2. *A typical grid on which the streamfunction vorticity equations are solved. The flow domain is the region exterior to the embedded objects. The objects are assumed to be stationary and rigid.*

We assume that the embedded objects shown in Figure 2, labeled $\Omega_j, j = 1, \ldots M_{bodies}$ are stationary and rigid, and so we must satisfy both a no-normal flow and a no-slip boundary condition on each object. On the boundary of the computational domain, we impose inflow conditions at the left edge, outflow conditions at the right edge, and free stream conditions on the top and bottom edges. The top and bottom act like free-slip boundaries parallel to the direction of the mean flow.

To solve the advection-diffusion equation in (7.6) for the vorticity, we use the fractional step approach described in Section 6. First, a given vorticity field is advected using a velocity field computed from the current streamfunction. The irregular geometry is handled using the capacity function, as described in Section 6. In this step, we impose no-flux boundary conditions on the embedded objects. The advection step is followed by the diffusion step described in Section 6.2. In this diffusion step, we impose a non-zero flux condition $\partial\omega/\partial n = g$ at the solid boundaries. This flux $g$ is determined by solving an auxiliary Stokes flow problem, described below. In the process of solving this Stokes flow problem, we also determine a streamfunction $\psi$ which we use to update our velocity field, and, because it satisfies the Stokes flow problem, satisfies $\psi_n = 0$.

The auxiliary Stokes flow problem is given in terms of a node-based quantity $\hat{\omega}$, obtained as a result of averaging the cell-averaged quantity $\omega$ onto grid nodes. The Stokes flow problem is

$$
\begin{aligned}
&\hat{\omega}_t = \nu \nabla^2 \hat{\omega}, && \partial\hat{\omega}/\partial n = g, && (x,y) \in \partial\Omega \\
&\nabla^2 \psi = -\hat{\omega}, && \psi = \overline{\psi}_j, && (x,y) \in \partial\Omega_j \\
&\psi_n = 0, && && (x,y) \in \partial\Omega \\
&\int_{\partial\Omega_j} \frac{\partial\hat{\omega}}{\partial n}\, ds = 0. && && j = 1, \ldots M_{bodies}
\end{aligned}
\tag{7.7}
$$

where the flux $g$ of vorticity along the boundaries of the objects is an unknown in the problem. Physically, the vorticity distribution $\hat{\omega}(x, y, t + \Delta t)$ that results from applying the flux $g$ over a single time step $\Delta t$ has the property that the solution to $\nabla^2 \psi = -\hat{\omega}$ satisfies $\psi_n = 0$, in addition to the Dirichlet condition $\psi = \overline{\psi}_j$.

The constants $\overline{\psi}_j, j = 1, \ldots M_{bodies}$ are also unknowns in the problem and are determined so that the circulation $\Gamma_j$

$$
\Gamma_j = \int_{\partial\Omega_j} \vec{u} \cdot \vec{\tau}\, ds = \int_{\partial\Omega_j} \psi_n\, ds = \int_{\Omega_j} \omega\, dA \tag{7.8}
$$

around each object $\Omega_j$ remains constant in time. This condition is satisfied by requiring that the net flux of vorticity into the bodies be zero for all time, or that the last equation in (7.7) hold. Note that we don't require that the circulation around the objects be zero; the circulation is determined initially by our choice of $\psi_n$ (or, equivalently by the initial vorticity distribution inside each object). Of course, when the bodies are stationary, then the no-slip condition is just $\psi_n = 0$ and inside of each body we have $\omega \equiv 0$, so that for each body, $\Gamma_j = 0$ for all time.

To discretize the equations for the Stokes flow problem in (7.7), we use a modified version of the Immersed Interface Method described in Section 5 of the sort developed in [9, 75, 72]. The IIM, as applied to these problems, gives second order accurate solutions to $\psi$ and $\hat{\omega}$, and $\psi_n = 0$ is satisfied to second order. Once we have discretized the equations, we can easily solve them by recognizing that we can set up a linear system for the flux function $g$ evaluated at discrete locations along the solid boundaries. This linear system can either be solved directly, by explicitly forming a matrix equation, or solved iteratively, using for example, the Generalized Minimum Residual algorithm (GMRES). Numerical tests demonstrate that this linear system is quite well conditioned [9].

**7.1. Numerical results for the streamfunction-vorticity equations.** Comparison of our results with those from the standard benchmark problem, flow around a cylinder, show that our algorithm yields results that are in excellent agreement with other published computational and experimental results. In [9], Calhoun compared the results of the above algorithm with the experimental results of Coutanceau and Bouard [16, 17] and the computational results of Fornberg [30] and found that velocity fields and vorticity wakes were quite well reproduced. For low Reynolds number flows ($Re < 40$), the flow remains steady and symmetric and any small perturbations in the wake of the cylinder are damped. At around Reynolds number 50, the streamline wake begins to oscillate and by Reynolds number 100, vortices are shed behind the cylinder forming the well known von Karman vortex street.

In the three sets of plots in Figures 3 to 5, we show results of our numerical algorithm for more complicated flows at Reynolds numbers varying from $Re = 20$ for Figure 3 to $Re \approx 100$ for Figure 5. In each case, the Reynolds number is based on the diameter of the object(s). In each set of plots, we show the streamfunction, advection of a tracer (grey shaded material), the horizontal velocity field and the vorticity distribution at some fixed time. For the calculation shown in Figure 3, the flow has reached a steady state, and there is no vortex shedding in the wake behind the bodies. From the plots of the horizontal velocities, one can see that the boundary layer around each object is fairly thick, and from the tracer plots, significantly more tracer is flowing around the objects than between them. In the calculation shown in Figure 4 at $Re = 40$, the shear gradient is steeper (and the boundary layer thinner) and more tracer material flows between the embedded objects. The last set of plots in Figure 5 provides an example of vortex shedding behind a complex bluff body. In this example, the boundary layers are fairly thin, although there are large areas where the velocity is essentially zero due to the shielding effect of the body on the flow.

Our goal in showing these plots is to demonstrate that our algorithm produces realistic results in some biologically relevant regimes. Our choice of examples was motivated to some extent by the work of Cheer and Koehl [12, 11], who considered flow around fixed objects. Wang [71] also considers flow around a fixed wing, and so in principle our scheme could be used for these calculations as well. However, these examples, as well as most other interesting examples in biofliud mechanics, really involve moving and deforming boundaries and more complicated boundary conditions. For these reasons, we are now considering methods for incorporating these features into our finite difference/finite volume methods.

Adding the more complicated boundary conditions should be a simple modification to the code. We would like to consider problems in which the boundaries of the embedded objects are absorbing or releasing some tracer material (with concentration $q$) into the flow. In general, the flux rate of this tracer will depend on the amount of tracer present near the boundary, and so this will lead to a boundary condition on the tracer that takes the form $\partial q/\partial n = f(q, \mathbf{x}, t)$, where $f$ may be a linear or non-linear function of $q$. This type of boundary condition appears, for example, in oxygenation of tissue, where the flux of oxygen through capillary walls is proportional to the partial pressure of oxygen in the tissue [67].

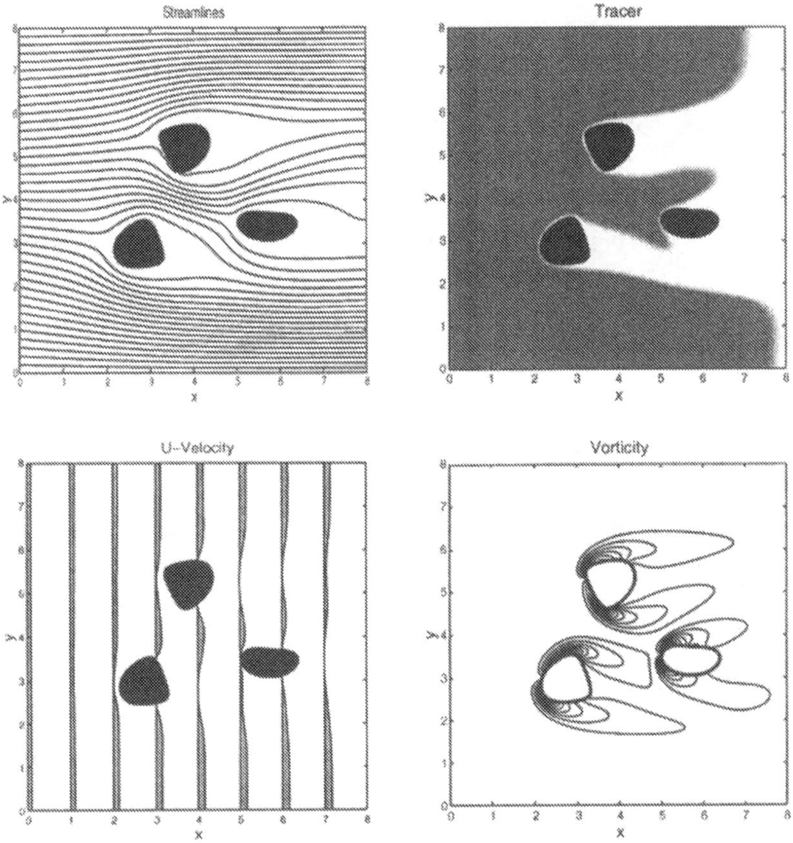


FIG. 3. *Flow past objects at Re* = 20 *Several quantities are illustrated at the same instant in time.*

We are currently working on extending our method to moving boundaries. Successful Cartesian-grid finite-volume methods for moving geometries have been developed for compressible flow, see [32] for one example. Our goal for incompressible flow is still to solve the streamfunction-vorticity equations for the fluid motion on a uniform grid, along with advection-diffusion equations for any tracer concentrations. The volume of the finite-volume cells will now be time-dependent, however, as the boundaries move through the grid. In the future we also hope to combine this algorithm with an immersed interface method for flexible membranes immersed in the fluid.

**8. Summary.** We have presented an overview of some recent work on Cartesian grid methods for solving fluid dynamics in complex geometry. The immersed boundary and immersed interface methods are finite-

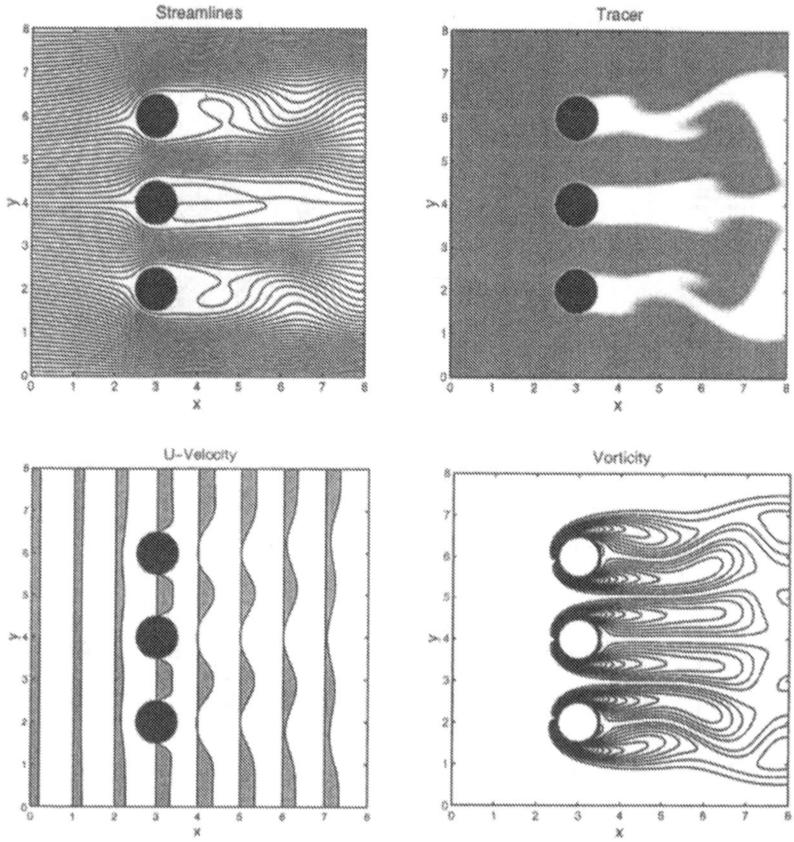


FIG. 4. *Flow past objects at Re = 40. Several quantities are illustrated at the same instant in time.*

difference methods which can deal with discontinuities in the solution across boundaries or interfaces. These methods are often used for problems with moving boundaries as well, in cases where the motion of the boundary is coupled to the fluid dynamics.

Finite-volume methods have some advantages for problems where conservation is important, and steep gradients in the solution can be well-resolved by high-resolution methods. The finite-volume method described here for advection-diffusion and the streamfunction-vorticity equations can deal with complex geometries. We have developed an approach to determine the flux of vorticity at a no-slip wall based on the behavior of the streamfunction at the wall (as computed by a finite-difference immersed interface method). This gives the required boundary condition for vorticity.

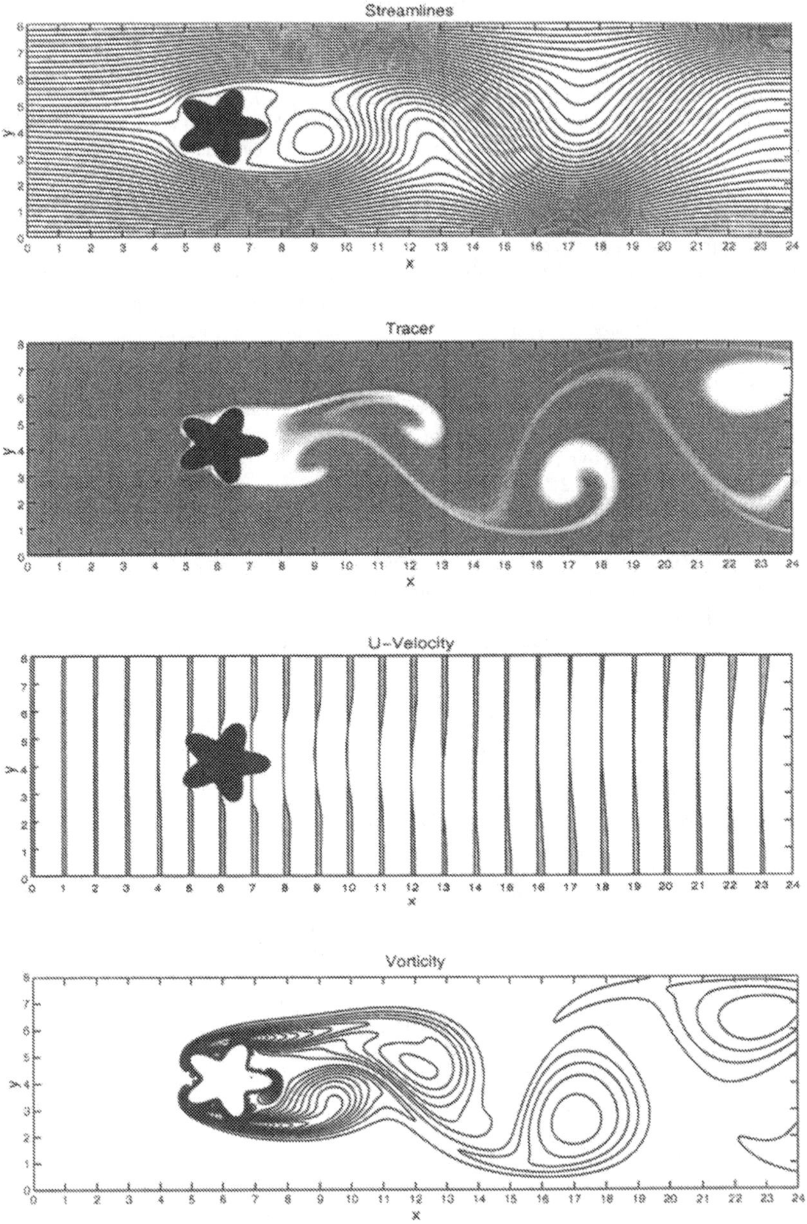


FIG. 5. *Flow at Re ≈ 100 with vortex shedding. Several quantities are illustrated at the same instant in time.*

## REFERENCES


[1] D.J. ACHESON, *Elementary Fluid Dynamics*, Oxford Applied Mathematics and Computing Science Series, Clarendon Press, 1990.

[2] A.S. ALMGREN, J.B. BELL, P. COLELLA, AND T. MARTHALER, *A Cartesian grid projection method for the incompressible Euler equations in complex geometries*, SIAM J. Sci. Comput., **18** (1997), pp. 1289–1309.

[3] C. ANDERSON, *Vorticity Boundary Conditions and Boundary Vorticity Generation for Two-dimensional Viscous Incompressible Flows*, J. Comput. Phys., **80** (1989), pp. 72–97.

[4] M. BERGER AND R.J. LEVEQUE, *An adaptive Cartesian mesh algorithm for the Euler equations in arbitrary geometries*. AIAA paper AIAA-89-1930, 1989.

[5] ———, *Stable boundary conditions for Cartesian grid calculations*, Computing Systems in Engineering, **1** (1990), pp. 305–311.

[6] R.P. BEYER, *A computational model of the cochlea using the immersed boundary method*, PhD thesis, University of Washington, 1989.

[7] ———, *A computational model of the cochlea using the immersed boundary method*, J. Comput. Phys., **98** (1992), pp. 145–162.

[8] R.P. BEYER AND R.J. LEVEQUE, *Analysis of a one-dimensional model for the immersed boundary method*, SIAM J. Num. Anal., **29** (1992), pp. 332–364.

[9] D. CALHOUN, *A Cartesian grid method for solving the streamfunction-vorticity equations in irregular geometries*, PhD thesis, University of Washington, 1999.

[10] D. CALHOUN AND R.J. LEVEQUE, *Solving the advection-diffusion equation in irregular geometries*, J. Comput. Phys., **156** (2000), pp. 1–38

[11] A. CHEER AND M. KOEHL, *Fluid flow through filtering appendages of insects*, IMA Journal of Mathematics Applied in Medicine and Biology, **4** (1987), pp. 185–199.

[12] ———, *Paddles and Rakes: Fluid Flow through ristled Appendages of Small Organisms*, J. Theor. Biol., **129** (1987), pp. 17–39.

[13] A.J. CHORIN, *Numerical solution of the Navier-Stokes equations*, Math. Comp., **22** (1968), pp. 745–762.

[14] ———, *Numerical study of slightly viscous flow*, J. Fluid Mech., **75** (1973), pp. 785–796.

[15] A.J. CHORIN AND J.E. MARSDEN, *A Mathematical Introduction to Fluid Mechanics*, Springer-Verlag, 1979.

[16] M. COUTANCEAU AND R. BOUARD, *Experimental determination of the main features of the vicous flow in the wake of a circular cylinder in uniform translation. Part I. Steady flow*, J. Fluid Mech., **79** (1977), pp. 231–256.

[17] ———, *Experimental determination of the main features of the vicous flow in the wake of a circular cylinder in uniform translation. Part II. Unsteady flow*, J. Fluid Mech., **79** (1977), pp. 257–272.

[18] M.S. DAY, P. COLELLA, M.J. LIJEWSKI, C.A. RENDLEMAN, AND D.L. MARCUS, *Embedded boundary algorithms for solving the Poisson equation on complex domains*. Preprint LBNL-41811, Lawrence Berkeley Lab, 1998.

[19] D. DE ZEEUW AND K. POWELL, *An adaptively-refined Cartesian mesh solver for the Euler equations*, J. Comput. Phys., **104** (1993), pp. 56–68.

[20] H.V. DER VORST, *Bi-CGSTAB: A fast and smoothly converging variant of Bi-CG for the solution of nonsymmetric linear systems*, SIAM J. Sci. Statist. Comput., **13** (1992), pp. 631–644.

[21] R. DILLON, L. FAUCI, A. FOGELSON, AND D. GAVER, *Modeling Biofilm Processes Using the Immersed Boundary Method*, J. Comput. Phys., **129** (1996), pp. 57–73.

[22] R. DILLON, L. FAUCI, AND D. GAVER, *A Microscale Model of Bacterial Swimming, Chemotaxis and Substrate Transport*, J. Theor. Biol., **177** (1995), pp. 325–340.

[23] W. E AND J. LIU, *Vorticity Boundary Condition and Related Issues for Finite Difference Schemes*, J. Comput. Phys., **124** (1996), pp. 368–382.

[24] L. FAUCI AND C.S. PESKIN, *A computational model of aquatic animal locomotion*, J. Comput. Phys., **77** (1988), pp. 85–108.

[25] L.J. FAUCI, *Interaction of oscillating filaments - a computational study*, J. Comput. Phys., **86** (1990), pp. 294–313.

[26] A. FOGELSON AND J. KEENER, *Immersed interface methods for Neumann and related problems in two and three dimensions*, to appear, SIAM J. Sci. Comput.

[27] A.L. FOGELSON, *A mathematical model and numerical method for studying platelet adhesion and aggregation during blood clotting*, J. Comput. Phys., **56** (1984), pp. 111–134.

[28] ———, *Mathematical and computational aspects of blood clotting*, in Proceedings of the 11th IMACS World Congress on System Simulation and Scientific Computation, Vol. 3, B. Wahlstrom, ed., North Holland, 1985, pp. 5–8.

[29] A.L. FOGELSON AND C.S. PESKIN, *Numerical solution of the three dimensional stokes equations in the presence of suspended particles*, in Proc. SIAM Conf. Multi-phase Flow, SIAM, June 1986.

[30] B. FORNBERG, *A numerical study of steady viscous flow past a circular cylinder*, J. Fluid Mech., **98** (1980), pp. 819–855.

[31] H. FORRER, *Boundary Treatments for Cartesian-Grid Methods*, PhD thesis, ETH-Zurich, 1997.

[32] H. FORRER AND M. BERGER, *Flow simulations on Cartesian grids involving complex moving geometries*, in Proc. 7'th Intl. Conf. on Hyperbolic Problems, R. Jeltsch, ed., Birkhäuser Verlag, 1998, pp. 315–324.

[33] H. FORRER AND R. JELTSCH, *A higher-order boundary treatment for Cartesian-grid methods*, J. Comput. Phys., **140** (1998), pp. 259–277.

[34] R. GLOWINSKI, T.-S. PAN, AND J. PERIAUX, *A fictitious domain method for Dirichlet problem and applications*, Comp. Meth. Appl. Mech. Eng., **111** (1994), pp. 283–303.

[35] ———, *A fictitious domain method for external incompressible viscous flow modeled by Navier-Stokes equations*, Comp. Meth. Appl. Mech. Eng., **112** (1994), pp. 133–148.

[36] T.Y. HOU, Z. LI, H. ZHAO, AND S. OSHER, *A hybrid method for moving interface problems with application to the Hele-Shaw flow*, J. Comput. Phys., **134** (1997), pp. 236–252.

[37] M. ISRAELI, *On the Evaluation of Iteration Parameters for the Boundary Vorticity*, Studies in Applied Mathematics, LI (1972), pp. 67–71.

[38] H. JOHANSEN AND P. COLELLA, *A Cartesian grid embedded boundary method for Poisson's equation on irregular domains*, J. Comput. Phys., **147** (1998), pp. 60–85.

[39] M. KOEHL, *Fluid flow through hair-bearing appendages: Feeding, smelling, and swimming at low and intermediate Reynolds numbers.*, in Biological Fluid Dynamics, C.P. Ellington and T.J. Pedley, eds., Vol. 49, Soc. Exp. Biol. Symp, 1995, pp. 157–82.

[40] R.J. LEVEQUE, CLAWPACK *software*. `http://www.amath.washington.edu/~rjl/clawpack.html`.

[41] ———, *Cartesian grid methods for flow in irregular regions*, in Num. Meth. Fl. Dyn. III, K.W. Morton and M.J. Baines, eds., Clarendon Press, 1988, pp. 375–382.

[42] ———, *High resolution finite volume methods on arbitrary grids via wave propagation*, J. Comput. Phys., **78** (1988), pp. 36–63.

[43] ———, *Numerical Methods for Conservation Laws*, Birkhäuser-Verlag, 1990.

[44] ———, *Wave propagation algorithms for multi-dimensional hyperbolic systems*, J. Comput. Phys., **131** (1997), pp. 327–353.

[45] R.J. LEVEQUE AND Z. LI, *The immersed interface method for elliptic equations with discontinuous coefficients and singular sources*, SIAM J. Numer. Anal., **31** (1994), pp. 1019–1044.

[46] ———, *Immersed interface methods for Stokes flow with elastic boundaries or surface tension*, SIAM J. Sci. Comput., **18** (1997), pp. 709–735.

[47] Z. Li, *The Immersed Interface Method — A Numerical Approach for Partial Differential Equations with Interfaces*, PhD thesis, University of Washington, 1994.

[48] ———, *A fast iterative algorithm for elliptic interface problems*, SIAM J. Numer. Anal., **35** (1998), pp. 230–254.

[49] ———, *The immersed interface method using a finite element formulation*, Applied Numer. Math., **27** (1998), pp. 253–267.

[50] X.-D. Liu, R.P. Fedkiw, and M. Kang, *A boundary condition capturing method for Poisson's equation on irregular domains*. CAM Report 99-15, UCLA Mathematics Department, 1999.

[51] A. Mayo, *The fast solution of Poisson's and the biharmonic equations on irregular regions*, SIAM J. Num. Anal., **21** (1984), pp. 285–299.

[52] A. Mayo and A. Greenbaum, *Fast parallel iterative solution of Poisson's and the biharmonic equations on irregular regions*, SIAM J. Sci. Stat. Comput., **13** (1992), pp. 101–118.

[53] A.A. Mayo and C.S. Peskin, *An implicit numerical method for fluid dynamics problems with immersed elastic boundaries*, Contemp. Math., **141** (1993), pp. 261–277.

[54] C.S. Peskin, *Numerical analysis of blood flow in the heart*, J. Comput. Phys., **25** (1977), pp. 220–252.

[55] ———, *Lectures on mathematical aspects of physiology*, Lectures in Appl. Math., **19** (1981), pp. 69–107.

[56] C.S. Peskin and D.M. McQueen, *Modeling prosthetic heart valves for numerical analysis of blood flow in the heart*, J. Comput. Phys., **37** (1980), pp. 113–132.

[57] ———, *Computer-assisted design of pivoting-disc prosthetic mitral valves*, J. Thorac. Cardiovasc. Surg., **86** (1983), pp. 126–135.

[58] ———, *Computer-assisted design of butterfly bileaflet valves for the mitral position*, Scand. J. Thorac. Cardiovasc. Surg., **19** (1985), pp. 139–148.

[59] K. Powell, *Solution of the Euler and Magnetohydrodynamic Equations on Solution-Adaptive Cartesian Grids*. Von Karman Institute for Fluid Dynamics Lecture Series, 1996.

[60] L. Quartapelle and F. Valz-griz, *Projection conditions on the vorticity in viscous incompressible flows*, Int. J. Numer. Methods. Fluids, **1** (1981), pp. 129–144.

[61] J.J. Quirk, *An alternative to unstructured grids for computing gas-dynamic flow around arbitrarily complex 2-dimensional bodies*, Comput. Fluids, **23** (1994), pp. 125–142.

[62] ———, *A Cartesian grid approach with hierarchical refinement for compressible flows*. ICASE Report No. TR-94-51, NASA Langley Research Center, 1994.

[63] J.M. Stockie and S.I. Green, *Simulating the motion of flexible pulp fibres using the immersed boundary method*, J. Comput. Phys., **147** (1998), pp. 147–165.

[64] J.M. Stockie and B.T.R. Wetton, *Stability analysis for the immersed fiber problem*, SIAM J. Appl. Math., **55** (1995), pp. 1577–1591.

[65] D. Sulsky and J.U. Brackbill, *A numerical method for suspension flow*, J. Comput. Phys., **96** (1991), pp. 339–368.

[66] A. Thom, *The flow past circular cylinders at low speeds*, Proc. Roy. Soc. A, **141** (1933), p. 651.

[67] M. Titcombe and M.J. Ward, *An asymptotic study of oxygen transport from multiple capillaries to skeletal muscle tissue*, to appear, SIAM J. Appl. Math., 2000.

[68] C. Tu and C.S. Peskin, *Stability and instability in the computation of flows with moving immersed boundaries: a comparison of three methods*, SIAM J. Sci. Stat. Comput., **13** (1992), pp. 1361–1376.

[69] S.O. Unverdi and G. Tryggvason, *Computations of multi-fluid flows*, Physica D, **60** (1992), pp. 70–83.
[70] ———, *A front-tracking method for viscous, incompressible, multi-fluid flows*, J. Comput. Phys., **100** (1992), pp. 25–37.
[71] Z.J. Wang, *Vortex shedding and frequency selection in flapping flight*. Submitted to the J. Fluid Mech., 1999.
[72] A. Wiegmann, *The Explicit Jump Immersed Interface Method and Interface Problems for Differential Equations*, PhD thesis, University of Washington, 1998.
[73] A. Wiegmann and K.P. Bube, *The immersed interface method for nonlinear differential equations with discontinuous coefficients and singular sources*, SIAM J. Numer. Anal., **35** (1998), pp. 177–200.
[74] ———, *The explicit jump immersed interface method: Finite difference methods for pde with piecewise smooth solutions*, SIAM J. Numer. Anal., **37** (2000), pp. 827–862.
[75] Z. Yang, *A Cartesian grid method for elliptic boundary value problems in irregular regions*, PhD thesis, University of Washington, 1996.

# COMPUTED SIMULATIONS OF CILIARY AND FLAGELLAR MOTILITY USING THE GEOMETRIC CLUTCH MODEL CAN REPLICATE A WIDE VARIETY OF EXPERIMENTAL CONDITIONS

CHARLES B. LINDEMANN*

**Abstract.** The Geometric Clutch hypothesis proposes that the strain that develops between the outer doublets of the flagellar/ciliary axoneme acts as the principal control to regulate the function of the dynein motor proteins. In this hypothetical scheme, the forces that develop transverse to the axis of the doublets ($t$-forces) act as a clutch to engage or disengage the dynein arms from their binding sites on adjacent doublets. These forces can be easily computed from the longitudinal tension, or compression, on a doublet and the local curvature. A computer model has been developed based on the Geometric Clutch principle. When the model is scaled as closely as possible to the physical dimensions and mechanical properties that have been measured in real cilia and flagella, the computed simulations successfully replicate the basic patterns of motility of the biological systems. Observed phenomena, such as the effective and recovery stroke of cilia, can be readily reproduced; and mechanical-sensitivity, a known property of cilia and flagella, is intrinsic to the computer simulation. Recently, the model has been further tested by comparing computed behavior and real behavior of bull sperm under identical conditions of mechanical restraint and dissection. The results of the real and computed experiments are in good agreement. The simulation accurately predicts the observed changes in the beating pattern, and the conditions that cause the beat to arrest.

**1. Introduction.** Cilia and flagella are organelles of living cells that are hair-like in appearance and move by propagating waves of bending along their length. The bending waves provide propulsion as the flagellum or cilium moves through the fluid surrounding the cell. The propulsive action provided by cilia or flagella is put to use in a great variety of applications in the biological world. It allows single-celled algae and protozoa to swim about in a pond or lake, sperm to swim in the fluid of the oviduct, and cells of the bronchial airways to move the mucous layer in order to clear debris from our lungs.

In all cases, movement of the flagellum or cilium is made possible by internal machinery composed of microtubules and molecular motors. This inner structure is called the axoneme and is highly conserved, having the same basic structure in both cilia and flagella (as shown in Figure 1). The arrangement of nine outer doublet microtubules and a centrally located pair of single microtubules is so universal that it is found in the flagella of primitive single-celled plants and in nearly identical form in the lung cilia of man. The basic structures of the flagellum were first described by the electron microscope studies of Afzelius (1959,1961), Gibbons (1961) and Gibbons and Grimstone (1960). Attached to the nine outer doublets are a series of projections, called the inner and outer dynein arms. These

*Department of Biological Sciences, Oakland University, Rochester, Michigan 48309; e-mail: lindeman@oakland.edu.

arm-like projections where shown to be composed of an ATPase enzyme (Gibbons and Rowe, 1965), and this enzyme, named dynein, was shown to be the motor that drives the bending process (Gibbons and Gibbons, 1973).

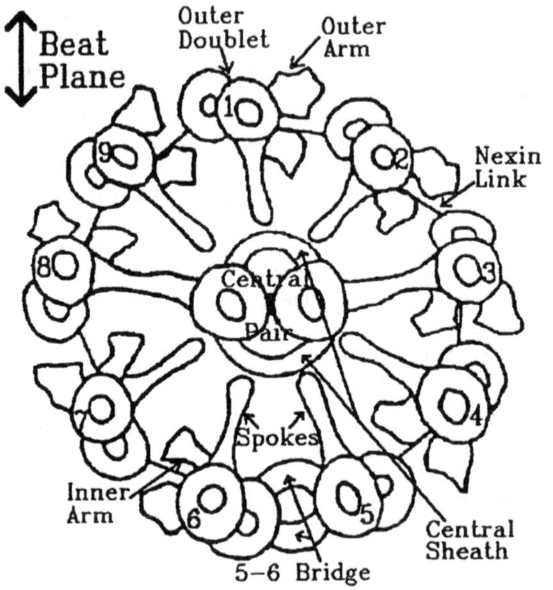


FIG. 1. *The eukaryotic flagellar axoneme. Structures commonly found in the typical axoneme are labeled on a silhouette diagram traced from an E.M. micrograph. The outer doublets, arranged in a ring of nine, each possess an inner and outer row of dynein arms. In many flagella, doublets 5 and 6 are permanently bridged, prohibiting interdoublet sliding between these doublets. Each of the outer doublets is also linked to its neighbors by nexin links. The outer doublets are connected to the central pair by a series of wagon wheel-like spokes that interact with the axonemal "hub". Some flagella have demonstrated the existence of a stable connection between outer doublets 3 and 8 that includes the central pair and roughly partitions the axoneme into two unequal "halves". The flagellar beat is in a plane perpendicular to this partition as indicated by the double headed arrow. Reprinted from Lindemann and Kanous, 1997 with permission.*

The axoneme converts the force developed by the dynein motors into the waves of bending that are responsible for the propulsive action of the flagella. The doublet microtubules slide relative to each other in the process of producing a bend, this sliding displacement was first demonstrated in cilia by Satir (1965, 1968). Later, it was shown that the doublets would actually slide apart in a telescoping fashion if protein connections between the outer doublets are cut by trypsin digestion (Summers and Gibbons,

1971). It also has been shown that the polarity of movement is for each doublet to slide base-ward on its neighbor (Sale and Satir, 1977). In reference to the numbering system given in Figure 1, this can be stated that each doublet ($n$) slides base-ward on the next higher numbered doublet ($n + 1$). Therefore in the orientation of Figure 1, where the dynein arms project clockwise, we would be looking toward the flagellar base. With this functional arrangement the motors on opposite sides of the axoneme bend the flagellum in opposite directions.

The doublet microtubules of the axoneme are organized in a circle; however, in actual practice all of the doublets are not equal in contributing to the bending. Early on it was recognized that in most cilia and flagella doublets 5 and 6 are permanently linked to each other, and therefore they limit the sliding that is possible in the axis defined by these two doublets (Afzelius, 1961; Gibbons, 1961). Furthermore, in some cilia and flagella there is a strong permanent linkage of doublets 3 and 8 through the central pair complex (Tamm and Tamm, 1984; Lindemann et al., 1992; Kanous et al., 1993). These two structural features make most cilia and flagella rather stiff and inflexible in the plane defined by the central pair. Consequently, bending waves are produced perpendicular to these structures. That means that sliding takes place mostly between the doublets 2, 3 and 4 on one side of the axoneme and doublets 7, 8, and 9 on the other side. Therefore, the dynein arms located on these two opposing sets of doublets do most of the actual work of bending the flagellum, and that the two opposite groups must work alternately to bend the flagellum first in one direction and then in the opposite direction.

A very interesting property of the axoneme is that it is a self-contained machine. It can use Mg-ATP as a chemical fuel to directly produce useful work. The ability of an isolated flagellum to produce swimming motion with externally supplied ATP was first demonstrated by Brokaw (1961). Subsequent studies also showed that the plasma membrane is unnecessary for the axoneme to work (Gibbons and Gibbons, 1972). Furthermore, it was shown that isolated pieces of the axoneme are able to propagate bends (Lindemann and Rikmenspoel, 1972a) and can continue to beat rhythmically (Lindemann and Rikmenspoel, 1972b). These observations established, beyond any doubt, that the axoneme is a self-contained, self-sufficient chemical oscillator with the ability to generate and propagate waves of bending using Mg-ATP as an energy source.

Another revealing property of the isolated flagellar axoneme is its response to imposed bending. It was found that isolated pieces of flagella incapable of initiating a beat on their own would non-the-less propagate a bend imposed by manipulation with a glass needle (Lindemann and Rikmenspoel, 1972a). This provided a clue to the underlying working mechanism of the axonemal structure. The process of imposing a bend was in some way able to recruit the force-producing action of the dynein motor proteins. A second important clue to understanding axonemal function was contributed by the work of Kamiya and Okagaki (1986). They showed that individual doublet pairs from the frayed ends of a Chlamydomonas flagellum could generate a kind of beat cycle where the sliding of one doublet on the next would continue until a sufficient amount of bending terminated the sliding by prying the doublets apart. The experimental evidence thus suggests a mechanism that might serve to terminate an episode of sliding between adjacent doublets.

How can it be that bending plays a role in both initiating and terminating the action of the dynein motors? The answer may lie in the evaluation of the forces that act transverse to the axis of the outer-doublet microtubules ($t$-forces for short).

Data from our lab indicated that in rat and bull sperm flagella the doublets numbered 3 and 8 appeared to be linked permanently to each other through the central pair (Lindemann et al., 1992, Kanous et al. 1993). Intrigued by this finding, I repaired to my workshop and made a crude balsa-wood model designed on the basis of the rat sperm results. I used strands of silicone adhesive to arrange nine balsa-wood strips in the pattern shown in Figure 2, and cemented the strips to wooden spacers at one end to construct a basal anchor. It was my hope that the wooden construction with plate-like elements, similar to the flattened outer dense fibers of a rat sperm flagellum, would provide some insight into how a structure with flattened elements and a central partition might move when force is applied to it. Much to my chagrin and initial disappointment the structure collapsed at every attempt to bend it. The collapse was such that the balsa wood strips flattened against the central partition. This would happen no matter how or where the bending was imposed. At first I assumed that this was a flaw of the simplified design of the wooden model. After further consideration, I began to realize that no design could get rid of the collapsing force, as this force originated from the elastic distention of the silicone linkers between the outer wooden elements.

I next explored the results of pushing one of the elements toward the base and pulling the adjacent element toward the tip. This produced an interesting result in that the force couple imposed on adjacent wooden elements did cause the whole structure bend, but to my surprise I saw how strongly the adjacent balsa strips pulled apart from one another while under this tension and compression. Figure 3 shows the result of passive bending and application of an active force couple on a three-element structure, with an assignment of the sign of the resulting $t$-force.

An analysis of the forces that pinch the linear elements together or spread them apart is presented in Lindemann (1994a). The transverse force originates from two sources; a local component due to the stretching of the elastic linkages between the linear elements, and a global component which is the product of the linear tension or compression multiplied by the local curvature. Of these two sources, the global component generally is the larger contributor to $t$-force everywhere except in the most distal part

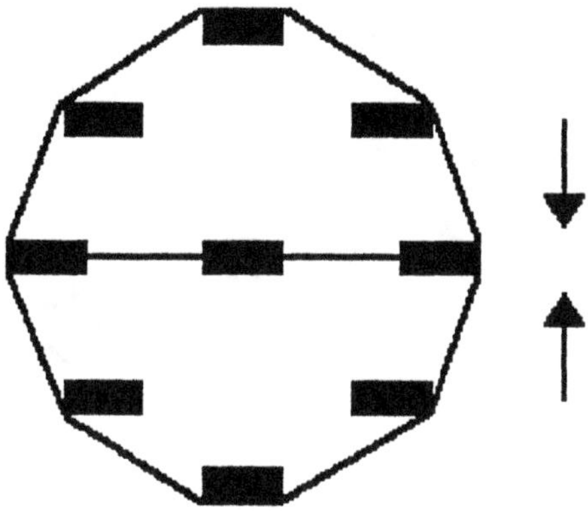

FIG. 2. *The balsa-wood model of the axoneme. The arrangement of the nine balsa-wood strips and silicone adhesive connecting strands of the axoneme model based on the partitioned-axoneme concept is shown (the concept is developed in Lindemann et al., 1992). All nine wooden elements were glued to spacers at one end to simulate a basal anchor. When the whole construction was subjected to external bending, it would only bend easily in the plane perpendicular to the linked central elements. When bent, the balsa wood strips always collapsed inward toward the partition as indicated by the arrows.*

of the structure. The local component is always positive (compressive), while the global component may vary between positive and negative. In a structure with the dimensions of a $10\mu$m cilium the global component is typically ten to twenty times larger than the local component at the end of the cilium near the basal anchor. It is therefore the global component that makes the $t$-force interesting as a potential regulator of dynein functioning.

**2. Hypothesis and model.** The Geometric Clutch hypothesis states that the $t$-force generated between the outer doublet microtubules of the axoneme regulates the initiation and termination of episodes of dynein action. As a pure principle, without attention to physical dimension and the details of a real flagellum, a model based on this hypothesis produced repetitive stable oscillations (Lindemann, 1994a). In this first and simplest computed model, the structure had to be initially bent to start activation of the dynein motors. Stable oscillations were only obtained if the force-producing bridges were harder to disengage (required more $t$-force to pry them off) than to engage. The need for some hysteresis in the cycle of events is not surprising, and it is quite reasonable that force-producing bridges would resist being pried loose once engaged at a binding site. The need for an initial bend is not in keeping with observation, as real cilia and flagella will start without prodding.

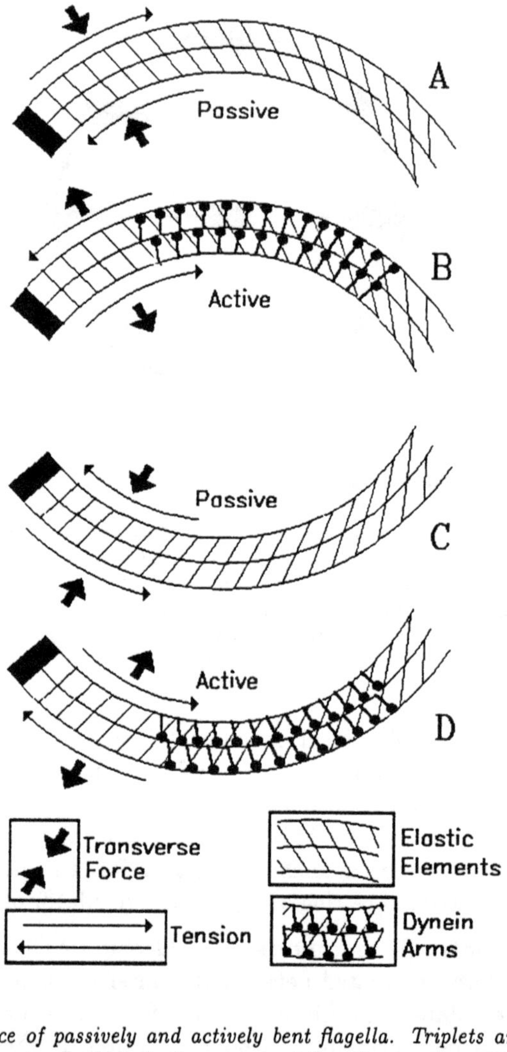


FIG. 3. *The t-force of passively and actively bent flagella. Triplets are displayed corresponding to three outer doublets in the process of flagellar bending. Both A and C depict doublets in passively bent flagella (bends imposed by an external force), while B and D portray doublets induced to bend by dynein action. The longitudinal force on the outer elements is displayed, along with the resultant t-force (bold arrows). While passive (imposed) bending compresses the axoneme in the plane of bending, active (dynein-induced) bending leads to axonemal distention. Inwardly directed t-forces (resulting in compression) are assigned a + (positive sign), while outwardly directed t-forces (causing distention) are designated by a − (negative sign). The curvature ($d\theta/ds$) multiplied by the tension yields the t-force. The corresponding signs of the curvature and tension are shown for each condition, to demonstrate the convention used in modeling the flagellum. Reprinted from Lindemann 1994a, with permission.*

A more sophisticated version of the model was developed that could initiate motility spontaneously, and could operate with the physical parameters of a real flagellum. In the second version each dynein was given a small initial probability to engage without the influence of $t$-force. The $t$-force and the adhesion force of the already attached dynein bridges each exerted a numerical influence on the probability of further bridge attachment or disengagement. Under this arrangement the adhesion of a few bridges at random, due to the base-level probability, precipitates a cascade of bridge attachment as illustrated in Figure 4. Ultimately a cascade of dynein bridge attachment produces enough force to initiate a bend and set things in motion. The growing bend and cumulative force from the activated bridges eventually creates sufficient negative $t$-force to break the bridges in the basal region of the simulated flagellum. This also fosters bridge attachment on the opposite side of the axoneme that, for all intents and purposes, is being passively bent and therefore is experiencing a positive $t$-force. The whole cycle of events thus created starts without an external push and repeats indefinitely.

The steps in the cycle on the two opposing sides of the axoneme are shown in Figure 5 (from Lindemann, 1994b). The key to this version of the model is the small base level of probability for spontaneous dynein engagement without $t$-force assistance. When the relative base level probability for spontaneous engagement is slightly higher on one side of the axoneme than the other it has a profound effect on the waveform of the resulting beat. The side of the axoneme with the higher base-level probability produces the principal bend stroke of the beat cycle and the ratio of the base level probability on the two sides of the axoneme determines the beat asymmetry. Varying these two parameters will automatically simulate a flagellar-like (more symmetric) or cilia-like (more asymmetric) beat cycle, with no need for further adjustments to the basic working mechanism (as shown in Figure 6).

Chlamydomonas has become the organism of choice for the study of the structure and genetics of the flagellum. This is owing to the fact that the organism is well suited to carry out molecular techniques, and an extensive collection of mutant strains is maintained in a central depository. When scaled to the size of a Chlamydomonas flagellum the Geometric Clutch simulation does a remarkable impersonation of the real thing (Figure 7). What is even more interesting is that it only requires the adjustment of a single parameter to alter the beat pattern from normal swimming (Figure 7A) to the reversal pattern the organism exhibits in response to a change in light intensity (Figure 7B). The parameter that must be changed is the rest-level probability of the $P$-bend dynein bridges. In nature, this would correspond to changing the length of the dynein stalk or changing the projection angle of the dynein arm from the outer doublet. This is an effect that could easily be accomplished by changing the conformation of a

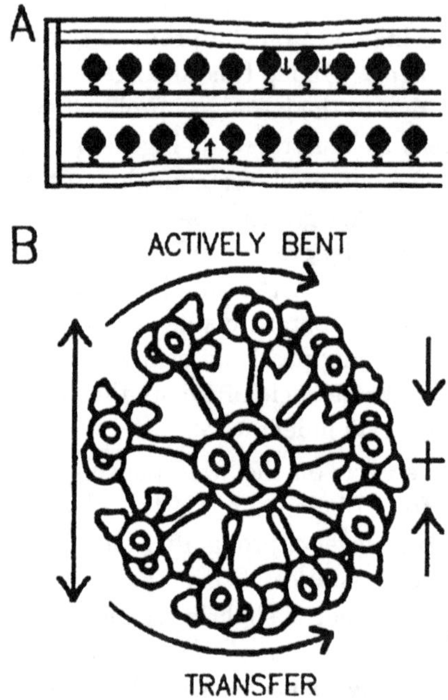


FIG. 4. *Adhesion Forces and Excitability.*
*A. When a dynein bridge forms in a resting axoneme, the interdoublet spacing is larger than the dynein length. Kinetic energy must initially contribute to stretching the dyneins, and permitting random bridge attachment. Once a dynein head with sufficient kinetic energy forms a bridge, the bridge is elastically stretched across the interdoublet space. Each dynein bridge attached in this manner contributes a force, pulling the doublets closer, which in turn increases the probability of additional bridge attachments. This makes a cascade of bridge attachment possible, and it also provides adhesion between the adjacent doublets. The resulting adhesion can only be overcome by a strong negative t-force.*
*B. Attachment of bridges on one side of the axoneme decreases the probability of a cascade of bridging on the opposite side. This is due to transfer of force through the interdoublet links. The transferred force increases interdoublet separation on the passive side. Reprinted from Lindemann, 1994b, with permission.*

single well-placed protein associated with the dynein arm. A prediction as straightforward as this should be possible to test.

The Geometric Clutch model with this stochastic bridge-cascade mechanism was also modified to see if the much larger and stiffer specifications of a bull sperm would also be compatible with the same underlying motor mechanism. To test this the length ($60\mu$m) and measured stiffness ($4 \times 10^{-12}$ dyne cm$^2$ from Lindemann et al., 1973) of a bull sperm flagellum were used in place of the values for a simple cilium. The axoneme of a bull sperm has an extra set of elements called the outer dense fibers that can be seen in Figure 8. The outer doublets of the bull sperm axoneme are attached to these outer dense fibers, which in turn are anchored in a structure called the connecting piece at the base of the flagellum (this arrangement is illustrated in Figure 9). In bull sperm, the doublets themselves are not anchored in a basal body as they are in other cilia and flagella. The attachment of the doublets to the outer dense fibers in a bull sperm was taken into account in the calculation of inter-doublet sliding and also in the working diameters used for torque and $t$-force calculation (as illustrated in Figure 9). The resulting modified model did not in any way modify the underlying axonemal motor or the $t$-force-switching algorithm. The results (reported in Lindemann, 1996) were remarkably consistent with the motility pattern that is seen in living bull sperm, as shown in Figure 10. The beat frequency, the waveform and the wave propagation were all fairly well duplicated. The most remarkable thing is that the same underlying motor can still work and work well in a situation where the forces and torque (see Figure 11) are an order of magnitude larger than they are in a $10\mu$m cilium.

The Geometric Clutch hypothesis seems to fit well with existing observations on cilia and flagella that are already a matter of record in the scientific literature (for a review see Lindemann and Kanous, 1997). A second, and more critical, challenge for the model is whether or not it can tell us anything new about cilia and flagella, which can help to guide development of the field. To do this a model must have predictive value. The model predicts that the beat cycle depends on an inter-doublet elastic component for the creation and distribution of the $t$-force. In a short monograph of predictions from the model, the existence of this component was predicted and the size was estimated to be between 0.01 and 0.03 dyne/cm (Lindemann and Kanous, 1995). Within a short time of this prediction, Yagi and Kamiya (1995) published an experimentally determined estimate of 0.02 dyne/cm for the nexin elasticity of a Chlamydomonas flagellum.

The most fundamental prediction of the Geometric Clutch model is that the $t$-force is responsible for switching the dyneins "off" at the appropriate step in the beat cycle, thereby allowing alternation of the action of the two opposing sets of dyneins (as illustrated in Figure 5). Therefore, any experimental procedure that reduces the development of $t$-force should result in an arrest of the beat cycle. The Geometric Clutch simulation of a bull sperm was used to devise two strategies that cause switching failure in the simulation. The first strategy was to block the formation of the basal bend in the flagellum by stopping the movement of the flagellum by imposing an equal and opposite torque at a point along the flagellum (blocking strategy). This prevents the $t$-force (product of linear tension on the doublets multiplied times the curvature) from reaching the $t$-force switching-threshold of the model program. The model was also modified so that the flagellum could be shortened while maintaining the original stiffness and inter-doublet working diameters over the shortened segment (clipping strategy). This strategy reduced the $t$-force by reducing the number of contributing dyneins while keeping the high basal stiffness. In both the blocking and clipping strategies, when $t$-force was sufficiently reduced, the beat cycle arrested. Figure 12 shows the result of computed blocking experiments in both $P$ and $R$ bend directions.

Both strategies, blocking and clipping, were readily tested in the laboratory on real bull sperm. The results are extremely encouraging. Real bull sperm flagella exhibit similar arrest behavior when subjected to either of the two procedures (Holcomb-Wygle et al., 1999).

**3. Future directions.** The Geometric Clutch computer model has been a useful template, which accurately mimics and predicts many specific behaviors seen in cilia and flagella. The greatly simplified representation of the axoneme as two sets of opposed dynein bridges controlled only by $t$-force and stochastic variability has performed well beyond expectation. It is now imperative to elaborate on the model to include more of the detailed properties of a complete axoneme. Some progress has already been made toward this goal.

A new version of the model is being developed that incorporates more of the properties of a complete axoneme. This newest version increases the number of independently regulated sets of dynein from two to four. In addition, each of the four sets has two individually specified populations of dyneins that can be assigned different characteristics. For the first time it is possible to specify inner and outer dynein arms with different properties. The first property to be explored with the new formulation is the effect of force-velocity characteristics on the performance of the beat cycle. The force exerted by each dynein in the new model is conditionally assigned in an algorithm that evaluates the sliding velocity at each location. The model has also been improved to include a primitive representation of the spoke apparatus. In the first attempt to model an action for the spokes, the spokes were made to bear a large share (80%) of the $t$-force across the axoneme until a threshold is reached at which the spokes detach from the central hub. Once this transition is reached, the $t$-force is redistributed to the active dyneins at the segment in question. In the electron micrographs of Warner and Satir (1974) it was demonstrated that at a critical point in bend development the spokes "jump" This is a clear indication that the spokes must be able to come loose from the hub as the cycle of bending develops. In the Geometric Clutch treatment of spokes they pop off the hub in response to reaching a threshold of $t$-force. This transition acts as a means of redistributing $t$-force to the dynein arms.

It is most encouraging that with these modifications in place the upgraded model produces life-like simulations of a cilium or flagellum as well as the earlier prototype. The upgraded version will be used to explore hypotheses related to differential inner and outer arm functions, a level of inquiry that was beyond the sophistication of the first model. Real cilia and flagella can produce a beat after removal of the outer dynein from the axoneme (Gibbons and Gibbons 1993). It will be very informative to see if the Geometric Clutch mechanism is compatible with this experimental data.

**4. Conclusions.** The Geometric Clutch hypothesis attempts to explain the organization of the beat cycle in cilia and flagella based on assigning a central role for the $t$-force. The $t$-force, which is the force acting to push the doublets together or apart, is postulated to determine when the dynein arms are activated and when their action is terminated. The computer model developed from this concept is very successful at producing cilia-like and flagella-like motion. The computed simulation can also reproduce many of the behaviors of the large flagellum of bull sperm, and has been able to accurately predict the results of laboratory experiments on bull sperm. The model may provide a viable framework with which to question and understand the workings of real cilia and flagella.

**Acknowledgements.** The author wishes to thank Ms. Kathleen A. Schmitz for assistance with the preparation of the manuscript and figures. Dana Holcomb-Wygle and Kathleen Schmitz carried out the experimental studies on bull sperm that are highlighted in the manuscript. This work was supported by grant # MCB-9722822 from the National Science Foundation.

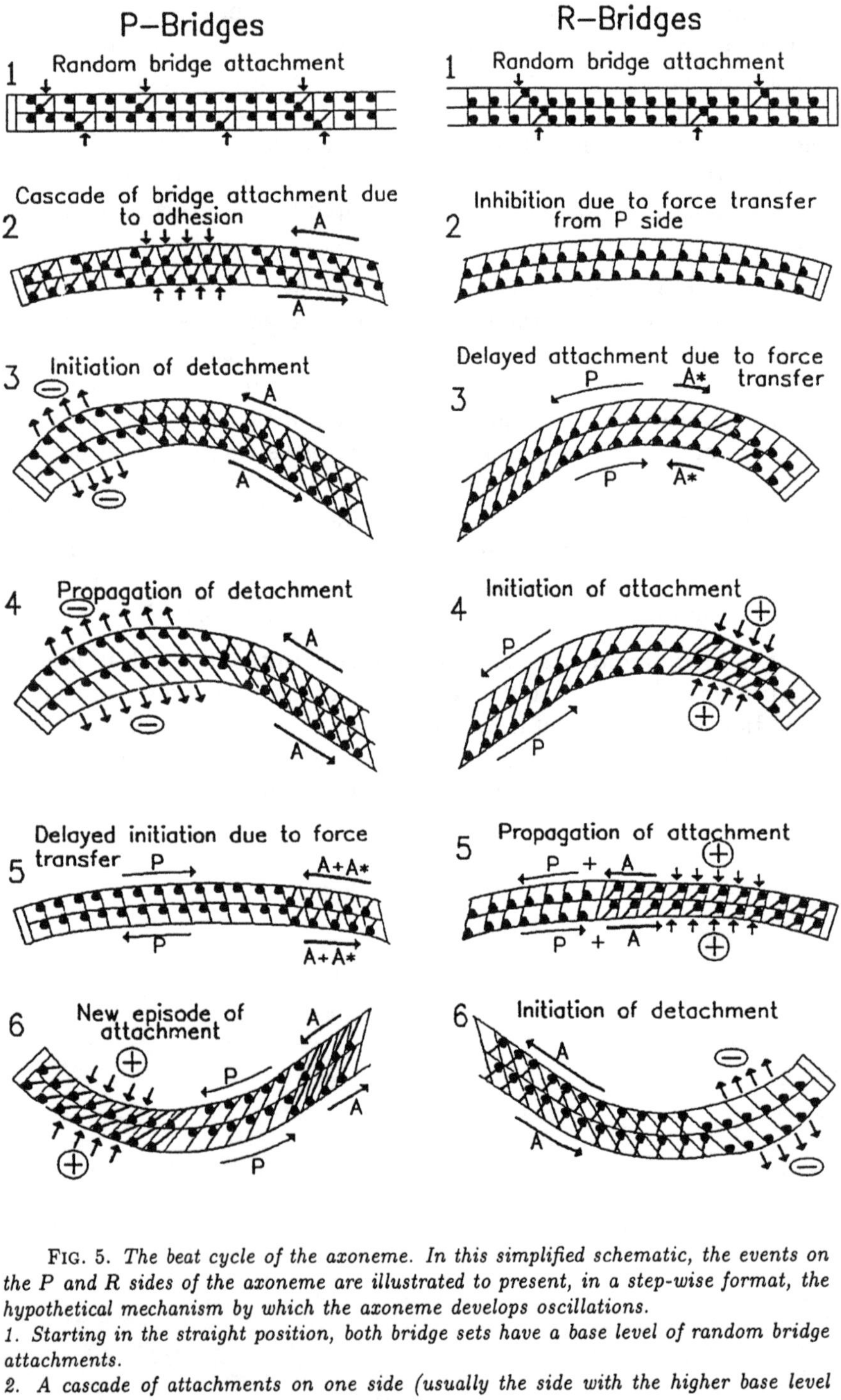


FIG. 5. *The beat cycle of the axoneme. In this simplified schematic, the events on the P and R sides of the axoneme are illustrated to present, in a step-wise format, the hypothetical mechanism by which the axoneme develops oscillations.*

*1. Starting in the straight position, both bridge sets have a base level of random bridge attachments.*

*2. A cascade of attachments on one side (usually the side with the higher base level*

*of attachment probability) begins an episode of sliding and (due to force transfer) simultaneously inhibits bridge attachment on the opposing side.*
*3. As bending increases, a negative t-force develops on the active side of the axoneme and is strongest near the base. Force transfer continues to inhibit the opposite side, but becomes weaker as detachment of dynein bridges proceeds.*
*4. A propagating area of bridge detachment on the P bridge side is accompanied by a positive t-force from passive bending on the R bridge side, and this ensures a cascade of bridge activation on the R side.*
*5. Curvature is now reversing due to the R bridge forces. P bridges are temporarily inhibited by force transfer from the R bridge side.*
*6: The curvature is now favorable for production of a negative t-force near the base on the R bridge side, and deactivation begins there. Meanwhile, a new cycle is beginning on the P bridge side, as positive t-force from passive links and residual active bridges contribute to activation. Arrows labeled P are passive force contributions originating from stretching the passive interdoublet links. Arrows labeled A are active forces from bridges, and A* indicates active force transferred from the opposite side. Reprinted from Lindemann 1994b, with permission.*

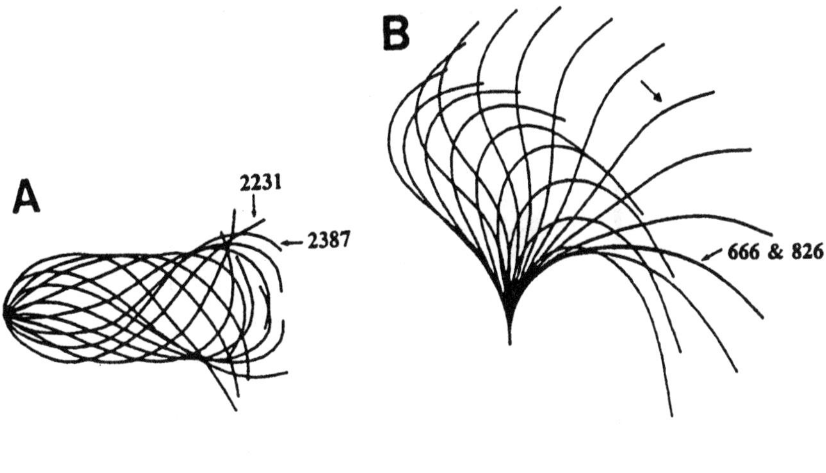


FIG. 6. *Computed simulations of a flagellum and a cilium. The output from the Geometric Clutch simulations of both a flagellar and a ciliary beat cycle are displayed. A 30μm long "flagellum" with a freely pivoting base is displayed in A, showing every 12th iteration of a cycle divided into intervals of 0.0001 seconds per iteration. A* 10μm *"cilium" with an anchored base is exhibited in B, showing every 8th iteration, utilizing the same iteration intervals as that used in the flagellar model. The numbers printed on the output denote the iteration number of the indicated beat position, corresponding to the first and last elements displayed. Note that the cilium clearly exhibits a two-phased beat cycle, with well-defined effective and recovery strokes. The flagellum shows wave propagation tip-ward from the base in both bending directions. The Geometric Clutch program was capable of producing both patterns of beating, without any fundamental change in the switching algorithm. The main determinants of the resultant beat included: base anchoring, axonemal length, and assignment of base level bridge attachment probability of the P and R bridges on opposing sides of the axoneme. The figure is reprinted from Lindemann, 1994b, with permission. The complete modeling parameters for the figure are given in that original report.*

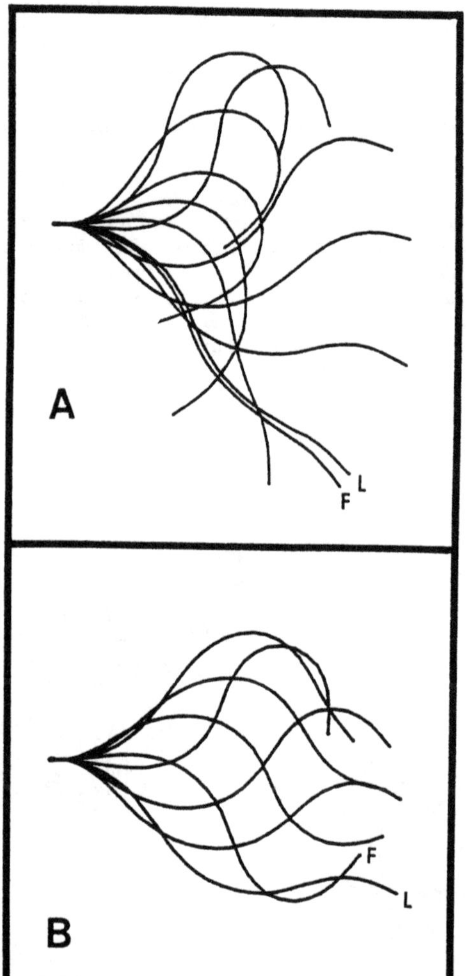


FIG. 7. *Simulation of a Chlamydomonas flagellum. Both A and B display one beat cycle of the Geometric Clutch model scaled to simulate the beat of a 12μm Chlamydomonas flagellum. The flagellum is fixed in position at the basal end, as is the case in Chlamydomonas, and a stiffness of* $1.0 \times 10^{-13}$ *dyne • cm*$^2$ *was used, which seems to provide the best fit to the Chlamydomonas waveform. In both A and B the output displays every 16*$^{th}$ *iteration at an iteration interval of 0.0001 second. The pattern in A resembles the normal forward beat of Chlamydomonas, and in B resembles the beat during a phobic response to light. The single model parameter that was changed to produce the two different patterns was the rest-level probability of engagement for the P-bend dyneins, which is set at 0.05 in A and 0.02 in B. The rest-level engagement probability of the R-bend dyneins is set at 0.005 in both.*

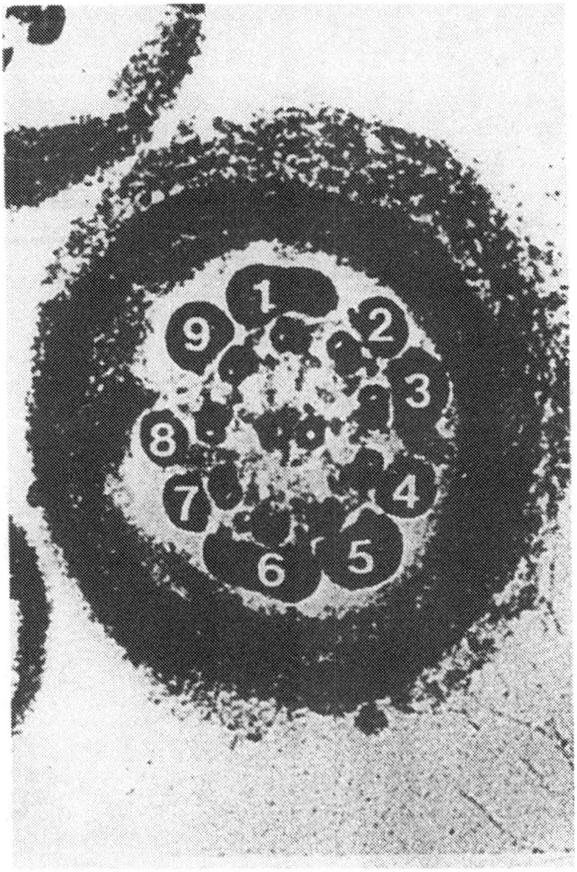


FIG. 8. *Special features of the mammalian sperm axoneme. A TEM cross-section of a bull sperm axoneme is displayed, with the outer doublets numbered in the same convention applied to other cilia and flagella. The presence of a fibrous (dense staining) outer sheath around the axoneme identifies the section as one from the flagellar principal piece. The outer dense fibers (ODFs) are attached to their respective doublet over much of their length, particularly in the principal piece region. These ODFs are largest at the flagellar base, and taper away to a termination point part way down the principal piece. Reprinted from Kanous et al. 1993, with permission.*

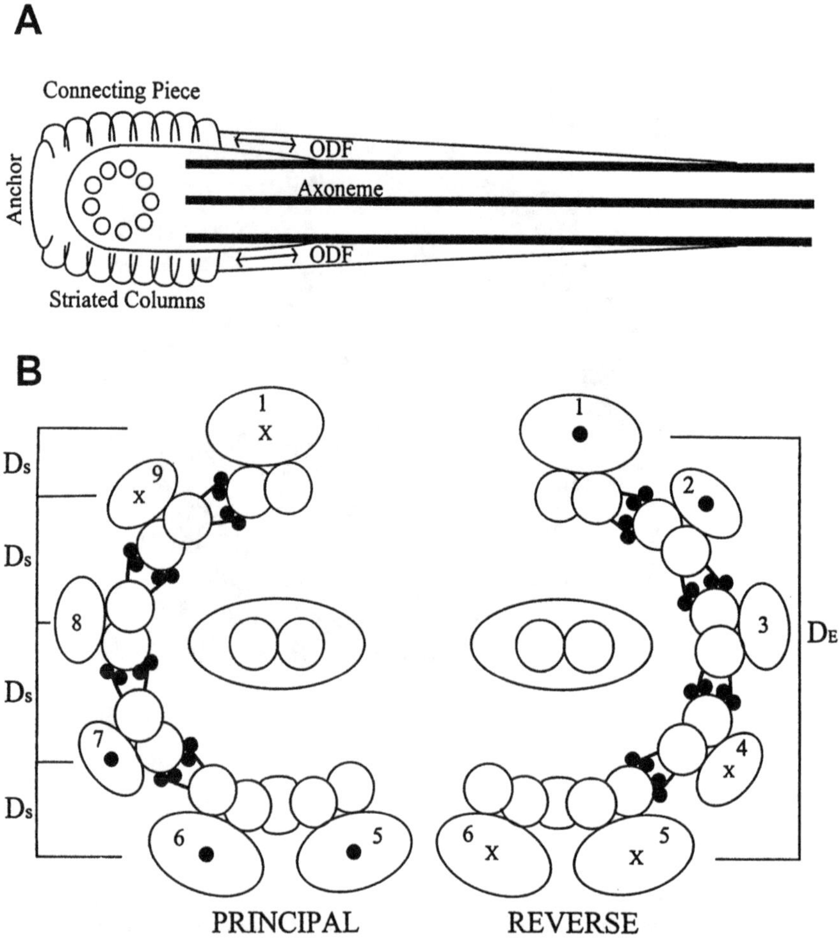


FIG. 9. *A functional schematic diagram of the mammalian sperm axoneme. A. In the mammalian sperm axoneme, the outer microtubule doublets are not anchored into a basal body at the flagellar base, but are attached to the ODFs along much of their length. However, the ODFs are affixed to the striated columns of the connecting piece. Consequently, when the doublets slide by the action of the dynein motors, the resultant force is transmitted through the ODFs to the connecting piece (arrows). The connecting piece assumes the role of a basal anchor replacing the basal body which disassembles during spermatogenesis. B. The sliding induced by the dynein-tubulin cross-bridge cycle during formation of principal and reverse bends results in maximal sliding displacement between elements #1 and #5-6. Unlike in simple flagella, the magnitude of interdoublet sliding is not determined by interdoublet spacing, but is dictated by the inter-ODF distances in the bending plane (identified as DS in the diagram). Force will also be transmitted to the ODFs, causing the greatest torque to develop from the force imparted to elements #1 and #5–6 because they exhibit the greatest separation (the effective diameter, $D_E$). •s and Xs indicate thrust in the directions into (•) and out of (X) the page if the axoneme were being viewed baseward. Reprinted from Lindemann 1996, with permission.*

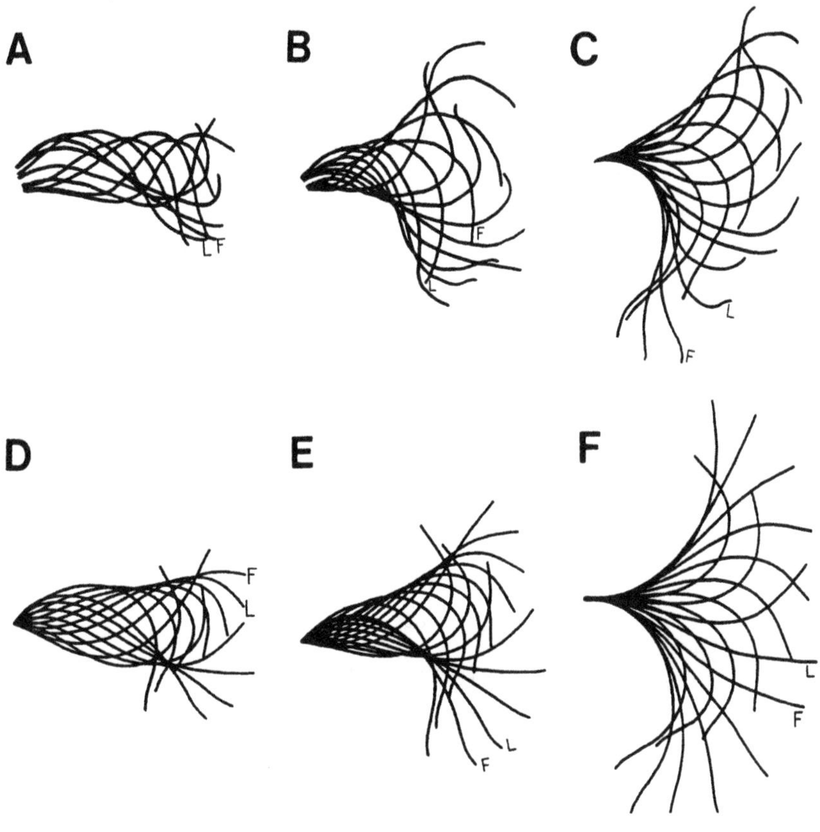


FIG. 10. *Tracings of live bull sperm and computed simulations. A, B, and C are traced positions of live bull sperm at 23° C from videotaped images using strobed illumination. Each tracing is at 0.0166 seconds. D, E, and F are computed simulations using the version of the Geometric Clutch Model that has been modified to use the dimensions and stiffness of a bull sperm. All three of the live sperm were partially immobilized by having the head stuck to a microscope slide. In A, the head is fairly free to pivot. In B, it is somewhat more restricted by surface friction with the slide while in C the head is firmly affixed to the glass slide. Mimicking the same three conditions in the computer model gave rise to D, E and F, using no other modifications. The computer output for A and B display iterations at 0.006 second intervals (every 12th iteration), while C is plotted at 0.0075 second intervals (every 15th iteration). The labels F and L designate the first and last traced positions in all figures. Reprinted from Lindemann 1996, with permission.*

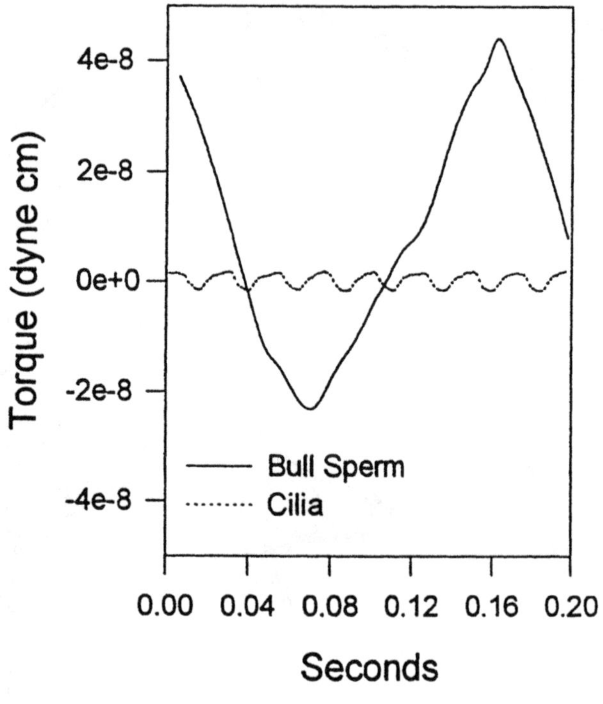


FIG. 11. *Comparison between the torque development of a bull sperm flagellum and that of a simple 10μm cilium. The bending torque at segment #2 (one segment from the base) is displayed for both a simulated bull sperm flagellum and a simulated simple cilium. However, since the simple cilium model used 100 nm as the functional diameter approximation, instead of the true spacing between doublets 5–6 and 1 (≡ 140 nm), the cilium values have been multiplied by 1.4 to provide a fairer comparison. Together, the greater number of contributing dynein bridges and the greater effective diameter contributed by the outer dense fibers in bull sperm result in approximately a 30× differential in the bending torque between a bull sperm flagellum and a simple 10μm cilium. Reprinted from Lindemann 1996, with permission.*

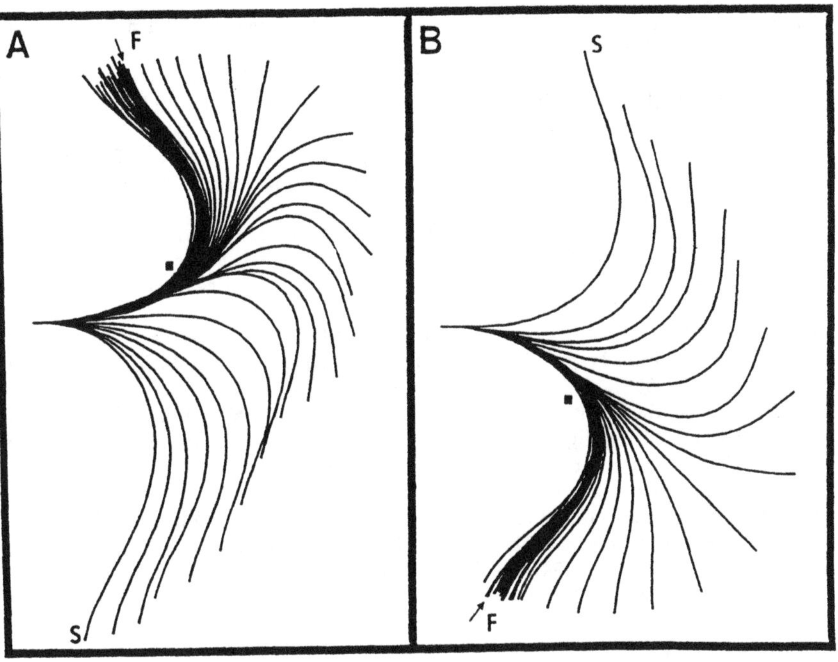


FIG. 12. *Computed blockade of the flagellar beat. The simulated flagellar beat of a bull sperm is shown in a computed blockade of the movement at the positions indicated by the squares. The blocking mechanism counters the driving torque at the segments between the obstacle and the flagellar base, but does not prevent the flagellum from moving or reversing direction if the internal torque balance changes to favor movement away from the blockage. By preventing the development of curvature in the basal portion of the flagellum, insufficient t-force is generated to allow the beat to reverse, and an arrest results. The sequences shown are plotted every 20 iterations (each iteration is 0.0005 second). Panel A shows a block in the R-bend direction for a total of 1000 iterations, and panel B shows a block in the P-bend direction for a total of 800 iterations. S denotes the first iteration of the display and F the final iteration.*

## REFERENCES

[1] AFZELIUS, B (1959). Electron Microscopy of the Sperm Tail. Results Obtained with a New Fixative. J. Biophys. Biochem. Cytol. **5**: 269–278.

[2] AFZELIUS, B (1961). The Fine Structure of the Cilia from Ctenophore Swimming Plates. J. Biophys. Biochem. Cytol. **9**: 383–394.

[3] BROKAW, C.J. (1961). Movement and Nucleoside Polyphosphatase Activity of Isolated Flagella from *Polytoma uvella*. Exp.Cell Res., **22**: 151–162.

[4] GIBBONS, I.R. (1961). The Relationship Between the Fine Structure and Direction of Beat in Gill Cilia of a Lamellibranch Mollusc. J. Biophys. Biochem. Cytol. **11**: 179–205.

[5] GIBBONS, I.R. AND GRIMSTONE, A.V. (1960). On Flagellar Structure in Certain Flagellates. J. Biophys. Biochem. Cytol. **7**: 697–716.

[6] GIBBONS, I.R. AND ROWE, A.J. (1965). Dynein: A Protein with Adenosine Triphosphatase Activity from Cilia. Science, **149**: 424–426.

[7] GIBBONS, I.R. AND GIBBONS, B.H. (1972). Flagellar Movement and Adenosine Triphosphatase Activity in Sea Urchin Sperm Extracted with Triton X-100. J. Cell Biol., **54**: 75–97.

[8] GIBBONS, I.R. AND GIBBONS, B.H. (1973). The Effect of Partial Extraction of Dynein Arms on the Movement of Reactivated Sea-urchin Sperm. J. Cell Sci., **13**: 337–357.

[9] HOLCOMB-WYGLE, D.L., SCHMITZ, K.A., AND LINDEMANN, C.B. (1999). Flagellar Arrest Behavior Predicted by the Geometric Clutch Model is Confirmed Experimentally by Micromanipulation Experiments on Reactivated Bull Sperm. Cell Motil. Cytoskeleton, **44**: 177–189.

[10] KAMIYA, R. AND OKAGAKI, T. (1986). Cyclical Bending of Two Outer-doublet Microtubules in Frayed Axonemes of Chlamydomonas. Cell Motil. Cytoskeleton, **6**: 580–585.

[11] KANOUS, K.S., CASEY, C., AND LINDEMANN, C.B. (1993). Inhibition of Microtubule Sliding by $Ni^{+2}$ and $Cd^{+2}$ Evidence for a Differential Response of Certain Microtubule Pairs Within the Bovine Sperm Axoneme. Cell Motil. Cytoskeleton, **26**: 66–76.

[12] LINDEMANN, C.B. (1994a). A Geometric Clutch Hypothesis to Explain Oscillations of the Axoneme of Cilia and Flagella. J. Theor. Biol., **168**: 175–189.

[13] LINDEMANN, C. B. (1994b). A Model of Flagellar and Ciliary Functioning Which Uses the Forces Transverse to the Axoneme as the Regulator of Dynein Activation. Cell Motil. Cytoskeleton, **29**: 141–154.

[14] LINDEMANN, C.B. (1996). The Functional Significance of the Outer Dense Fibers of Mammalian Sperm Examined by Computer Simulations with the Geometric Clutch model. Cell Motil. Cytoskeleton, **34**: 258–270.

[15] LINDEMANN, C.B. AND KANOUS, K.S. (1995). "Geometric Clutch" hypothesis of Axonemal Function: Key Issues and Testable Predictions. Cell Motil. Cytoskeleton, **31**: 1–8.

[16] LINDEMANN, C.B. AND KANOUS, K.S. (1997). A Model for Flagellar Motility. Int. Rev. of Cytol., **173**: 1–72.

[17] LINDEMANN, C.B., ORLANDO, A. AND KANOUS, K.S. (1992). The Flagellar Beat of Rat Sperm is Organized by the Interaction of Two Functionally Distinct Populations of Dynein Bridges with a Stable Central Axonemal Partition. J. Cell Sci., **102**: 249–260.

[18] LINDEMANN, C.B. AND RIKMENSPOEL, R. (1972a). Sperm Flagella: Autonomous Oscillations of the Contractile System. Science, **175**: 337–338.

[19] LINDEMANN, C.B. AND RIKMENSPOEL, R. (1972b). Sperm Flagellar Motion Maintained by ADP. Exp. Cell Res., **73**: 255–259.

[20] SALE, W.S. AND SATIR, P. (1977). Direction of Active Sliding of Microtubules in Tetrahymena Cilia. Proc. Natl. Acad. Sci. USA, **74**: 2045–2049.

[21] SATIR, P. (1965). Studies in Cilia. II. Examination of the Distal Region of the Ciliary Shaft and the Role of the Filaments in Motility. J. Cell Biol., **26**: 805–834.

[22] SATIR, P. (1968). Studies on Cilia. III. Further Studies on the Cilium Tip and a "Sliding Filament" Model of Ciliary Motility. J. Cell Biol., **39**: 77–94.

[23] SUMMERS, K.E. AND GIBBONS, I.R. (1971). Adenosine Triphosphate- induced Sliding of tubules in Trypsin-treated Flagella of Sea Urchin Sperm. Proc. Natl. Acad. Sci. USA, **68**: 3092–3096.

[24] Tamm, S.L. and Tamm, S. (1984). Alternate Patterns of Doublet Microtubule Sliding in ATP-disintegrated Macrocilia of the Ctenophore Beroe. J. Cell Biol., **99**: 1364–1371.

[25] Warner, F.D. and Satir, P. (1974). The Structural Basis of Ciliary Bend Formation. Radial Spoke Position Changes Accompanying Microtubule Sliding. J. Cell Biol., **63**: 35–63.

[26] Yagi, T. and Kamiya, R. (1995). Novel Mode of Hyper-oscillation in the Paralyzed Axoneme of a Chlamydomonas Mutant Lacking the Central-pair Microtubules. Cell Motil. Cytoskeleton, **31**: 207–214.

# A ONE-DIMENSIONAL FLUID DYNAMIC MODEL OF THE SYSTEMIC ARTERIES*

METTE S. OLUFSEN[†]

**Abstract.** The systemic arteries are modeled as a bifurcating tree of compliant and tapering vessels. Blood flow and pressure in the vessels are determined by solving the axisymmetric Navier-Stokes equations. The arterial tree ranging from the aorta to the arterioles consists of a tree with more than 20 generations. Computing blood flow and pressure for all vessels requires a prohibitive amount of time. To avoid using too much time, we have truncated the arterial tree after a limited number of generations and applied a suitable outflow boundary condition. To this end we propose a structured tree model in which a root impedance is determined using a semi-analytical approach. The structured tree is a binary asymmetric tree in which the radii of the daughter vessels are scaled linearly with the radius of the parent vessel. The root impedance of the structured tree is found by propagating solutions of a wave equation from the terminals to the root of the structured tree. The wave equation is derived by linearizing the axisymmetric Navier-Stokes equations together with applying a long-wave approximation. The root impedance of the structured tree provides a dynamical outflow boundary condition, which is computationally feasible. The structured tree outflow boundary condition is based on physiologic principles and it accommodates wave propagation effects for the entire systemic arterial tree. Blood flow in the large systemic arteries is verified by comparing simulations with data obtained from magnetic resonance measurements. The outflow boundary condition is verified by comparisons with literature data and with a standard model (the three-element windkessel model).

**Key words.** Arterial blood flow, Arterial modeling, Blood flow modeling, Systemic arteries, Arterial outflow conditions, Biofluid Dynamics.

**AMS(MOS) subject classifications.** Primary 76Z05, 92C35, 65M06, 35L60.

## 1. Introduction.

The systemic arteries consist of a sophisticated network of compliant and tapering vessels, which transport blood from the heart to the capillaries in order to supply all muscles and tissue with nutrients and oxygen. Blood flow and pressure in the systemic arteries can be found from solving the incompressible axisymmetric Navier-Stokes equations combined with a state equation relating pressure to cross-sectional area [2, 5, 19–22, 25, 26, 30, 31]. The basis for the approach used in this paper follows the ideas outlined in the above referenced work.

The incompressible axisymmetric Navier-Stokes equations are nonlinear and cannot be solved analytically. Applying the Navier-Stokes equations for the entire arterial tree would be too elaborate because the network becomes too large for any practical purposes. The complexity of the arterial tree made us interested in studying how the small arteries, in a

*This research was supported by "Group Infrastructure Grant" DMS-9631755 from the National Science Foundation, the Danish Academy of Technical Sciences grant EF501, and by the Danish Eureka project 1063.

[†]Center for BioDynamics & Department of Mathematics, Boston University, Boston, MA 02215.

physiological way, can be lumped to yield a computationally feasible outflow boundary condition for the large arteries. The purpose of the small arteries is to provide the capillary network, distributed evenly in muscles and tissue, with sufficient blood. Studying the distribution of the small arteries gave rise to the idea of modeling them as asymmetrical and binary structured trees in which the radii of the daughter vessels are scaled linearly with the radius of the parent vessel. In order to make it feasible to solve the Navier-Stokes equations for the vessels of the structured trees the equations are simplified (by linearization and a long-wave approximation). The simplified equations yield a wave equation which for each vessel can be solved analytically. Propagating the solution of the wave equations from the terminals to the root of the structured tree yields a root impedance which can be applied as an outflow boundary condition for the large arteries.

Previous modeling studies did not model the network of small arteries but used simple models comprising a constant resistor (a constant ratio between flow and pressure) [2, 12, 26] or an impedance based on the windkessel model [28, 30, 31]. The constant resistance condition has the disadvantage that it forces the flow and pressure to be in phase and it is very sensitive to small changes in peripheral resistance [19]. The resulting propagation of the pulse-wave is not physiological since the "shape" is changed and artificial reflections are invoked when the wave is propagated back through the arterial tree [28]. The windkessel model does not have these problems. However, the windkessel model is a lumped model which does not include the spatial distribution of the small arteries. Hence, it cannot include wave-propagation effects in the part of the arterial system that it models. On the other hand, windkessel models have been successfully applied as boundary conditions for the large arteries [31].

In this paper our arterial model [20–22] is derived and validated by comparing computed and measured flows. The measured flows are obtained by magnetic resonance techniques [22]. In addition, the structured tree model is validated by comparing it with the windkessel model and with impedances found in the literature. Section 2 describes the model of the large arteries, Section 3 describes the structured tree and the windkessel models, Section 4 presents our results and compares our computed flows with flows obtained using magnetic resonance techniques, and Section 5 compares our model with the windkessel model and with impedance data obtained from the literature.

**2. The large systemic arteries.** The incompressible axisymmetric Navier-Stokes equations combined with a state equation relating pressure to the cross-sectional area can be used to predict the blood flow and pressure in the large systemic arteries [20, 26]. The large arteries are modeled as a binary tree (all junctions have two daughters) with vessels being cylindrical, compliant and tapering (see Figure 1). The geometry of the large arteries is faithful to the human arteries with a number of exceptions: A number of small vessels branching from the large arteries are left out and arteries present on both sides of the body are modeled as identical.

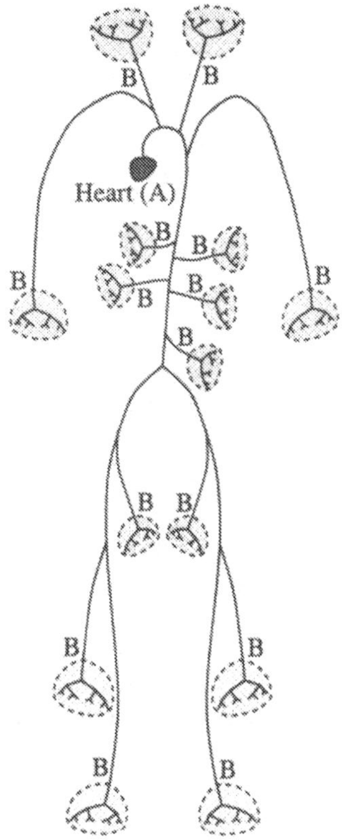


FIG. 1. *The systemic arterial tree. The tree consists of the large arteries originating at the heart (marked by A) and terminating at the points marked by B's. The small arteries are modeled as structured trees attached at the terminals of the large arteries. (The figure is adapted from [21] and reprinted with permision from American Physiological Society.)*

The incompressible axisymmetric Navier-Stokes equations give rise to two equations: Conservation of mass and $x$-momentum [1, 9]:

$$\frac{\partial u_x}{\partial x} + \frac{1}{r}\frac{\partial (r u_r)}{\partial t} = 0 \tag{2.1}$$

$$\frac{\partial u_x}{\partial t} + u_x \frac{\partial u_x}{\partial x} + u_r \frac{\partial u_x}{\partial r} + \frac{1}{\rho}\frac{\partial p}{\partial x} = \nu \left( \frac{1}{r}\frac{\partial}{\partial r}\left( r \frac{\partial u_x}{\partial r} \right) + \frac{\partial^2 u_x}{\partial x^2} \right) \tag{2.2}$$

where $\rho$ is the fluid density (constant), $\nu = \mu/\rho$ is the fluid kinematic viscosity (constant), $\mathbf{u} = [u_r(r,x,t), u_x(r,x,t)]$ is the fluid velocity, and $p(x,t)$ is the fluid pressure.

A boundary layer is applied to account for the no slip condition. We assume that the longitudinal velocity can be approximated using the following velocity profile which,except for a thin boundary layer,is flat across the vessel diameter:

$$u_x = \begin{cases} u, & r \leq R - \delta \\ u(R-r)/\delta, & R - \delta < r \leq R \end{cases}$$

where $\delta$ (constant) is the thickness of the boundary layer. $\delta$ can be estimated from $(\nu T)^{1/2} \approx 0.2$ cm, where $T$ is the duration of one cardiac cycle [14, 24]. Introducing the flow

$$q(x,t) = \int_A u_x dx$$

the one-dimensional equations can be derived by integrating over the vessel cross-sectional area and disregarding terms of order $\delta$ and smaller [22].

$$\frac{\partial A}{\partial t} + \frac{\partial q}{\partial x} = 0 \tag{2.3}$$

$$\frac{\partial q}{\partial t} + \frac{\partial}{\partial x}\left(\frac{q^2}{A}\right) + \frac{A}{\rho}\frac{\partial p}{\partial x} = -\frac{2\pi\nu R q}{\delta A}. \tag{2.4}$$

Here $A(x,t) = \pi R(x,t)^2$ is the cross-sectional area and $R(x,t)$ is the radius of the vessels. Assuming that the deformation of the arterial wall is purely elastic, a relationship between pressure and cross-sectional area can be expressed as [13, 20]:

$$p(x,t) = \frac{4}{3}\frac{Eh}{r_0(x)}\left(1 - \sqrt{\frac{A_0(x)}{A(x,t)}}\right) \tag{2.5}$$

where $A_0(x) = \pi r_0(x)^2$ is the cross-sectional area when the pressure of the vessel equals the pressure of the surrounding tissue. Using estimates from Segers et al. [30, 31, 34] it is possible to estimate $Eh/r_0$ as a function of $r_0$.

$$\frac{Eh}{r_0} = k_1 \exp(k_2 r_0) + k_3 \tag{2.6}$$

where $k_1, k_2$, and $k_3$ are constants (see Figure 2).

In order to solve equations (2.3–2.5) two boundary conditions are needed for each vessel; one at the inlet and one at the outlet. For the arterial tree, boundary conditions are needed at the aortic root (at A in Figure 1), at the terminals of the large arteries (at B in Figure 1), and at each bifurcation (an outlet condition from the parent vessel and inlet conditions into each of the daughter vessels).

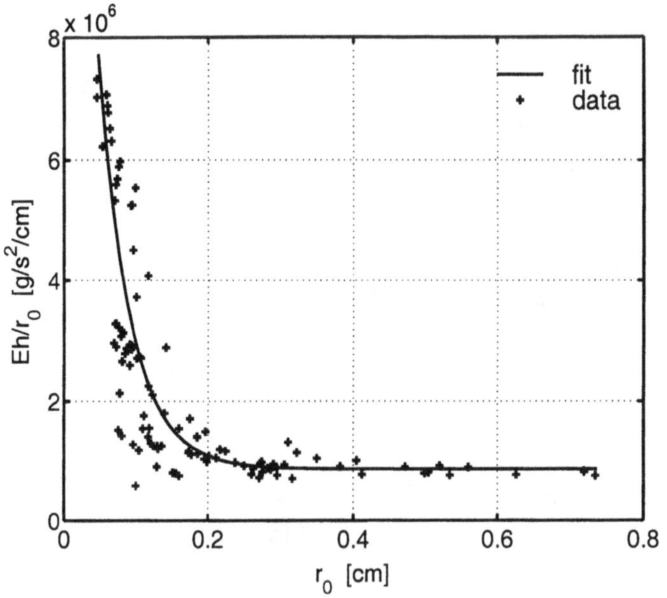


FIG. 2. *Fit to data for $Eh/r_0$, where $E$ is Young's modulus, $h$ is the wall-thickness and $r_0$ is the radius. The fitted function has the form $k_1 \exp(k_2 r_0) + k_3$, where $k_1 = 2.00 \times 10^7 g/(s^2 cm)$, $k_2 = -22.53\ cm^{-1}$, and $k_3 = 8.65 \times 10^5 g/(s^2 cm)$. (The figure is reprinted from [21] with permision from American Physiological Society.)*

The boundary condition at the aortic root can either be estimated from studying the characteristics of the pulse-wave or from measurements of either blood flow or pressure. The pulse-wave is periodic and has a skew symmetry. Consequently, the maximum of the pulse-wave occurs during the first third of the cardiac-cycle. The flow into the aorta can be approximated using the following periodic function.

$$
\begin{aligned}
q(0,t) &= \frac{\overline{q}_0 t}{\tau^2} \exp(-t^2/(2\tau^2)) \qquad && 0 \le t < T \\
q(0,t+jT) &= q(0,t) && j = 1,2,3,\ldots
\end{aligned}
\tag{2.7}
$$

where $\overline{q}_0$ is the cardiac output and $\tau$ is the time at which the maximal cardiac output is reached. The above expression yields a generic boundary condition which enable qualitative studies of the model. However, the pulse-wave varies significantly among human subjects. Hence, using a measured flow/pressure profile makes it possible to study the quantitative behavior of the model. The simulations presented in this paper are all based on magnetic resonance measurements of the flow in the ascending aorta (see Figure 3).

The boundary conditions at the terminals of the large arteries and at the bifurcations must also be based on estimates for either flow, pressure, or a relationship between them. In this paper, the outflows at the terminals of the large arteries are determined using a functional dependence between

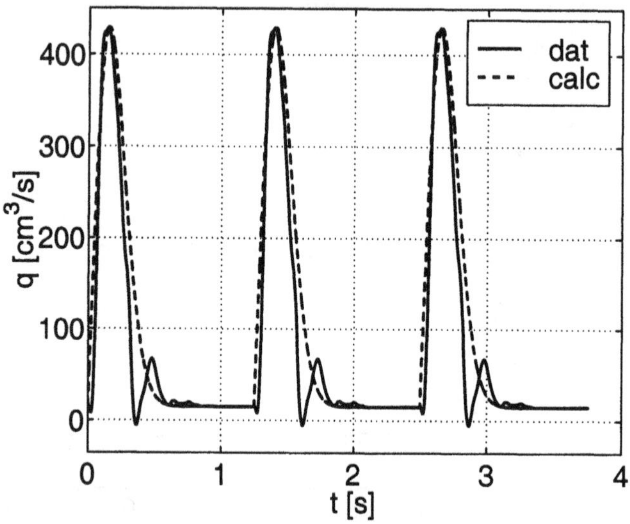


FIG. 3. *The inflow (at A in Figure 9) is a periodic repetition of data measured during one cardiac-cycle. The two curves show the measured data and the inflow approximated using (2.7).*

flow and pressure. A good approach is to determine an expression for the impedance based on the viscous flow in the network of small arteries and arterioles. Assume that the impedance can be defined by:

$$Z(x,\omega) = \frac{P(x,\omega)}{Q(x,\omega)} \tag{2.8}$$

where

$$p(x,t) = \sum_{k=-\infty}^{\infty} P(x,\omega_k)e^{i\omega_k t}, \quad q(x,t) = \sum_{k=-\infty}^{\infty} Q(x,\omega_k)e^{i\omega_k t}$$

with $\omega_k = \frac{2\pi k}{T}$ being the angular frequency, and

$$P(x,\omega_k) = \frac{1}{T}\int_{-T/2}^{T/2} p(x,t)e^{-i\omega_k t}dt, \quad Q(x,\omega_k) = \frac{1}{T}\int_{-T/2}^{T/2} q(x,t)e^{-i\omega_k t}dt.$$

Using the convolution theorem, an analytic relation between $p(x,t)$ and $q(x,t)$ can be determined:

$$p(x,t) = \frac{1}{T}\int_{-T/2}^{T/2} z(x,t-\tau)q(x,\tau)d\tau. \tag{2.9}$$

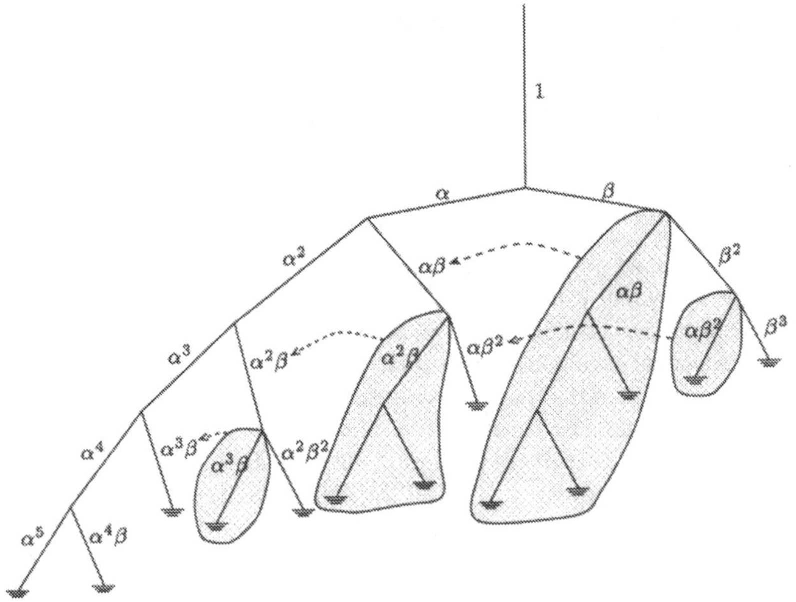

FIG. 4. *A structured tree, at each bifurcation the radii of the daughter vessels are scaled linearly by factors α and β. The shaded parts of the figure correspond to subtrees which are represented more than once. Assuming identical conditions at all leaves of the structured tree, the impedances arising from the shaded subtrees have already been evaluated and does not need to be reevaluated. (The figure is adapted from [21] and reprinted with permision from American Physiological Society.)*

The remaining boundary conditions are the bifurcation conditions. Assume that there is no leakage and that the bifurcation occurs at a point. Hence, the flow over a bifurcation must add:

$$q_{pa} = q_{d_1} + q_{d_2}. \tag{2.10}$$

The subscript $pa$ refers to the parent vessel and the subscripts $d_i$ refer to the daughter vessels. Furthermore, assume that pressure is continuous. Consequently the remaining two boundary conditions can be obtained by [22],

$$p_{pa} = p_{d_1} = p_{d_2}. \tag{2.11}$$

**3. The small arteries.** Since blood flow through the capillaries is almost steady and non-oscillatory, the arteries should cushion the fluctuations resulting from the pulsating ejection of blood from the heart. The pulse-wave is damped because of the vast branching of the vessels combined with an increase in total cross-sectional area. Even though each vessel tapers and the cross-sectional area of each daughter vessel is smaller than that of the parent vessel, the combined cross-sectional area of the two daughter vessels is larger than that of the parent vessel.

The trees comprising the smaller vessels have more than 20 generations. At this point the capillary level is reached and all vessels have approximately the same diameter. At the capillary level loops are formed and the structure transforms from a tree-like structure into a "swamp" of vessels. Due to the formation of loops and because the majority of the peripheral resistance originates in the arterioles the capillaries are not included in our model.

**3.1. The structured tree model.** In fact, there is evidence [6, 29, 35], that the small arteries are distributed in an optimal and structured way. Using the ideas outlined in above cited work, we have constructed a one-dimensional asymmetrical structured tree model for the small arteries, in which the radii of the daughter vessels are scaled linearly from the radius of the parent vessels by factors $\alpha$ and $\beta$, which are both smaller than one.

Computing the full axisymmetric Navier-Stokes equations for such a tree would not be practical. In addition, in the small arteries viscosity becomes more prominent than inertia. Consequently the non-linear effects becomes less pronounced [8]. Such a model was derived by Atabek et al. [4, 3, 25]. Their model corresponds to linearization of the momentum equation (2.2) and neglection of terms of order $\partial^2 u_x/\partial x^2$ (eliminating these terms corresponds to the long-wave approximation suggested by Atabek et al.). The resulting momentum equation becomes:

$$\frac{\partial u_x}{\partial t} + \frac{1}{\rho}\frac{\partial p}{\partial x} = \frac{\mu}{r}\frac{\partial}{\partial r}\left(r\frac{\partial u_x}{\partial r}\right). \tag{3.1}$$

Assuming that all variables are periodic, they can be written on the harmonic form $u_x(r,x,t) = U_x(r,x)e^{i\omega t}$, $p = P(x)e^{i\omega t}$. Using these harmonic variables (3.1) can be rewritten as a Bessel equation with the following analytic solution:

$$i\omega Q + K\frac{A_0}{\rho}\frac{\partial P}{\partial x} = 0 \tag{3.2}$$

where

$$K = \begin{cases} 1 - 2i^{-1/2}\mathrm{w}^{-1}, & \mathrm{w} > 3 \\ (4/3 - 8i\mathrm{w}^{-2})^{-1}, & \mathrm{w} \le 2 \end{cases}$$

and $\mathrm{w}^2 = r_0^2\omega/\nu$ is the Womersley number. It should be noted that $K$ is not continuous, but derived from asymptotic expansions for $\mathrm{w} \to 0$ and $\mathrm{w} \to \infty$. However, the equation can be made continuous using linear-interpolation between $\mathrm{w} = 2$ and $\mathrm{w} = 3$.

Finally, using the above periodic variables the continuity equation (2.3) can be written as:

$$i\omega C P + \frac{\partial Q}{\partial x} = 0 \tag{3.3}$$

where $C$ is the compliance

$$C = \frac{dA}{dp} = \frac{3A_0 r_0}{2Eh}\left(1 - \frac{3pr_0}{4Eh}\right)^{-3} \approx \frac{3A_0 r_0}{2Eh}\,.$$

The latter approximation applies since $Eh \gg pr_0$. Combining (3.2) and (3.3) and yields the following wave equation:

$$\omega^2 Q + c^2 \frac{\partial^2 Q}{\partial x^2} = 0, \qquad c^2 = \frac{A_0 K}{\rho C}. \tag{3.4}$$

For each of the small vessels the flow and pressure can be found by solving (3.4) for $Q$ and using this solution together with (3.2) or (3.3) to get $P$. The resulting impedance is given by:

$$Z(x,\omega) = \frac{P(x,\omega)}{Q(x,\omega)} = -\frac{b\cos(\omega x/c) - a\sin(\omega x/c)}{ig(a\cos(\omega x/c) + b\sin(\omega x/c))} \tag{3.5}$$

where $a$ and $b$ are arbitrary constants and $g = cC = \sqrt{CA_0K/\rho}$. Assuming that $Z(L,\omega)$ is known, it is possible to find an expression for $b/a$ using (3.5). The input impedance (at $x = 0$) for any given vessel can be found by evaluating (3.5) at $x = 0$ and inserting the expression for $b/a$.

$$Z(0,\omega) = \frac{ig^{-1}\sin(\omega L/c) + Z(L,\omega)\cos(\omega L/c)}{\cos(\omega L/c) + igZ(L,\omega)\sin(\omega L/c)}. \tag{3.6}$$

In order to extend the above analysis to the more general case of a tree a bifurcation must be determined.

As for the large arteries (equations (2.10) and (2.11)) we assume that the pressures are continuous and that the flows, and hence the admittances, add:

$$\frac{1}{Z_{pa}} = \frac{1}{Z_{d_1}} + \frac{1}{Z_{d_2}}. \tag{3.7}$$

If the geometry of the tree and the impedances at the distal terminals (at the leaves of the tree) are known equations (3.6) and (3.7) can be used recursively to find the impedance at the root of the structured trees (for details see [22]). Because of the viscous treatment of the blood in the small vessels, the structured tree provides the peripheral resistance for the large arteries.

By inverse Fourier transform of the root impedance $Z(x,\omega)$ the response function $z(x,t)$ can be found. $z(x,t)$ is needed when evaluating the convolution integral in (2.9). The convolution integral spans over one period so it requires knowledge of the solution at all times in order to be evaluated. Equations (2.3) and (2.4) for the large arteries are solved using an explicit method (Richtmeyer's two-step version of Lax-Wendroff's method) and consequently, the solution will not be computed simultaneously at all times during the period. However, because propagation of the pulse-wave is periodic this difficulty can be overcome using iteration: At any time $t'$ the convolution integral (2.9) can be evaluated using values from the previous period for $t \in [t' : T]$ and the ones already computed for $t \in [0 : t']$. Numerical experiments suggest that a stable solution can be reached after 3–4 periods (see Figure 5).

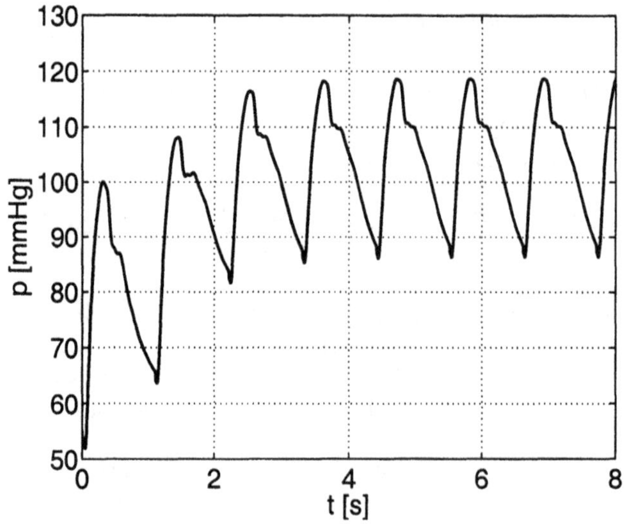


FIG. 5. *Aortic pressure 8 cm after the aortic valve. The plot shows that $p(x, t)$ converges after three periods.*

The structured tree can be characterized by the following parameters:

- Scaling ratios $\alpha$ and $\beta$.
- Order of the structured tree.
- Parameters for the vessel length, radius, wall-thickness, and Young's Modulus.

The scaling parameters $\alpha$ and $\beta$ can be determined by studying how the radii of daughter vessels are related to the radius of the parent vessel and relationships for the area- and asymmetry ratios. Ulyings et al. [33] suggested that the change in radii over a bifurcation is related by the following power law.

$$r_{pa}^{\xi} = r_{d_1}^{\xi} + r_{d_2}^{\xi} \tag{3.8}$$

where the exponent $\xi$ is constant [32, 27]. This condition is derived from the principle of minimum work. The area– and asymmetry ratios can be defined by [35]:

$$\eta = \frac{r_{d_1}^2 + r_{d_2}^2}{r_{pa}^2}, \qquad \gamma = \left(r_{d_2}/r_{d_1}\right)^2. \tag{3.9}$$

It has been shown [10, 23, 33] that $\eta$ and $\gamma$ are approximately constant throughout the small arteries. Assuming that the linear scaling at each bifurcation has the form:

$$r_{d_1} = \alpha r_{pa}, \qquad r_{d_2} = \beta r_{pa}.$$

The scaling parameters $\alpha$ and $\beta$ can be found from (3.8) and (3.9) to be:

$$\alpha = \left(1 + \gamma^{\xi/2}\right)^{-1/\xi}, \qquad \beta = \alpha\sqrt{\gamma}.$$

The small arteries and arterioles consist of a tree with more than 20 generations and at the capillary level all vessels have approximately the same radius. The structured tree is binary and asymmetric. Terminating the branching of the structured tree after a fixed number of generations would yield terminal vessels with significant variations in their diameters. Therefore, we have chosen to terminate the structured tree when the radii of the terminal vessels become smaller than some given minimum value, depending on the muscle or tissue in question [21]. For example, tissue with a low peripheral resistance (e.g. the renal or carotid arteries) can be represented by structured trees with a large minimum radius and tissue with a high peripheral resistance (e.g. the femoral or brachial arteries) can be represented by structured trees with a small minimum radius.

Finally, the length-to-radius ratio can be approximated as $l_{rr} = L/r \approx 50 \pm 10$ [11, 32, 7, 6], the wall-thickness-to-radius relationship can be approximated by $h = kr_0$ where $k \approx 0.16$ [14, 18], and Young's modulus can (similar to the large arteries) be approximated using (2.6).

**3.2. The three-element windkessel model.** A much simpler approach is to lump the small arteries into a windkessel model from which the impedance can be predicted at each of the terminals of the large arteries. The windkessel model consists of a resistance in series with a parallel combination of a resistance and a capacitor (see Figure 6). The latter simulates the compliance of the terminal [31].

The impedance of the windkessel model is given by:

$$Z(0,\omega) = \frac{R_1 + R_2 + i\omega C_T R_1 R_2}{1 + i\omega C_T R_2}.$$

The above equation can be used as a boundary condition instead of the convolution integral in (2.9) or it can be transformed into the time-domain yielding a first-order ordinary differential equation.

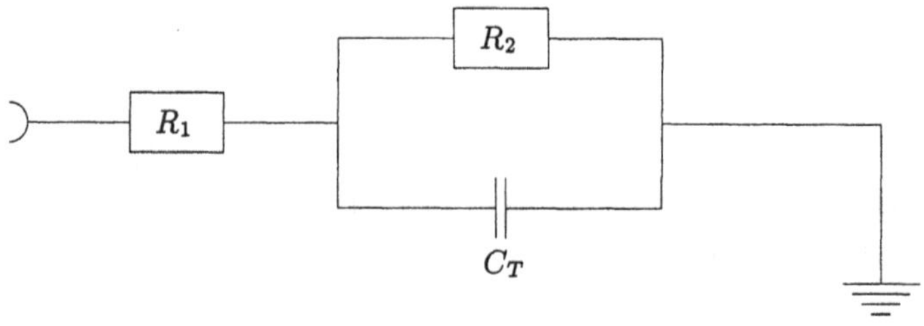


FIG. 6. *The three-element windkessel model used for predicting the impedance at the terminals of the large arteries. The resistances $R_1$ and $R_2$ and the capacitance $C_T$ must be estimated for each of the terminal vessels.*

**4. Results.** Our main aim was to show that our structured tree model provides a feasible outflow boundary condition for determining blood flow and pressure in the large arteries. To do so, we showed that our model can reproduce the essential characteristics featuring the arterial pulse both qualitatively and quantitatively. This was done by comparing our model with magnetic resonance data [22].

All simulations in this paper were based on the same set of parameters. Most lengths and diameters were based on magnetic resonance measurements [22]. At locations where measured data was not available, lengths and diameters were estimated from combining literature data with measured and computed flows. The parameters representing $Eh/r_0$ were chosen as shown in Figure 2, while density and viscosity were kept constant. The parameters for the structured trees followed the choices discussed in Section 3. In order to take the variations in peripheral resistance into account the minimum radius was varied for the various structured trees.

The total cross-sectional area of the systemic arteries increases from approximately 5 $cm^2$ at the root of the aorta to 400 $cm^2$ at the arterioles. The maximum (systolic) pressure of the large arteries increases away from the heart, towards the periphery, while the mean pressure decreases [17]. The increase of systolic pressure is mainly due to tapering and branching of the large arteries and the peripheral resistance in the arterioles. As a result of these features the pulse-wave is reflected. In the large arteries, the reflected wave is superimposed on the pulse-wave, increasing the systolic pressure. At the same time the large increase in cross-sectional area causes a decrease in mean pressure (see Figure 7). Furthermore, the velocity of the wave propagation is larger than the velocity of the pulse-wave and as a result the reflected wave separates from the incoming pressure wave and becomes more prominent at peripheral locations than at proximal locations. Finally, the steepness of the incoming pressure and flow profiles increases towards the periphery (see Figure 8). The increase of the incoming waves is due to pressure dependence of the wave propagation velocity $c(p)$, which causes wave components with higher pressure to travel faster than those with lower pressure.

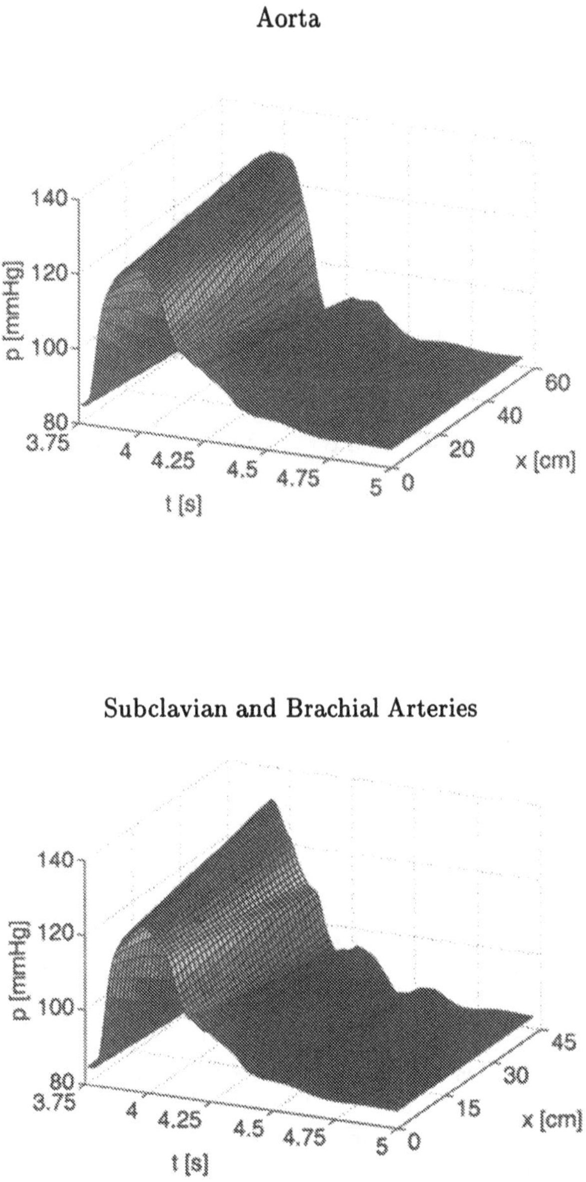


FIG. 7. *The graphs show pressure (mmHg) as functions of space $x$ and time $t$ for the aorta and the subclavian and brachial arteries. (The figure is reprinted from [21] with permision from American Physiological Society.)*

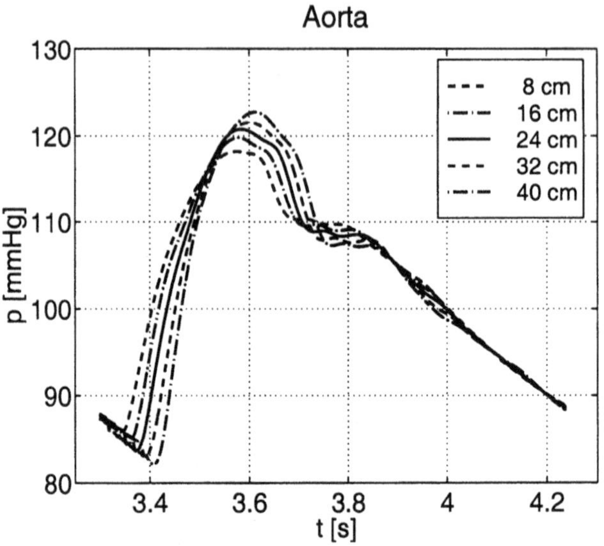


FIG. 8. *Aortic pressure at five locations along the aorta.*

The flow-data were obtained using magnetic resonance measurements in a young man. His average heart rate at rest was 55 beats per minute and his systolic and diastolic blood pressures were 120/80 mmHg (measured with a cuff) [22]. These mean measurements corresponded well to the pressures computed at the left subclavian artery (see Figure 11). Both measured and computed flows are determined at the locations shown in Figure 9. The agreement we have obtained with the measured data are shown in Figure 6.

**5. Verification of the structured tree model.** Modeling of the impedances at the terminals of the large arteries have been described in two ways, by a structured tree model and a windkessel model. The windkessel model is a lumped model predicting the impedances based on the resistive and compliant behavior of the small arteries. This model has the disadvantage that it is not able to capture the wave-propagation phenomena on the part of the arterial system that it models. Furthermore, the model needs estimations of three parameters for each terminal segment. However, assuming that the parameters can be estimated, the model is still able to capture the overall behavior of the impedances.

Analysis of the structured tree and the windkessel models shows that they both have two essential parameters. Elasticity (Young's modulus or total compliance) and peripheral resistance (the minimal radius applied at the leaves of the structured tree or the total resistance). In order to achieve agreement between the structured tree and the windkessel model parameters for the windkessel model can be fitted from the structured

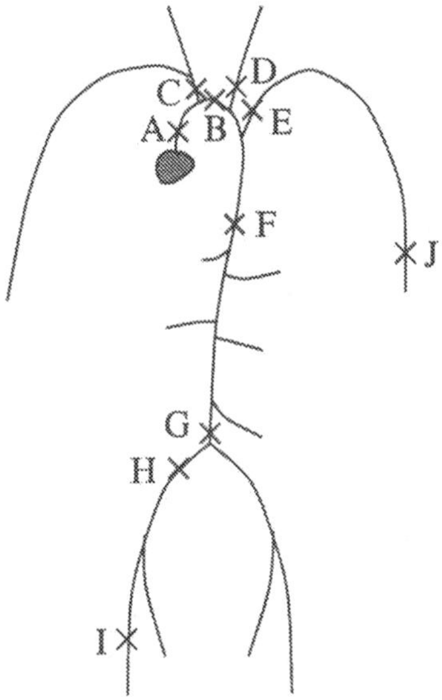


FIG. 9. *The flow is measured in the ascending aorta and at 9 peripheral locations. These are marked with crosses on the figure. In order to avoid too many artifacts due to entry region flows, were possible, all flows are measured 2 cm beyond the bifurcations.*

tree model. The total resistance $R_1 + R_2 = Z(0)$ (the DC-term), $R_1 = \lim_{\omega\to\infty} Z(\omega)$, and the total compliance $C_T$ can be determined from the descent-ratio of $Z(\omega)$. Figure 12 shows Bode plots of the two models and compares them with impedance data obtained from the literature [15, 16]. The differences between the models become evident in that the windkessel model cannot include high-frequency oscillations present in the structured tree model and in the measured data. Furthermore, the windkessel model cannot include the large dip seen in the phase. However, it should be noted that the structured tree model is not designed to be valid for the large arteries and as a result one should not assume perfect matches without some adjustments of the parameters. For example, the brachiocephalic artery shown in Figure 12 the length to radius relationship is modified from 50-100 [21].

**6. Conclusion.** Our study has shown that it is possible to construct a model for the systemic arteries including both large and small arteries. The small arteries are modeled as asymmetrical binary structured trees. The pressure and flow profiles of the large arteries all have the right characteristics. The pressure increases away from the heart, the wave-speed of the reflected dicrotic-wave is faster than that of the main wave, the dicrotic-wave becomes more prominent away from the heart, and the steepness of the pressure wave increases away from the heart (see Figures 7 and 8).

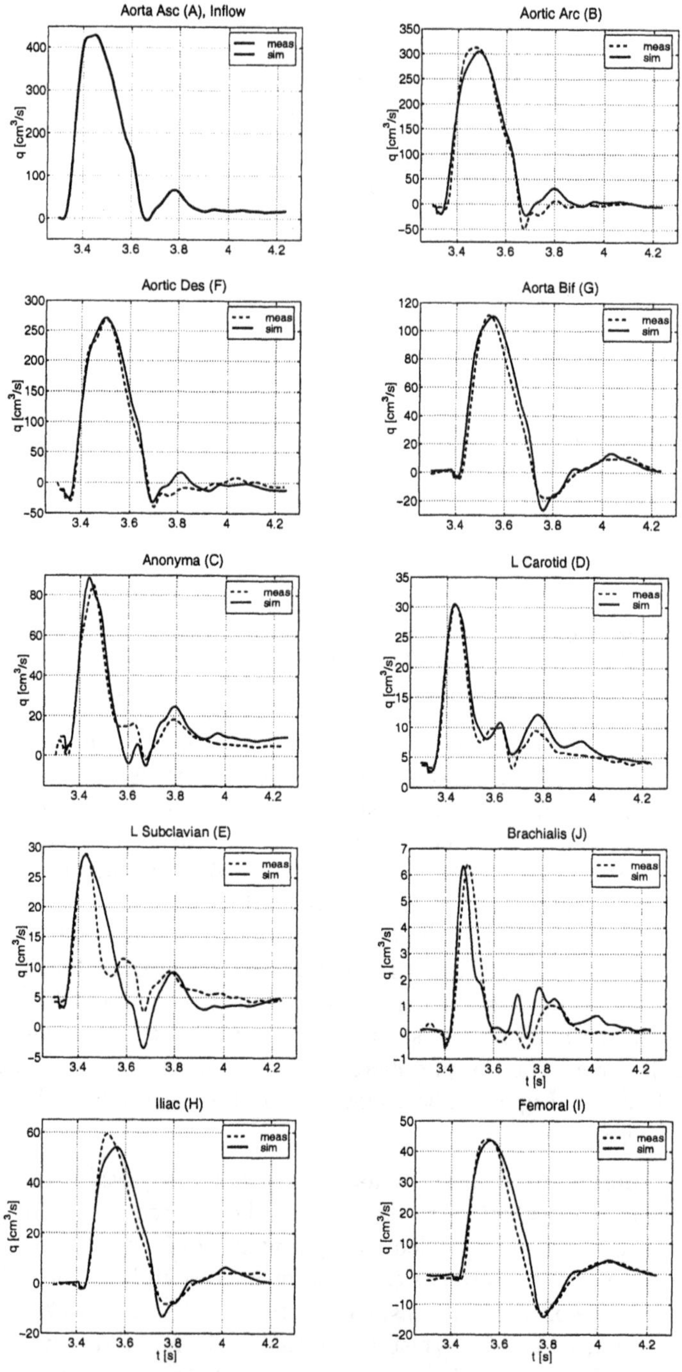


FIG. 10. *Measured and simulated flows. The letter in parenthesis indicate where the flow is measured as shown in Figure 9.*

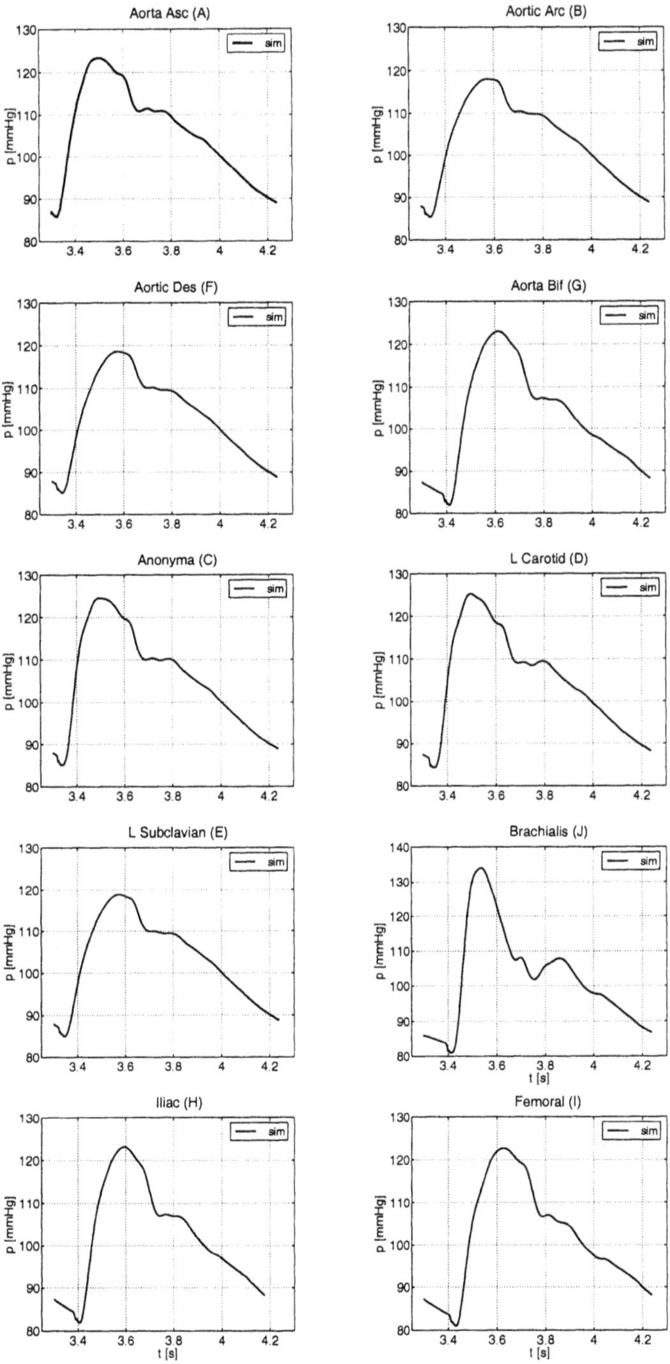


FIG. 11. *Simulated pressures corresponding to the flows shown in Figure 6.*

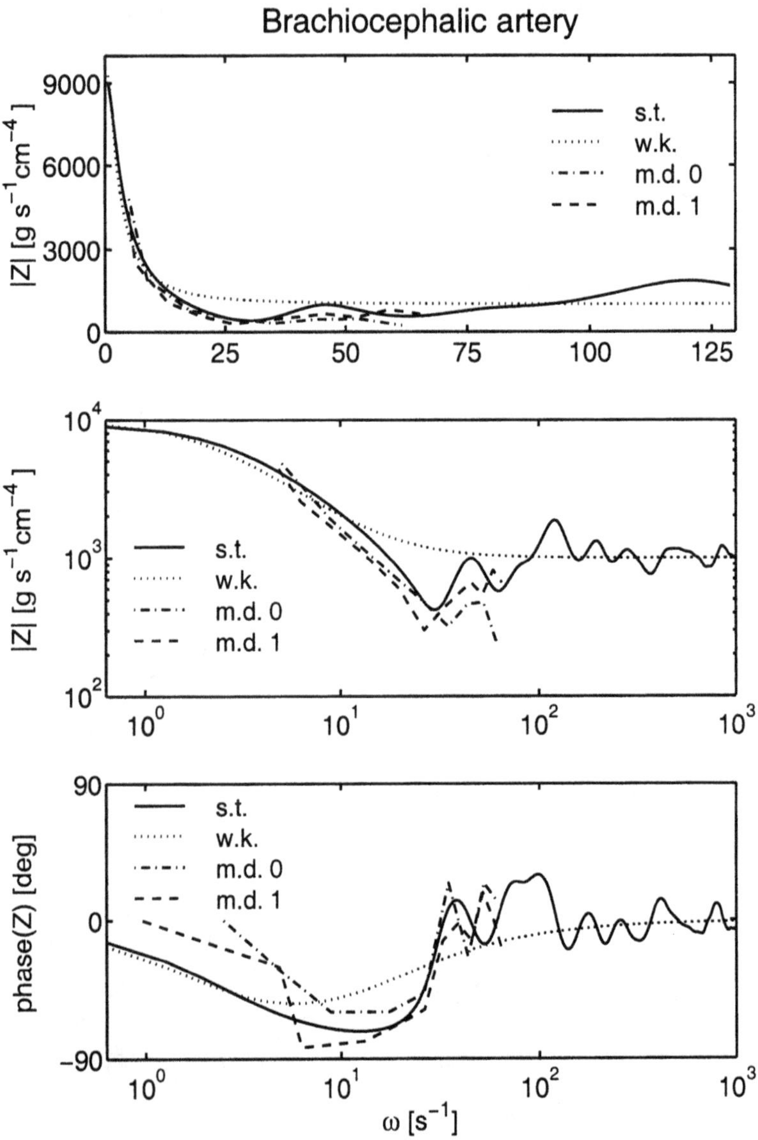


FIG. 12. *The top graph shows the impedance* $|Z(0,\omega)|$ *as a function of the angular frequency* $\omega = 2\pi f$ *for* $\omega \in [0 : 125]$ *Hz. The next two graphs show Bode plots, a double logarithmic plot of the modulus of the impedance* $|Z(0,\omega)|$ *and a logarithmic plot of the phase. Both are plotted as a function of frequency. The dotted lines signify results from the windkessel model (w.k.), the solid lines signify results from the structured tree model (s.t.), and the dashed lines signify measured results (m.d. 0) and (m.d. 1). The measured data are reproduced from [15]. (The figure is reprinted from [21] with permision from American Physiological Society.)*

The advantage of our arterial model is that it can give us the flow and pressure for both large and small arteries. Furthermore, the boundary condition (the structured tree outflow condition) reflects the physical domain and includes the wave-propagation effects.

Several modifications can be made to the structured tree model. The most obvious is the following modification: Instead of varying the minimum radius of the structured tree depending on the tissue it supplies, the minimum radius could be constant for all structured trees. Thus, the peripheral resistance could be varied by addition of a pure resistance at the leaves of the structured trees. Adding a pure resistance at the bottom makes it easier to adapt the model to various physiological conditions without changing the order of the structured trees. For example, during exercise the peripheral resistance of the various tissues decreases, but the order of the trees of small arteries do not change. However, the resistance at the bottom of the large arteries cannot be modeled as a pure resistance. Hence, it is advantagous to insert the structured trees and then apply a pure resistance at the bottom of the structured tree.

Finally, it possible to argue that the structured model is too complicated and that the much simpler windkessel model is adequate enough since it too provides a dynamic boundary condition. Replacing the windkessel model with a structured tree model comes at the cost of replacing an ordinary differential equation by a convolution integral. However, we find the structured tree model relevant, regardless, since it is based on the underlying physiology and includes wave-propagation effects in the part of the arterial system that it models.

## REFERENCES


[1] D.J. Acheson, *Elementary Fluid Dynamics*, Oxford University Press, Oxford U.K., 1990.

[2] M. Anliker, *Current and Future Aspects of Biomedical Engineering*, in Triangle, **16** (1977), pp. 129–140.

[3] H.B. Atabek and H.S. Lew, *Wave Propagation Through a Viscous Incompressible Fluid Contained in an Initially Stressed Elastic Tube*, in Biophys J., **6** (1966), pp. 481–503.

[4] H.B. Atabek, *Wave Propagation Through a Viscous Fluid Contained in a Tethered, Initially Stressed, Orthotropic Elastic Tube*, in Biophys J., **8** (1968), pp. 626–649.

[5] A.C.L. Barnard and W.A. Hunt and W.P. Timlake and E. Varley, *A Theory of Fluid Flow in Compliant Tubes*", in Biophys J., **6** (1966), pp. 717–724.

[6] J.B. Bassingthwaighte and L.S. Liebovitch and B.J. West, *Fractal Physiology*, Oxford University Press, NY, 1994.

[7] D.H. Bergel, *Cardiovascular Fluid Dynamics*, Volume 1, Academic Press, WHERE, 1972.

[8] C.G. Caro and T.J. Pedley and R.C. Schroter and W.A. Seed, *The Mechanics of the Circulation*, Oxford University Press, Oxford, U.K., 1972.

[9] A.J. Chorin and J.E. Marsden, *A Mathematical Introduction to Fluid Mechanics*, Third, corrected printing, Springer Verlag, WHERRE, 1998.

[10] K. Horsfield and M.J. Woldenberg, *Diameters and Cross-Sectional Areas of Branches in the Human Pulmonary Arterial Tree*, in Anat. Rec., **223** (1989), pp. 245–251.

[11] A.S. Iberall, *Anatomy and Steady Flow Characteristics of the Arterial System with an Introduction to its Pulsatile Characteristics*, in Math. Biosci., **1** (1967), pp. 375–385.

[12] J. Lighthill, *Mathematical Biofluiddynamics*, Third printing, SIAM, Philadelphia, 1989.

[13] J. Mazumdar, *Biofluid Mechanics*, World Scientific Co., Singapore, 1992.

[14] D.A. McDonald, *Blood Flow in Arteries*, Edward Arnold, London, U.K., 1974.

[15] C.J. Mills and I.T. Gabe and J.H. Gault, *Pressure-flow relationships and vascular impedance in man*, in Cardiovasc. Res., **4** (1970), pp. 405–417.

[16] W.W. Nichols and M.F. O'Rourke, *McDonald's Blood Flow in Arteries*, Fourth edition, Edward Arnold, London, U.K., 1998.

[17] A. Noordergraaf, *Circulatory System Dynamics*, Academic Press, San Diego, 1978.

[18] S. Oka and M. Nakai, *Optimality Principle in Vascular Bifurcation*, in Biorheology, **24** (1987), pp. 737–751.

[19] M. Olufsen and J. Ottesen, *Outflow Conditions in Human Arterial Flow*, in Computer Simulations in Biomedicine, H. Power and R.T. Hart (eds.), Computational Mechanics Publications, Southhampton, U.K. 1995, pp. 249–256.

[20] M. Olufsen, *Modeling the Arterial System with Reference to an Anesthesia Simulator*, Ph.D. Thesis, Text No. 345, IMFUFA, Roskilde University, Denmark, 1998.

[21] M. Olufsen, *Structured Tree Outflow Condition for Blood Flow in Larger Systemic Arteries*, in Am. J. Physiol., **276** (1999), pp. H257–H268.

[22] M. Olufsen and C.S. Peskin and W.Y. Kim and E.M. Pedersen and A. Nadim and J. Larsen, *Numerical Simulation and Experimental Validation of Blood Flow in Arteries with Structured-Tree Outflow Conditions*, Ann. of Biomed. Eng., **28**(11), 2000 (in press).

[23] G.L. Papageorgiou and B.N. Jones and V.J. Redding, *The Area Ratio of Normal Arterial Junctions and its Implications in Pulse Wave Reflections*, in Cardiovasc. Res., **24** (1990), pp. 478–484.

[24] E.M. Pedersen and H-W. Sung and A.C. Burlson and A.P Yoganathan, *Two-Dimensional Velocity Measurements in a Pulsatile Flow Model of the Abdominal Aorta Simulating Different Hemodynamic Conditions*, in J. Biomech., **26** (1993), pp. 1237–1247.

[25] T.J. Pedley, *The Fluid Mechanics of Large Blood Vessels*, Cambridge University Press, Cambridge, U.K., 1980.

[26] T.J. Pedley, *Partial Differential Equations in Biology*, Courant Institute of Mathematical Sciences, New York University, NY, 1976.

[27] M.S. Pollanen, *Dimensional Optimization at Different Levels at the Arterial Hierarchy*, in J. Theor. Biol., **159** (1992), pp. 267–270.

[28] J.K Raines and M.Y. Jaffrin and A.H. Shapiro, *A Computer Simulation of Arterial Dynamics in the Human Leg*, in J. Biomech., **7** (1974), pp. 77–91.

[29] W. Schreiner, *Computer Generation of Complex Arterial Tree Models*, in J. Biomech. Eng., **15** (1993), pp. 148–150.

[30] P. Segers and F. Dubois and D. DeWachter and P. Verdonck, *Role and Relevancy of a Cardiovascular Simulator*, J. Cardiovasc. Eng., **3** (1998), pp. 48–56.

[31] N. Stergiopulos and D.F. Young and T.R. Rogge, *Computer Simulation of Arterial Flow with Applications to Arterial and Aortic Stenosis*, in J. Biomech., **25** (1992), pp. 1477–1488.

[32] N. Suwa and T. Niwa and H. Fukasawa and Y. Sasaki, *Estimation of Intravascular Blood Pressure Gradients by Mathematical Analysis of Arterial Casts*, in Tohoku J. Exp. Med., **79** (1963), pp. 168–198.

[33] H.B.M. UYLINGS *Optimization of Diameters and Bifurcation Angles in Lung and Vascular Tree Structures*, in Bull. Math. Biol., **39** (1977), pp. 509–520.

[34] N. WESTERHOF AND F. BOSMAN AND C.J. DEVRIES AND A. NOORDERGRAAF, *Analog Studies of the Human Systemic Arterial Tree*, in J. Biomech., **2** (1969), pp. 121–143.

[35] M. ZAMIR, *Nonsymmetrical Bifurcations in Arterial Branching*, in J. Gen. Physiol., **72** (1978), pp. 837–845.

# HYDRODYNAMICS OF LIQUID CAPSULES ENCLOSED BY ELASTIC MEMBRANES

C. POZRIKIDIS*

**Abstract.** The flow-induced deformation of liquid capsules enclosed by elastic membranes is discussed with reference to the biofluid-dynamics of vesicles and biological cells. The deformation of a membrane from a reference configuration causes the development of non-isotropic in-plane elastic tensions, transverse shear tensions, and bending moments according to constitutive laws that reflect the membrane constitution. The type and degree of capsule deformation, the internal and external structure of the flow, and the macroscopic rheological properties of a suspension are determined by the magnitude of the developing elastic tensions and bending moments relative to the strength of the imposed flow. Integral representations of the flow past a collection of capsules are reviewed, and integral equations for the velocity distribution over the membranes are presented for two-dimensional, three-dimensional, and axisymmetric flow at vanishing Reynolds number. The velocity field due to the presence of the capsules is expressed in terms of distributions of Stokes flow singularities over the membranes, represented by the single-layer and double-layer hydrodynamic potential of Stokes flow. The strength of the distribution of the single-layer potential expressing interfacial distributions of point forces is identified with jump in the hydrodynamic traction across a membrane. Expressions for this jump are reviewed and derived in terms of the developing elastic tensions and bending moments under the formalism of the theory of thin shells.

**1. Introduction.** Considerable progress has been made in the last two decades in describing the fluid mechanics of deformable liquid particles and in quantifying the rheological and statistical properties of their suspensions. Much of this progress has become possible thanks to the development of efficient numerical algorithms that have permitted the simulation of highly deformed particles with a variety of interfacial properties, and the simulation of the flow of suspensions of a multitude of particles in homogeneous and wall-bounded flows. The state of the art for suspensions of liquid drops whose interfaces exhibit constant surface tension is reviewed by Li et al. (1996), Charles & Pozrikidis (1998), and Li & Pozrikidis (2000).

In practice, the interfaces of deformable particles show a variety of behaviors that are not adequately described by uniform surface tension. First, surfactant molecules alter the intensity of the forces between the molecules of two immiscible species on either side of an interface, and render the surface tension a function of the local surfactant concentration. The established tangential surface-tension gradients generate interfacial stresses that may have a significant influence on the flow inside and around the particle and on the particle deformation. Second, impurities and the growth of interfacial layers due to chemical reactions generate interfaces that behave like two-dimensional Newtonian or viscoelastic Boussinesq fluids (e.g. Scriven 1960, Waxman 1984, Edwards et al. 1991). Polymerized mem-

*Department of Mechanical and Aerospace Engineering, University of California, San Diego, La Jolla, California 92093-0411, USA; e-mail: cpozrikidis@ucsd.edu; Internet URL: http://stokes.ucsd.edu/c_pozrikidis.

branes develop stress resultants recognized as elastic tensions and bending moments similar to those exhibited by thin elastic shells. Third, the interfaces of biological cells and laboratory vesicles consisting of symmetric lipid bilayers resist changes in area and favor flat shapes (e.g. Lipowsky 1991). A certain class of cells are enclosed by fibrous and laminated membranes that are responsible for viscoelastic behavior (Zinemanas & Nir 1988).

An example of an interface that is significantly more complex than the usual interface between two immiscible fluids is the membrane of a red blood cell. Laboratory studies have shown that the membrane of the red cell is composed of a lipid bilayer resting on a network of, and transversed by, an assortment of proteins (e.g. Evans & Skalak 1980, Mohandas & Evans 1994). The bilayer is responsible for incompressible behavior that requires the surface area of any infinitesimal or finite portion of the membrane to be conserved during deformation; thus, a spherical cell may not deform under constant volume and develops infinite tensions when subjected to flow. The cytoskeleton is responsible for elastic behavior that causes the cell to return to the resting shape of the biconcave disk in hydrostatics. When subjected to hydrodynamic tractions, the membrane develops anisotropic elastic tensions and a position-dependent isotropic tension that ensures incompressible deformation. The excess area of a healthy membrane, combined with its low modulus of elasticity, allows the red blood cells to readily deform and squeeze through the microcapillaries. Under pathological conditions, the membrane of a certain class of diseased cells loses its ability to deform, and blood is no longer able to reach the outer branches of the microcirculation.

Membranes consisting of lipid bilayers exhibit bending elasticity, that is, resistance to bending from an equilibrium configuration (Lipowsky 1991). If the bilayer is symmetric, the equilibrium shape has zero mean curvature. More generally, bending elasticity allows the membranes to obtain a great variety of shapes and to exhibit different modes of oscillatory behavior in excitation (Seifert 1997). The surface tension incorporates two contributions: One associated with a surface energy that is proportional to the membrane surface area, and an entropic tension associated with the membrane's wrinkled shape due to thermal fluctuations. Applying an external tension to counteract the entropic tension smoothes out the corrugated shapes and reveals an unfolded area that increases at a logarithmic rate with tension. To a first approximation, the energy stored in a membrane consisting of a symmetric bilayer may be set equal to the sum of the total surface area multiplied by the surface tension, and the integral of the square of the mean curvature multiplied by twice the modules of the membrane bending stiffness. The structure and dynamics of biological membranes is discussed in two illuminating volumes edited by Lipowsky & Sackmann (1995).

The mathematical modeling of tensions and bending moments developing in elastic membranes draws heavily from the theory of thin shells (e.g. Naghdi 1972, Libai & Simmonds 1998). In this theory, the membrane is described as a curved two-dimensional medium of small or zero thickness, and its mid-surface is described in parametric form using two surface curvilinear coordinates. Three approaches are available for describing the membrane deformation, for deriving equilibrium conditions, and for computing the elastic tensions and bending moments developing due to deformation. In the first approach, the membrane is viewed as a thin sheet of a three-dimensional material, and asymptotic equations originating from three-dimensional continuum mechanics are derived in the limit of zero thickness. In the second approach, special assumptions are made regarding the deformation of fibers that are normal to the mid-surface of the membrane (e.g. Le Dret & Raoult 1995). In the third approach, the third dimension is abandoned at the outset, and the membrane is regarded as a purely two-dimensional medium. The third approach has significant advantages: It circumvents certain inconsistencies encountered in the first two approaches (e.g. Budiansky & Sanders 1963), and it is appropriate for molecular membranes for which the assumption of continuum in the normal direction is not meaningful.

A number of analytical and computational studies have addressed the deformation of elastic capsules in hydrodynamics and hydrostatics. Zarda et al. (1977) computed equilibrium shapes of inflated red blood cells using a model that includes the membrane elasticity under tension in its own plane, as well as flexural stiffness. Secomb et al. (1986) used lubrication theory to describe the motion of tightly-fitting red blood cells traveling through narrow capillaries. Barthès-Biesel (1980), Barthès-Biesel & Rallison (1981), and Barthès-Biesel & Sgaier (1985) used asymptotic methods to study the steady and transient small deformation of capsules with spherical unstressed shapes enclosed by elastic or viscoelastic membranes. Li, Barthès-Biesel & Helmy (1988) and Diaz et al. (2000) used a boundary-integral method to study large axisymmetric deformations in uniaxial extensional flow. Barthès-Biesel (1991) discussed critical shear rates above which a capsule exhibits continued deformation leading to membrane failure and breakup. Leyrat-Maurin & Barthès-Biesel (1994) simulated the passage of capsules through axisymmetric constrictions with applications to the filtration of red blood cells. Pozrikidis (1995), Ramanujan & Pozrikidis (1998), Navot (1998), and Eggleton & Popel (1998) simulated the deformation of elastic capsules with various unstressed shapes in simple shear flow. Breyiannis & Pozrikidis (1999) studied the deformation of two-dimensional solitary capsules and the rheological and statistical properties of their suspensions in shear flow. More recently, Pozrikidis (2000a) and Kwak & Pozrikidis (2000) investigated the effect of membrane bending stiffness on capsule deformation in simple shear flow and uniaxial extensional flow. Previously, Kraus et al. (1996) had accounted for the effect of bending stiffness by setting the jump in the interfacial hydrodynamic traction across the membrane equal to the gradient of an appropriate membrane energy function, as will be discussed in Section 6. Other authors have studied the deformation of capsules enclosed by incompressible membranes developing a position-dependent tension (Zarda et al. 1977, Pozrikidis 1990, Zhou & Pozrikidis 1995, Kraus et al. 1996).

In this article, we review the equations governing the motion of the fluid inside and outside a solitary capsule or a collection of suspended capsules, in the limit of vanishing Reynolds numbers. In Section 2, we present boundary-integral formulations for two-dimensional, axisymmetric, and three-dimensional flow. Coupling the fluid mechanics with the membrane mechanics is effected by means of interfacial force and torque balances that relate the jump in the hydrodynamic traction across the interface to the membrane tensions and bending moments. In sections 3–5, we present the specific forms of the balances for two-dimensional, three-dimensional, and axisymmetric flow. Finally, in Section 6 we discuss extensions and complementary formulations.

**2. Integral formulation for Stokes flow.** Consider the motion of a collection of neutrally buoyant capsules enclosed by elastic membranes, consisting of the same or different Newtonian fluids, and suspended in another Newtonian ambient fluid. In the limit of vanishing Reynolds number *Re*, where *Re* is defined with respect to the typical capsule size, inertial forces are insignificant, and the motion of the fluid inside and outside the capsules is governed by the linear equations of Stokes flow (e.g. Pozrikidis 1992, 1997).

In this section, we review boundary integral formulations that provide us with a basis for computing the instantaneous distribution of the velocity over the membranes, for simulating the motion of the capsules, and for evaluating flow variables at any point in the flow.

**2.1. Two-dimensional flow.** To begin, we consider a suspension of $N$ two-dimensional, neutrally-buoyant capsules evolving under the influence of an imposed flow, as illustrated in Figure 1(a). Requiring that the velocity is continuous across the interfaces, and following a standard methodology (e.g. Pozrikidis 1992), we find that the velocity at a point $\mathbf{x}_0$ that lies within the suspending fluid is given by the integral representation

$$
\begin{aligned}
u_j(\mathbf{x}_0) = u_j^\infty(\mathbf{x}_0) - \frac{1}{4\pi\mu_s}\sum_{m=1}^{N}\int_{C_m} \Delta f_i(\mathbf{x})\, G_{ij}(\mathbf{x},\mathbf{x}_0)\, dl(\mathbf{x}) \\
+\frac{1}{4\pi}\sum_{m=1}^{N}(1-\lambda_m)\int_{C_m} u_i(\mathbf{x})\, T_{ijk}(\mathbf{x},\mathbf{x}_0)\, n_k(\mathbf{x})\, dl(\mathbf{x}),
\end{aligned}
\tag{2.1}
$$

where $\mathbf{u}^\infty$ is the imposed velocity prevailing in the absence of the capsules, $C_m$ stands for the interface of the $m$th capsule, $l$ is the arc length along $C_m$, $\mu_s$ is the viscosity of the ambient suspending fluid, $\mu_m$ is the viscosity of the $m$th capsule, $\lambda_m = \mu_m/\mu_s$ is the corresponding viscosity ratio, and $\mathbf{n}$ is the unit vector normal to the interfaces pointing into the ambient fluid.

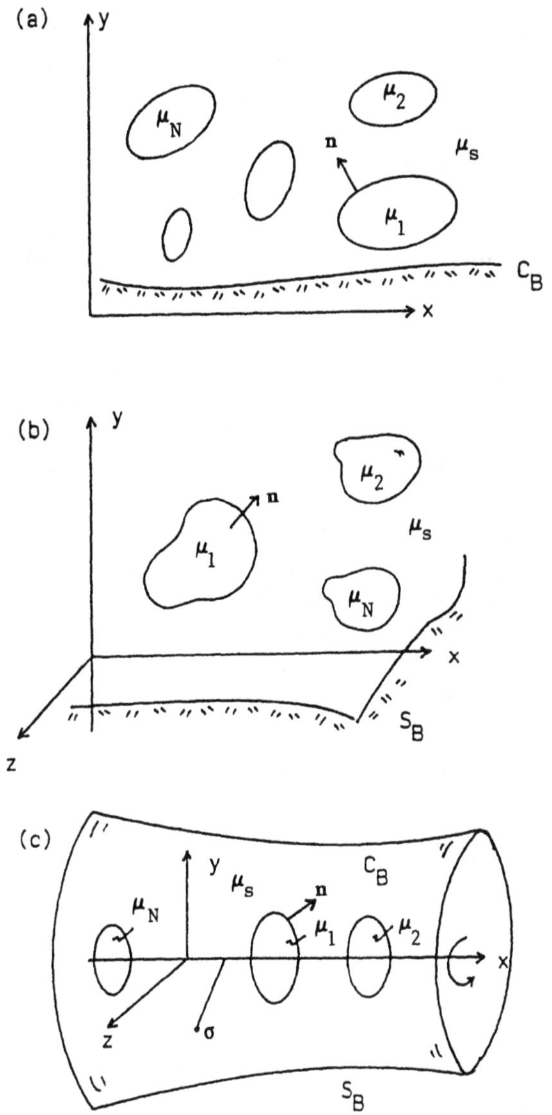


FIG. 1. *Schematic illustration of a suspension of capsules in (a) two-dimensional, (b) three-dimensional, and (c) axisymmetric flow.*

The first integral on the right-hand side of (2.1) is the single-layer potential, and the second integral is the double-layer potential of Stokes flow. The kernels **G** and **T** are the Green's functions of Stokes flow representing, respectively, the velocity and stress field due to a point force in solitary or periodic configuration. We have assumed that the Green's function conforms with the specified periodicity of the flow, and observes the geometry of the boundaries of the flow. For example, if the flow is bounded by the rigid boundary $C_B$, then the velocity Green's function **G** is required to vanish when either the point force or the field point is positioned on $C_B$.

The density of the distribution of the single-layer potential, denoted by $\Delta\mathbf{f}$, is the jump in the traction across the capsule interface, defined as

$$\Delta\mathbf{f} \equiv \mathbf{f}^{(s)} - \mathbf{f}^{(c)} = (\sigma^{(s)} - \sigma^{(c)}) \cdot \mathbf{n}, \tag{2.2}$$

where $\sigma$ is the Newtonian stress tensor (e.g. Pozrikidis 1997), the superscript $(s)$ denotes the ambient suspending fluid, and the superscript $(c)$ denotes a capsule. The computation of $\Delta\mathbf{f}$ will be discussed in Section 3.

For a point $\mathbf{x}_0$ that lies inside the $q$th capsule, where $q = 1, \ldots, N$, the velocity is given by the integral representation (2.1), except that all terms on the right-hand side are divided by the viscosity ratio $\lambda_q$. Taking the limit as the point $\mathbf{x}_0$ approaches the interface of the $q$th capsule, and expressing the limit of the double-layer potential in terms of its principal value, we derive the following Fredholm integral equation of the second kind for the interfacial velocity, also known as the boundary integral equation,

$$\begin{aligned} u_j(\mathbf{x}_0) = \frac{2}{1+\lambda_q} \Big[ & u_j^\infty(\mathbf{x}_0) - \frac{1}{4\pi\mu_s} \sum_{m=1}^{N} \int_{C_m} \Delta f_i(\mathbf{x})\, G_{ij}(\mathbf{x}, \mathbf{x})\, dl(\mathbf{x}) \\ & + \frac{1}{4\pi} \sum_{m=1}^{N} (1-\lambda_m) \int_{C_m}^{PV} u_i(\mathbf{x})\, T_{ijk}(\mathbf{x}, \mathbf{x}_0)\, n_k(\mathbf{x})\, dl(\mathbf{x}) \Big], \end{aligned} \tag{2.3}$$

where $PV$ denotes the principal value (Pozrikidis 1992). The numerical solution of this integral equation is the cornerstone of computational procedures for simulating the capsule motion and the deformation of the membrane.

**2.2. Three-dimensional flow.** Next, we consider a suspension of $N$ three-dimensional, neutrally buoyant capsules evolving under the influence of an imposed flow, as illustrated in Figure 1(b). The counterpart of the integral representation (2.1) is

$$\begin{aligned} u_j(\mathbf{x}_0) = u_j^\infty(\mathbf{x}_0) & - \frac{1}{8\pi\mu_s} \sum_{m=1}^{N} \int_{D_m} \Delta f_i(\mathbf{x})\, G_{ij}(\mathbf{x}, \mathbf{x}_0)\, dS(\mathbf{x}) \\ & + \frac{1}{8\pi} \sum_{m=1}^{N} (1-\lambda_m) \int_{D_m} u_i(\mathbf{x})\, T_{ijk}(\mathbf{x}, \mathbf{x}_0)\, n_k(\mathbf{x})\, dS(\mathbf{x}), \end{aligned} \tag{2.4}$$

where $D_m$ stands for the interface of the $m$th capsule, $dS$ is an infinitesimal surface area of $D_m$, and the rest of the symbols were defined in Section 2.1. For a point $\mathbf{x}_0$ that lies inside the $q$th capsule, where $q = 1, \ldots, N$, the velocity is given by the integral representation (2.4) except that all terms on the right-hand side are divided by the viscosity ratio $\lambda_q$. The counterpart of the integral equation (2.3) is

$$(2.5)\quad \begin{aligned} u_j(\mathbf{x}_0) = \frac{2}{1+\lambda_q} [u_j^\infty(\mathbf{x}_0) &- \frac{1}{8\pi\mu_s} \sum_{m=1}^{N} \int_{S_m} \Delta f_i(\mathbf{x})\, G_{ij}(\mathbf{x}, \mathbf{x}_0)\, dS(\mathbf{x}) \\ &+ \frac{1}{8\pi} \sum_{m=1}^{N} (1-\lambda_m) \int_{S_m}^{PV} u_i(\mathbf{x})\, T_{ijk}(\mathbf{x}, \mathbf{x}_0)\, n_k(\mathbf{x})\, dS(\mathbf{x})], \end{aligned}$$

where the point $\mathbf{x}_0$ lies at the interface of the $q$th capsule, $q = 1, \ldots, N$, and PV denotes the principal value of the double-layer integral.

**2.3. Axisymmetric flow.** As a special case of three-dimensional flow, we consider a suspension of $N$ axisymmetric, coaxial, neutrally buoyant capsules evolving under the influence of an imposed axisymmetric flow, possibly in the presence of axisymmetric boundaries, as illustrated in Figure 1(c). The counterpart of the representation (2.1) is

$$(2.6)\quad \begin{aligned} u_\alpha(\mathbf{x}_0) = u_\alpha^\infty(\mathbf{x}_0) &- \frac{1}{8\pi\mu_s} \sum_{m=1}^{N} \int_{C_m} G_{\alpha\beta}(\mathbf{x}, \mathbf{x}_0)\, \Delta f_\beta(\mathbf{x})\, dl(\mathbf{x}) \\ &+ \frac{1}{8\pi} \sum_{m=1}^{N} (1-\lambda_m) \int_{C_m} u_\beta(\mathbf{x})\, T_{\alpha\beta\gamma}(\mathbf{x}, \mathbf{x}_0)\, n_\gamma(\mathbf{x})\, dl(\mathbf{x}), \end{aligned}$$

where Greek symbols stand for $x$ or $\sigma$ denoting, respectively, the $x$ position and distance from the $x$ axis, as illustrated in Figure 1(c), $C_m$ stands for the trace of the interface of the $m$th capsule in a meridional plane of constant $\varphi$, and the rest of the symbols were defined in Section 2.1. The axisymmetric Green's functions of Stokes flow are discussed by Pozrikidis (1992).

For a point $\mathbf{x}_0$ that lies inside the $q$th capsule, where $q = 1, \ldots, N$, the velocity is given by the integral representation (2.6) except that all terms on the right-hand side are divided by the viscosity ratio $\lambda_q$. The counterpart of the integral equation (2.5) is

$$(2.7)\quad \begin{aligned} u_\alpha(\mathbf{x}_0) = \frac{2}{1+\lambda_q} [u_\alpha^\infty(\mathbf{x}_0) &- \frac{1}{8\pi\mu_s} \sum_{m=1}^{N} \int_{C_m} G_{\alpha\beta}(\mathbf{x}, \mathbf{x}_0) \Delta f_\beta(\mathbf{x})\, dl(\mathbf{x}) \\ &+ \frac{1}{8\pi} \sum_{m=1}^{N} (1-\lambda_m) \int_{C_m}^{PV} u_\alpha(\mathbf{x})\, T_{\alpha\beta\gamma}(\mathbf{x}, \mathbf{x}_0)\, n_\gamma(\mathbf{x})\, dl(\mathbf{x})], \end{aligned}$$

where the point $\mathbf{x}_0$ lies at the interface of the $q$th capsule, $q = 1, \ldots, N$.

**3. Two-dimensional membranes.** To compute the traction discontinuity $\Delta\mathbf{f}$ defined in equation (2.2), we perform force and torque balances over an infinitesimal section of the membrane. To illustrate the procedure, in this section we consider a two-dimensional membrane in the $xy$ plane whose generators are parallel to the $z$ axis. The unit vector $\mathbf{n}$ is normal to the membrane and points into the ambient suspending fluid, as illustrated in Figure 2. The unit tangential vector $\mathbf{t}$ is oriented such that the pair of vectors $\mathbf{n}$ and $\mathbf{t}$, considered in this particular order, define a right-handed system of local orthogonal coordinates.

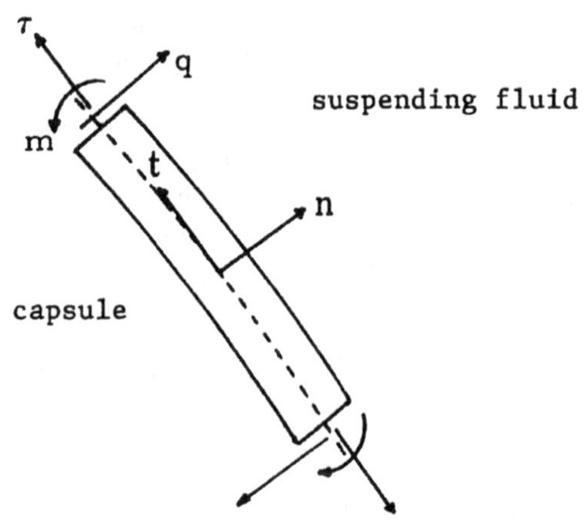


FIG. 2. *Tensions (stress resultants) and bending moment developing over a cross-section of a two-dimensional membrane.*

Following the formalism of shell theory (e.g. Fung 1965, Green & Zerna 1968, Green & Adkins 1970, Flügge 1973, Møllmann 1981, Ogden 1984, Beatty 1987, McDonald 1996), we consider the mid-line of the trace of the membrane in the $xy$ plane, drawn with the dashed line in Figure 2, and introduce the in-plane tension or stress resultant denoted by $\tau$, the transverse shear tension denoted by $q$, and the bending moment $m$. Considering now a small section of the membrane of infinitesimal arc length $dl$, where the arc length $l$ is measured in the direction of the tangent vector $\mathbf{t}$ from an arbitrary origin, we assume that the mass and thus the inertia of the membrane is negligible, and balance the hydrodynamic tractions and the membrane tensions to find that the jump in the hydrodynamic traction across the membrane is given by

$$\Delta \mathbf{f} \equiv (\sigma^{(s)} - \sigma^{(c)}) \cdot \mathbf{n} = -\frac{\partial}{\partial l}(\tau\, \mathbf{t} + q\, \mathbf{n}), \tag{3.1}$$

where the superscipr ts $(s)$ and $(c)$ denote, respectively, the suspending fluid and the fluid enclosed by the capsule. Expanding out the derivative on the right-hand side of (3.1), and using the Frenet relations

$$\frac{\partial \mathbf{n}}{\partial l} = \kappa\, \mathbf{t}, \qquad \frac{\partial \mathbf{t}}{\partial l} = -\kappa\, \mathbf{n}, \tag{3.2}$$

where $\kappa = \nabla \cdot \mathbf{n}$ is the curvature of the membrane in the $xy$ plane, we find

$$\Delta \mathbf{f}^{(m)} \equiv (\kappa\, \tau - \frac{\partial q}{\partial l})\, \mathbf{n} - (\frac{\partial \tau}{\partial l} + \kappa\, q)\, \mathbf{t}, \tag{3.3}$$

Performing an analogous torque balance, we derive a relation between the transverse shear tension and the bending moment,

$$q = \frac{\partial m}{\partial l}, \tag{3.4}$$

Replacing now the transverse shear tension on the right-hand side of (3.3) with the right-hand side of (3.4), we arrive at the final form

$$\Delta \mathbf{f} \equiv (\kappa\, \tau - \frac{\partial^2 m}{\partial l^2})\, \mathbf{n} - (\frac{\partial \tau}{\partial l} + \kappa\, \frac{\partial m}{\partial l})\, \mathbf{t}. \tag{3.5}$$

To evaluate the right-hand side of (3.5), we require constitutive relations for the elastic tension $\tau$ and bending moment $m$. Considering the elastic tension, we introduce the stretch or extension ratio

$$\lambda \equiv \frac{\partial l}{\partial l^R}, \tag{3.6}$$

where $l^R$ is the arc length along the membrane at the position of a material point particle, measured in a reference state where the membrane is unstressed, and $l$ is the corresponding arc length in the deformed state; the origin of arc length is located at the same material point particle in both states. Moreover, we introduce the surface strain-energy function $W^S(\lambda, K)$, where $K \equiv \kappa - \kappa^R$ is a measure of bending and $\kappa^R$ is the curvature of the membrane in a reference state where the bending moments vanish, and write

$$\tau = \frac{\partial W^S}{\partial \lambda}, \qquad m = \frac{\partial W^S}{\partial k}, \tag{3.7}$$

(Steigmann & Ogden 1997). For example, if the membrane consists of a symmetric lipid bilayer, then the reference configuration is flat, and the reference curvature vanishes, $\kappa^R = 0$. For small deformations, the strain-energy function may be approximated with the quadratic form

$$W^S = \frac{1}{2} E_S (\lambda - 1)^2 + \frac{1}{2} \kappa_B K^2, \tag{3.8}$$

where $E_S$ is the surface modulus of elasticity, and $\kappa_B$ is the modulus of bending stiffness. Subject to (3.8), the in-plane tension and bending moment are given by the relations

$$\tau = E_S (\lambda - 1), \qquad m = \kappa_B K, \tag{3.9}$$

which provide us with linear constitutive equations.

**4. Three-dimensional membranes.** The theory of thin shells provides us with a natural framework for establishing equilibrium equations and for developing constitutive relations relating the tensions and bending moments to the membrane deformation (e.g. Fung 1965, Green & Zerna 1968, Green & Adkins 1970, Simonds & Danielson 1972, Flügge 1973, Gurtin & Murdoch 1975, Møllmann 1981, Ogden 1984, Beatty 1987, McDonald 1996; Steigmann & Ogden 1997, 1999).

**4.1. Stress resultants and bending moments.** Consider a membrane at a specified reference configuration, and label the point particles comprising it using the convected surface curvilinear coordinates $(\xi, \eta)$, so that a line of constant $\eta$, a line of constant $\xi$, and a line along the unit normal vector $\mathbf{n}^R$ define a system of right-handed but not necessarily orthogonal coordinates, as depicted at the top drawing of Figure 3. The positions of point particles in the reference state are denoted by $\mathbf{x}^R(\xi, \eta)$. Assume now that the membrane deforms, and denote the new positions of the point particles by $\mathbf{x}(\xi, \eta)$. The developing in-plane stress resultants or elastic tensions $\tau_{\alpha\beta}$, tansverse shear tensions $q_\alpha$, and bending moments $m_{\alpha\beta}$, are illustrated in Figure 4. In the "membrane approximation" of thin-shell theory, the transverse tensions and bending moments are neglected, and only the in-plane stress resultants are considered (Corneliussen & Shield 1961). This approximation, however, is not appropriate for biological membranes where bending moments make an important, if not essential, contribution.

**4.2. Interface force and torque balances in Cartesian coordinates.** Describing the membrane tensions and bending moments in global Cartesian coordinates facilitates the interfacing of the membrane mechanics and the hydrodynamics on either side of an interface. In the Cartesian formulation, the domain of definition of the membrane tensions and bending moments is extended into the whole three-dimensional space subject to appropriate constraints, as follows (Gurtin & Murdoch 1975; Libai & Simonds 1998, Chapter VIII; Valid 1995, Pozrikidis 2000a).

The in-plane tensions are described in terms of the Cartesian tensor $\tau$, so that the in-plane tension exerted on a cross-section of the membrane that

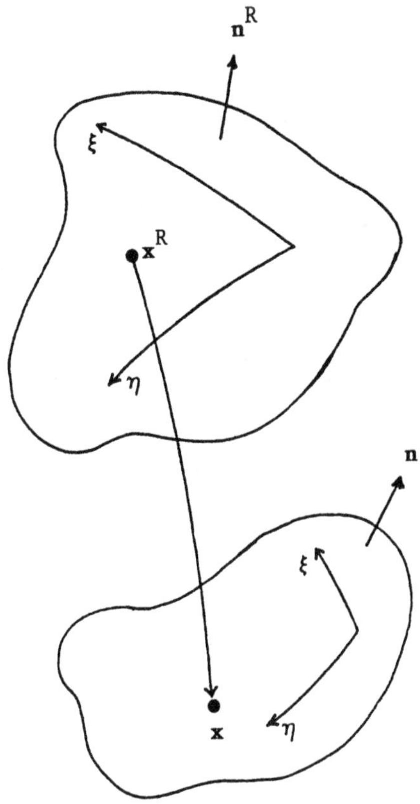


FIG. 3. *Schematic illustration of a three-dimensional membrane at the reference and at a deformed state. Point particles distributed over the membrane are marked permanently by two convected surface curvilinear coordinates.*

is normal to the tangential unit vector $\mathbf{b}$ is given by $\mathbf{b} \cdot \tau$ and, furthermore, $\mathbf{n} \cdot \tau = \mathbf{0}$ and $\tau \cdot \mathbf{n} = \mathbf{0}$; the last restriction ensures that the tension lies in the tangential plane. For example, if an interface exhibits isotropic tension $\gamma$, then $\tau = \gamma\, \mathbf{P}$, where $\mathbf{P} = \mathbf{I} - \mathbf{nn}$ is the tangential projection operator. The transverse shear tension is described in terms of the Cartesian vector $\mathbf{q}$, so that the transverse shear tension exerted on a cross-section of the membrane that is normal to the tangential unit vector $\mathbf{b}$ in the direction of the unit normal vector $\mathbf{n}$ is given by $\mathbf{b} \cdot \mathbf{q}$ and, furthermore, $\mathbf{n} \cdot \mathbf{q} = 0$. The bending moments are expressed in terms of the Cartesian tensor $\mathbf{m}$, so that the bending moment vector exerted on a cross-section of the membrane that is normal to the tangential unit vector $\mathbf{b}$ is given by $\mathbf{n} \times (\mathbf{b} \cdot \mathbf{m})$ and, furthermore, $\mathbf{n} \cdot \mathbf{m} = \mathbf{0}$ and $\mathbf{m} \cdot \mathbf{n} = \mathbf{0}$; the last restriction ensures that the

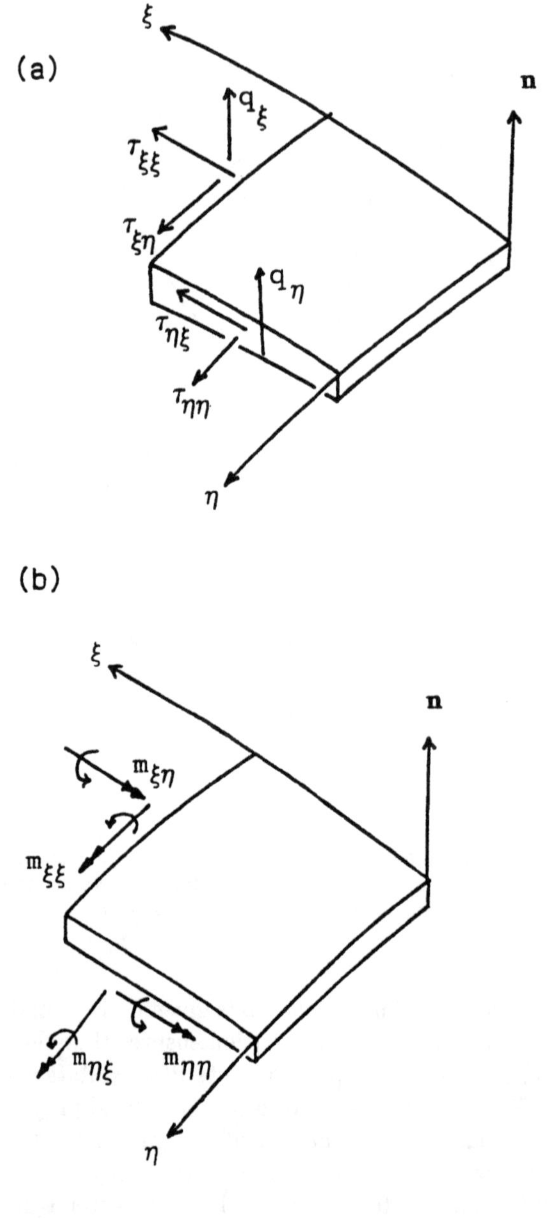


FIG. 4. *Depiction of (a) tensions (stress resultants), and (b) bending moments developing around the edges of a patch on a three-dimensional membrane.*

moment vector lies in the tangential plane.

Consider now a membrane patch enclosed by the contour $C$. Assuming that the mass and thus the inertia of the membrane is negligible, we perform a force balance over the patch and obtain

$$\int_{patch} (\sigma^{(s)} - \sigma^{(c)}) \cdot \mathbf{n}\, dS + \int_C \mathbf{b} \cdot (\tau + \mathbf{q}\,\mathbf{n})\, dl = \mathbf{0}, \tag{4.1}$$

where $\mathbf{t}$ is the unit vector tangential to $C$, $\mathbf{b} = \mathbf{t} \times \mathbf{n}$ is the unit vector that is tangential to the membrane and lies in a plane that is normal to the contour $C$, and $l$ is the arc length along $C$. Using the surface divergence theorem to convert the contour integral into a surface integral on the right-hand side of (4.1), and then taking the limit as the size of the patch becomes infinitesimal, we find that the jump in the hydrodynamic traction across the membrane is given by

$$\begin{aligned} \Delta\mathbf{f} \equiv (\sigma^{(s)} - \sigma^{(c)}) \cdot \mathbf{n} &= -(\mathbf{P} \cdot \nabla) \cdot (\tau + \mathbf{q}\,\mathbf{n}) \\ &= -\mathrm{Trace}[(\mathbf{P} \cdot \nabla)(\tau + \mathbf{q}\,\mathbf{n})], \end{aligned} \tag{4.2}$$

where $\mathbf{P} = \mathbf{I} - \mathbf{nn}$ is the tangential projection operator. The right-hand side of (4.2) expresses the surface divergence of the generalized elastic tension tensor in Cartesian coordinates. The tangential derivatives are taken with respect to two isometric orthogonal rectilinear coordinates that are tangential to the membrane at the point where the divergence is evaluated.

An analogous torque balance with respect to the arbitrary point $\mathbf{x}_c$ requires

$$\begin{aligned} &\int_{patch} (\mathbf{x} - \mathbf{x}_c) \times [(\sigma^{(s)} - \sigma^{(c)}) \cdot \mathbf{n}]\, dS \\ &\quad + \int_C (\mathbf{x} - \mathbf{x}_c) \times [\mathbf{b} \cdot (\tau + \mathbf{q}\,\mathbf{n})]\, dl + \int_C \mathbf{n} \times (\mathbf{b} \cdot \mathbf{m})\, dl = \mathbf{0}. \end{aligned} \tag{4.3}$$

Using the divergence theorem to convert the contour integral into a surface integral on the right-hand side of (4.3), taking the limit as the size of the patch becomes infinitesimal, and then using the force balance (4.2), we derive an expression for the transverse shear tension,

$$\mathbf{q} = [(\mathbf{P} \cdot \nabla) \cdot \mathbf{m}] \cdot \mathbf{P} = \mathrm{Trace}[(\mathbf{P} \cdot \nabla)\mathbf{m}] \cdot \mathbf{P}, \tag{4.4}$$

and another expression for the antisymmetric part of the in-plane tension tensor,

$$\tau - \tau^T = \mathbf{B} \cdot \mathbf{m} - \mathbf{m}^T \cdot \mathbf{B}, \tag{4.5}$$

where the superscript $T$ denotes the matrix transpose, and $\mathbf{B} = \nabla\mathbf{n}$ is the symmetric Cartesian curvature tensor (Pozrikidis 2000a).

**4.3. Interface force and torque balances in surface curvilinear coordinates.** The Cartesian formulation described in Section 4.2 requires that the membrane tensions and bending moments be extended into the whole space in an appropriate fashion. This extension can be avoided by working in surface curvilinear coordinates (e.g. Barthès-Biesel 1980). To set up this formulation, we introduce the generally non-unit tangential vectors

$$\mathbf{t}_\xi \equiv \frac{\partial \mathbf{x}}{\partial \xi}, \qquad \mathbf{t}_\eta \equiv \frac{\partial \mathbf{x}}{\partial \eta}, \tag{4.6}$$

and the corresponding arc length metric coefficients

$$h_\xi = |\mathbf{t}_\xi|, \qquad h_\eta = |\mathbf{t}_\eta|. \tag{4.7}$$

The *first fundamental form* of the surface is defined as the square of the length of an infinitesimal fiber whose end-points are separated by a vector corresponding to the infinitesimal coordinates $d\xi$ and $d\eta$,

$$(dl)^2 = a_{\xi\xi}\,(d\xi)^2 + 2\,a_{\xi\eta}\,d\xi\,d\eta + a_{\eta\eta}\,(d\eta)^2, \tag{4.8}$$

where

$$a_{\xi\xi} = h_\xi^2, \qquad a_{\xi\eta} = a_{\eta\xi} = \mathbf{t}_\xi \cdot \mathbf{t}_\eta, \qquad a_{\eta\eta} = h_\eta^2, \tag{4.9}$$

The surface area of a patch confined between two $\xi$ or $\eta$ segments with infinitesimal spans $d\xi$ and $d\eta$ is equal to $dS = \sqrt{a}\,d\xi\,d\eta$, where $a = h_\xi^2\,h_\eta^2 - a_{\xi\eta}^2$.

The *second fundamental form* of the surface is defined as the quadratic form

$$Q = -b_{\xi\xi}\,(d\xi)^2 - 2\,b_{\xi\eta}\,d\xi\,d\eta - b_{\eta\eta}\,(d\eta)^2, \tag{4.10}$$

where

$$b_{\xi\xi} = \mathbf{n} \cdot \frac{\partial \mathbf{t}_\xi}{\partial \xi}, \qquad b_{\xi\eta} = b_{\eta\xi} = \mathbf{n} \cdot \frac{\partial \mathbf{t}_\xi}{\partial \eta} = \mathbf{n} \cdot \frac{\partial \mathbf{t}_\eta}{\partial \xi}, \qquad b_{\eta\eta} = \mathbf{n} \cdot \frac{\partial \mathbf{t}_\eta}{\partial \eta}, \tag{4.11}$$

and $\mathbf{n} = \mathbf{t}_\xi \times \mathbf{t}_\eta / |\mathbf{t}_\xi \times \mathbf{t}_\eta|$ is the unit normal vector. The normal curvature of the surface in the direction of an infinitesimal vector whose end-points correspond to the infinitesimal increments $d\xi$ and $d\eta$ is equal to the ratio of the second to the first fundamental form of the surface.

Next, we introduce the surface contravariant components of the tension tensor $\tau$ denoted by $\tau^{\alpha\beta}$, the surface contravariant components of the transverse shear tension vector denoted by $q^\xi$ and $q^\eta$, and the surface contravariant components of the bending moment tensor $\mathbf{m}$ denoted by $m^{\alpha\beta}$. Greek superscripts and subscripts stand for $\xi$ or $\eta$. Subject to these definitions, the force equilibrium equation (4.2) takes the form

$$\Delta\mathbf{f} \equiv (\sigma^{(s)} - \sigma^{(c)}) \cdot \mathbf{n} = \Delta f^n\,\mathbf{n} + \Delta f^\xi\,\mathbf{t}_\xi + \Delta f^\eta\,\mathbf{t}_\eta \tag{4.12}$$

where

$$\Delta f^n = -b_{\alpha\beta}\,\tau^{\alpha\beta} - q^\beta|_\beta, \qquad \Delta f^\xi = -\tau^{\beta\xi}|_\beta + b^\xi_\alpha\, q^\alpha,$$
$$\Delta f^\eta = -\tau^{\beta\eta}|_\beta + b^\eta_\alpha\, q^\alpha, \tag{4.13}$$

(Møllmann 1981, p.165; Waxman 1984, eq. (3.5)). Correspondingly, the torque equilibrium equations (4.4) and (4.5) take the form

$$q^\xi = m^{\alpha\xi}|_\alpha, \qquad q^\eta = m^{\alpha\xi}|_\alpha, \qquad \tau^{\xi\eta} - \tau^{\eta\xi} = b^\xi_\alpha\, m^{\alpha\eta} - b^\eta_\alpha\, m^{\alpha\xi} \tag{4.14}$$

where the mixed components $b^\beta_\alpha$ are related to the pure components $b_{\alpha\beta}$ by the relation $b_{\beta\alpha} = a_{\beta\gamma}\, b^\gamma_\alpha$ (Møllmann 1981, p. 165; Waxman 1984, eq. (3.9)). A vertical bar signifies the covariant derivative taken with respect to the subscribed variable and defined in terms of the Christoffel symbols (e.g. Aris 1962).

**4.4. Interface force and torque balance in lines of principal curvature.** Considerable simplifications occur by referring to surface curvilinear coordinates whose tangential vector at every point is oriented in the direction of the principal curvature, defined as the *lines of principal curvature*. A line of constant $\eta$, a line of constant $\xi$, and a line directed along the unit normal vector $\mathbf{n}$ define a right-handed system of orthogonal curvilinear coordinates. The directions of the stress resultants and bending moments are defined in Figure 4. In the case of an axisymmetric interface supporting axisymmetric tensions, to be discussed in Section 5, the lines of principal curvature are and remain the traces of the membrane in meridional and azimuthal planes.

In surface curvilinear coordinates that are lines of principal curvature, the decomposition (4.12) takes the form

$$\Delta \mathbf{f} \equiv (\sigma^{(s)} - \sigma^{(c)}) \cdot \mathbf{n} = \Delta f^n\, \mathbf{n} + \Delta f'^\xi\, \mathbf{e}_\xi + \Delta f'^\eta\, \mathbf{e}_\eta, \tag{4.15}$$

where $\mathbf{n}$ is the unit normal vector pointing into the ambient fluid, and $\mathbf{e}_\xi \equiv \mathbf{t}_\xi/|\mathbf{t}_\xi|$, $\mathbf{e}_\eta \equiv \mathbf{t}_\eta/|\mathbf{t}_\eta|$ are unit tangential vectors. The normal and tangential components of the traction discontinuity are given by

$$\Delta f^n = \kappa_\xi\, \tau_{\xi\xi} + \kappa_\eta\, \tau_{\eta\eta} - \frac{1}{h_\xi h_\eta}\left[\frac{\partial}{\partial \xi}(h_\eta q_\xi) + \frac{\partial}{\partial \eta}(h_\xi q_\eta)\right], \tag{4.16}$$

$$\Delta f'^\xi = -\frac{1}{h_\xi h_\eta}\left[\frac{\partial}{\partial \xi}(h_\eta \tau_{\xi\xi}) + \frac{\partial}{\partial \eta}(h_\xi \tau_{\eta\xi}) + \frac{\partial h_\xi}{\partial \eta}\,\tau_{\xi\eta} - \frac{\partial h_\eta}{\partial \xi}\,\tau_{\eta\eta}\right] - \kappa_\xi\, q_\xi, \tag{4.17}$$

$$\Delta f'^\eta = -\frac{1}{h_\xi h_\eta}\left[\frac{\partial}{\partial \xi}(h_\eta \tau_{\xi\eta}) + \frac{\partial}{\partial \eta}(h_\xi \tau_{\eta\eta}) - \frac{\partial h_\xi}{\partial \eta}\,\tau_{\xi\xi} + \frac{\partial h_\eta}{\partial \xi}\,\tau_{\eta\xi}\right] - \kappa_\eta\, q_\eta, \tag{4.18}$$

where $\kappa_\xi = -b_{\xi\xi}/h_\xi^2$ and $\kappa_\eta = -b_{\eta\eta}/h_\eta^2$ are the principal curvatures. Expressions (4.14) simplify to

$$q_\xi = \frac{1}{h_\xi h_\eta}\left[\frac{\partial}{\partial \xi}(h_\eta m_{\xi\xi}) + \frac{\partial}{\partial \eta}(h_\xi m_{\eta\xi}) + \frac{\partial h_\xi}{\partial \eta}\, m_{\xi\eta} - \frac{\partial h_\eta}{\partial \xi}\, m_{\eta\eta}\right], \tag{4.19}$$

$$q_\eta = \frac{1}{h_\xi h_\eta}\left[\frac{\partial}{\partial \xi}(h_\eta m_{\xi\eta}) + \frac{\partial}{\partial \eta}(h_\xi m_{\eta\eta}) - \frac{\partial h_\xi}{\partial \eta}\, m_{\xi\xi} + \frac{\partial h_\eta}{\partial \xi}\, m_{\eta\xi}\right], \tag{4.20}$$

$$\tau_{\xi\eta} - \tau_{\eta\xi} = -\kappa_\xi\, m_{\xi\eta} + \kappa_\eta\, m_{\eta\xi}, \tag{4.21}$$

(e.g. Møllmann 1981, p. 33). In Section 5, we shall present the specific forms of these expressions for axisymmetric membranes in cylindrical polar coordinates.

**4.5. Surface deformation.** As a prelude to evaluating the elastic tensions, we refer to Figure 3 and introduce the three-dimensional Cartesian relative deformation gradient tensor $\mathbf{F}$ with components

$$F_{ij} = (\frac{\partial x_i}{\partial x_j^R})_{\xi,\eta}. \tag{4.22}$$

Let the infinitesimal vector $d\mathbf{l}^R$ describe a small fiber that is either tangential or normal to the membrane at the reference state. After deformation, the fiber has rotated and stretched or compressed to its image described by

$$d\mathbf{l} = \mathbf{F} \cdot d\mathbf{l}^R. \tag{4.23}$$

The nine components of the relative deformation gradient tensor $\mathbf{F}$ may be evaluated from knowledge of the images of two fibers that are tangential to the membrane at a point, and the image of a fiber that is normal to the membrane at that point. In the present formulation, the image of a fiber that is normal to the membrane is assumed to vanish, so that the deformation of this fiber does not enter the computation of the elastic tensions explicitly, but only implicitly by means of the deformation of the tangential fibers, and according to constitutive laws expressing membrane material properties.

For the purpose of computing the elastic tensions, equation (4.23) is replaced by the equation

$$d\mathbf{l} = \mathbf{F}^S \cdot d\mathbf{l}^R, \tag{4.24}$$

where

$$\mathbf{F}^S \equiv \mathbf{F} \cdot (\mathbf{I} - \mathbf{n}^R\mathbf{n}^R) \tag{4.25}$$

is the relative *surface* deformation gradient, and the superscript S stands for "surface". Clearly, $\mathbf{n}^R$ is an eigenvector of $\mathbf{F}^S$ corresponding to a vanishing eigenvalue, which shows that $\mathbf{F}^S$ is singular.

If $d\mathbf{l}^R$ is a tangential fiber in the reference state, then $d\mathbf{l}$ is also a tangential fiber in the deformed state, and this requires $\mathbf{n}\cdot d\mathbf{l} = \mathbf{n}\cdot\mathbf{F}^S\cdot d\mathbf{l}^R = 0$. Since, however, the orientation of $d\mathbf{l}^R$ is arbitrary, it must be $\mathbf{n}\cdot\mathbf{F}^S = \mathbf{0}$, which suggests that $\mathbf{F}^S = (\mathbf{I} - \mathbf{n}\mathbf{n})\cdot\mathbf{F}^S$ or

$$\mathbf{F}^S \equiv (\mathbf{I} - \mathbf{n}\mathbf{n}) \cdot \mathbf{F} \cdot (\mathbf{I} - \mathbf{n}^R\mathbf{n}^R). \tag{4.26}$$

Thus, $\mathbf{n}$ is an eigenvector of the transpose of $\mathbf{F}^S$ corresponding to the vanishing eigenvalue. The polar decomposition theorem allows us to write $\mathbf{F}^S = \mathbf{R}\cdot\mathbf{U} = \mathbf{V}\cdot\mathbf{R}$, where $\mathbf{R}$ is an orthogonal matrix expressing plane rotation, and $\mathbf{U}$ and $\mathbf{V}$ are the positive-definite and symmetric right and left stretch tensors expressing pure deformation.

Following standard procedure in the theory of elasticity (e.g. Green & Zerna 1968, Green & Adkins 1970, Beatty 1987), we introduce the positive-definite and symmetric surface left Cauchy-Green deformation tensor

$$\mathbf{V}^2 \equiv \mathbf{F}^S \cdot \mathbf{F}^{S^T}, \tag{4.27}$$

where the superscript T denotes the matrix transpose. The eigenvalues of $\mathbf{V}^2$ are equal to $\lambda_1^2$, $\lambda_2^2$, and 0, corresponding to the orthogonal tangential eigenvectors $\mathbf{v}_1$, $\mathbf{v}_2$ and to the normal vector $\mathbf{n}$. The eigenvectors of $\mathbf{V}^2$ are also eigenvectors of the tension tensor $\tau$.

In terms of the principal elastic tensions $\tau_1^P$ and $\tau_2^P$ and the unit tangential eigenvectors $\mathbf{e}_1 = \mathbf{v}_1/|\mathbf{v}_1|$ and $\mathbf{e}_2 = \mathbf{v}_2/|\mathbf{v}_2|$, the symmetric tension tensor in the absence of bending moments is given by the spectral decomposition

$$\tau = \tau_1^P\, \mathbf{e}_1\, \mathbf{e}_1 + \tau_2^P\, \mathbf{e}_2\, \mathbf{e}_2. \tag{4.28}$$

**4.6. Constitutive equations for the elastic tensions.** Next, we proceed to relate the tensions to the surface strains by means of a constitutive equation. As a prelude, we consider a three-dimensional elastic medium and express the force exerted on a small material patch of surface area $dS$ that is perpendicular to the unit normal vector $\mathbf{n}$ in terms of the Eulerian stress tensor $\sigma$, in the familiar form

$$d\mathbf{f} = \mathbf{n}\cdot\sigma\, dS. \tag{4.29}$$

Furthermore, we introduce the *first Piola-Kirchhoff tensor* $\mathbf{T}$, also called the *Lagrange or nominal stress tensor*, and the *Piola-Kirchhoff tensor* $\mathbf{S}$, defined by the relations

$$d\mathbf{f} = \mathbf{n}\cdot\sigma\, dS = \mathbf{n}^R\cdot\mathbf{T}\, dS^R = \mathbf{n}^R\cdot\mathbf{S}\cdot\mathbf{F}^T\, dS^R, \tag{4.30}$$

where $\mathbf{n}^R$ is the unit vector normal to the patch in a reference state, $dS^R$ is the corresponding surface area, and the superscript T denotes the matrix transpose (e.g. Fung 1965, p. 438; Ogden 1984, p. 152). The Eulerian stress tensor $\sigma$ is related to the first Piola-Kirchhoff tensor $\mathbf{T}$ and to the Piola-Kirchhoff tensor $\mathbf{S}$ by the equation

$$\sigma = \frac{1}{J}\mathbf{F}\cdot\mathbf{T} = \frac{1}{J}\mathbf{F}\cdot\mathbf{S}\cdot\mathbf{F}^T \tag{4.31}$$

where $J = \mathrm{Det}(\mathbf{F})$ is the fractional volume of an infinitesimal volume element after the deformation; for an incompresible material, $J = 1$. For a Green-elastic or hyperelastic three-dimensional medium, the first Piola-Kirchhoff tensor and the Piola-Kirchhoff tensor derive from a strain-energy function $W(\mathbf{F})$ by means of the relations

$$T_{ij} = \frac{\partial W}{\partial F_{ij}}, \qquad S_{ij} = \frac{\partial W}{\partial E_{ij}}, \tag{4.32}$$

where $\mathbf{E} = \frac{1}{2}(\mathbf{F}^T \cdot \mathbf{F} - \mathbf{I})$ is the Green (material or Lagrangean) strain tensor (e.g. Fung 1965, p. 449; Green & Adkins 1970, p. 7; Ogden 1984, pp. 204-209; Beatty, 1987).

Considering now the two-dimensional analog of the preceding equations over the curved surface of a membrane in the absence of bending moments, we replace equation (4.31) by

$$\tau = \frac{1}{J^S}\,\mathbf{F}^S \cdot \mathbf{T}^S = \frac{1}{J^S}\,\mathbf{F}^S \cdot \mathbf{S}^S \cdot \mathbf{F}^{S^T} \tag{4.33}$$

where $J^S = \lambda_1\lambda_2$ is the fractional surface area of a material membrane patch after deformation, and $\mathbf{T}^S$ and $\mathbf{S}^S$ are the surface Piola-Kirchhoff tensors. The counterparts of relations (4.32) are

$$T^S_{ij} = \frac{\partial W^S}{\partial F^S_{ij}}, \qquad S^S_{ij} = \frac{\partial W^S}{\partial E^S_{ij}} \tag{4.34}$$

where $\mathbf{E}^S = \frac{1}{2}(\mathbf{F}^{S^T} \cdot \mathbf{F}^S - \mathbf{I})$ is the surface Green (material or Lagrangean) strain tensor.

Referring to local Cartesian coordinates with two axes parallel to the principal directions of the tension tensor at a point, and using equations (4.33) and (4.34), we find that the principal tensions are given by

$$\tau_1^P = \frac{1}{\lambda_2}\frac{\partial W^S}{\partial \lambda_1}, \qquad \tau_2^P = \frac{1}{\lambda_1}\frac{\partial W^S}{\partial \lambda_2}. \tag{4.35}$$

Expression (4.28) combined with equations (4.35) provides us with a complete description of the elastic tensions.

Kinematic constraints require that the surface strain-energy function $W^S$ depend on the surface deformation gradient only through strain invariants. Skalak et al. (1973) introduced the invariants

$$I_1^S = \lambda_1^2 + \lambda_2^2 - 2, \qquad I_2^S = \lambda_2^2\lambda_2^2 - 1, \tag{4.36}$$

in terms of which expressions (4.35) take the form

$$\begin{aligned} \tau_1^P &= 2\,\frac{\lambda_1}{\lambda_2}\frac{\partial W^S}{\partial I_1^S} + 2\,\lambda_1\lambda_2\,\frac{\partial W^S}{\partial I_2^S}, \\ \tau_2^P &= 2\,\frac{\lambda_2}{\lambda_1}\frac{\partial W^S}{\partial I_1^S} + 2\,\lambda_1\lambda_2\,\frac{\partial W^S}{\partial I_2^S}, \end{aligned} \tag{4.37}$$

Substituting expressions (4.37) into (4.28), we find

$$\tau = \frac{2}{\lambda_1\lambda_2}\,(\lambda_1^2\,\mathbf{e}_1\,\mathbf{e}_1 + \lambda_2^2\,\mathbf{e}_2\,\mathbf{e}_2)\frac{\partial W^S}{\partial I_1^S} + 2\lambda_1\lambda_2\,(\mathbf{e}_1\,\mathbf{e}_1 + \mathbf{e}_2\,\mathbf{e}_2)\frac{\partial W^S}{\partial I_2^S}, \tag{4.38}$$

or

$$\tau = \frac{2}{\lambda_1\lambda_2}\,\frac{\partial W^S}{\partial I_1^S}\,\mathbf{V}^2 + 2\,\lambda_1\lambda_2\frac{\partial W^S}{\partial I_2^S}\,(\mathbf{I} - \mathbf{nn}) \tag{4.39}$$

Note that when $\partial W^S/\partial I_2^S = 0$, the tensions are isotropic.

Skalak et al. (1973) proposed the following strain energy function for the membrane of a red blood cell,

$$W^S = \frac{B}{4}\left(\frac{1}{2}\,I_1^{S^2} + I_1^S - I_2^S\right) + \frac{C}{8}\,I_2^{S^2} \tag{4.40}$$

where $B$ and $C$ are physical constants with estimated values on the order of $B = 0.005$ dyn/cm and $C = 100$ dyn/cm. The large magnitude of the constant $C$ compared to the magnitude of $B$ ensures that the membrane is nearly incompressible: a small deviation of $I_2^S$ from unity generates large elastic tensions.

Barthès-Biesel & Rallison (1981) introduced the alternative strain invariants

$$\Lambda_1 = \ln(\lambda_1\lambda_2) = \frac{1}{2}\ln(I_2^S + 1), \qquad \Lambda_2 = \frac{1}{2}(\lambda_1^2 + \lambda_2^2) - 1 = \frac{1}{2}\,I_1^S, \tag{4.41}$$

in terms of which expressions (4.35) take the form

$$\tau_1^P = \frac{1}{\lambda_1\lambda_2}\,\left(\frac{\partial W^S}{\partial \Lambda_1} + \lambda_1^2\,\frac{\partial W^S}{\partial \Lambda_2}\right), \qquad \tau_2^P = \frac{1}{\lambda_1\lambda_2}\,\left(\frac{\partial W^S}{\partial \Lambda_1} + \lambda_2^2\,\frac{\partial W^S}{\partial \Lambda_2}\right), \tag{4.42}$$

Substituting expressions (4.42) into (4.28), we find

$$\tau = \frac{1}{\lambda_1\lambda_2}\,(\mathbf{e}_1\,\mathbf{e}_1 + \mathbf{e}_2\,\mathbf{e}_2)\frac{\partial W^S}{\partial \Lambda_1} + (\lambda_1^2\,\mathbf{e}_1\,\mathbf{e}_1 + \lambda_2^2\,\mathbf{e}_2\,\mathbf{e}_2)\frac{\partial W^S}{\partial \Lambda_2}, \tag{4.43}$$

or

$$\tau = \frac{1}{\lambda_1\lambda_2}\,\left[\frac{\partial W^S}{\partial \Lambda_1}\,(\mathbf{I} - \mathbf{nn}) + \frac{\partial W^S}{\partial \Lambda_2}\,\mathbf{V}^2\right]. \tag{4.44}$$

Note that when $\partial W^S/\partial \Lambda_2 = 0$, the tensions are isotropic. In the limit of small deformations, the strain energy function obtains the standard Mooney-Rivlin form

$$W^s = \alpha_1\,\Lambda_1 + \frac{1}{2}\,(\alpha_1 + \alpha_2)\,\Lambda_1^2 + \alpha_3(\Lambda_2 - \Lambda_1), \tag{4.45}$$

where $\alpha_1$, $\alpha_2$, and $\alpha_3$ are material constants (see Section 5).

**4.7. Tensions in a thin elastic shell.** It is instructive to compare the results derived in Section 4.6 with corresponding results for the tensions developing in a thin shell of a three-dimensional incompressible elastic material with uniform thickness $h$ (e.g. Green & Adkins 1970, pp. 156–159; McDonald 1996, p. 399). For this purpose, we introduce the *volume* strain invariants

$$I_1^V \equiv \lambda_1^2 + \lambda_2^2 + \frac{1}{\lambda_1^2\lambda_2^2}, \qquad I_2^V \equiv \lambda_1^2 + \lambda_2^2 + \lambda_1^2\lambda_2^2, \tag{4.46}$$

and express the principal elastic tensions in terms of the volume strain energy function $W^V$ as

$$\begin{aligned} \tau_1^P &= \frac{2h}{\lambda_1\lambda_2}(\lambda_1^2 - \frac{1}{\lambda_1^2\lambda_2^2})\,(\frac{\partial W^V}{\partial I_1^V} + \lambda_2^2\,\frac{\partial W^V}{\partial I_2^V}), \\ \tau_2^P &= \frac{2h}{\lambda_1\lambda_2}(\lambda_2^2 - \frac{1}{\lambda_1^2\lambda_2^2})\,(\frac{\partial W^V}{\partial I_1^V} + \lambda_1^2\,\frac{\partial W^V}{\partial I_2^V}) \end{aligned} \tag{4.47}$$

The Mooney-Rivlin strain-energy function is given by

$$W^V = \frac{E}{6h}\,[(1-\alpha)(I_1^V - 3) + \alpha\,(I_2^V - 3)], \tag{4.48}$$

where $E$ is the volume modulus of elasticity and $\alpha$ is a material parameter varying between 0 and 1; $\alpha = 0$ corresponds to a linear, neo-Hookean medium (e.g. Ogden 1984, p. 221). In the limit of small deformations, expression (4.48) reduces to (4.45) with $\alpha_1 = 0, \alpha_2 = \frac{2}{3}\,E$, and $\alpha_3 = \frac{1}{3}\,E$, yielding $W^S = \frac{E}{3}(\Lambda_1^2 + \Lambda_2 - \Lambda_1)$, as discussed by Barthès-Biesel & Rallison (1981).

**4.8. Constitutive equations for bending moments.** The bending moments developing in a hyperelastic membrane derive from a strain energy function of appropriate strain and bending measures. Nonlinear theories of shells applicable for finite deformations have been developed and reviewed by several authors including Sanders (1963), Budiansky & Sanders (1963), Budiansky (1968), Naghdi (1972), Simmonds & Danielson (1972), Libai & Simmonds (1998), and more recently by Steigmann & Ogden (1997, 1999) and Pozrikidis (2000a).

To illustrate the methods, we restrict our attention to the most tractable case of infinitesimal displacements, and refer to orthogonal curvilinear coordinates that are lines of principal curvature (e.g. Møllmann 1981, p. 17). Considering the displacement of a material point particle over the membrane, denoted by $\mathbf{v}$, we introduce the strain measures

$$\begin{aligned} \epsilon_{\xi\xi} &= \frac{1}{h_\xi}\,\frac{\partial \mathbf{x}}{\partial \xi}\cdot\frac{\partial \mathbf{v}}{\partial \xi}, \qquad \epsilon_{\eta\eta} = \frac{1}{h_\eta}\,\frac{\partial \mathbf{x}}{\partial \eta}\cdot\frac{\partial \mathbf{v}}{\partial \eta}, \\ \epsilon_{\xi\eta} = \epsilon_{\eta\xi} &= \frac{1}{2h_\xi h_\eta}\,(\frac{\partial \mathbf{x}}{\partial \eta}\cdot\frac{\partial \mathbf{v}}{\partial \xi} + \frac{\partial \mathbf{x}}{\partial \xi}\cdot\frac{\partial \mathbf{v}}{\partial \eta}), \end{aligned} \tag{4.49}$$

The measure $\epsilon_{\xi\xi}$ expresses the elongation of a fiber in the direction of the $\xi$ axis, the measure $\epsilon_{\eta\eta}$ expresses the elongation of a fiber in the direction of the $\eta$ axis, and $\epsilon_{\xi\eta}$ is a measure of the deformation of an infinitesimal patch. Three corresponding measures of bending, $\kappa_{\xi\xi}$, $\kappa_{\eta\eta}$, and $\kappa_{\xi\eta}$ may be defined in terms of the rotation vector $\omega$ of a surface patch due to the deformation (e.g. Møllmann 1981, pp. 21, 25). The strain and bending measures may now be used to define the vectorial strain measure

$$\mathbf{h} \equiv (\epsilon_{\xi\xi}, \epsilon_{\xi\eta}, \epsilon_{\eta\eta}, \kappa_{\xi\xi}, \kappa_{\xi\eta}, \kappa_{\eta\eta}), \tag{4.50}$$

and the surface strain energy function

$$W^S = \frac{1}{2}\mathbf{h}\cdot\mathbf{D}\cdot\mathbf{h} \tag{4.51}$$

where $\mathbf{D}$ is a positive-definite matrix expressing membrane material properties (e.g. Møllmann 1981, p. 45). For example, if the membrane is a thin shell of a three-dimensional isotropic elastic material, the strain energy function may be approximated with *Love's first approximation* describing the infinitesimal displacement of a thin plate of thickness $h$,

$$\begin{aligned} W^S = \frac{1}{2}\frac{Eh}{1-\nu^2}\left[(1-\nu)\,(\epsilon_{\xi\xi}^2 + 2\,\epsilon_{\xi\eta}^2 + \epsilon_{\eta\eta}^2) + \nu(\epsilon_{\xi\xi}+\epsilon_{\eta\eta})^2\right] \\ + \frac{E_B}{2}\left[(1-\nu)\,(\kappa_{\xi\xi}^2 + 2\,\kappa_{\xi\eta}^2 + \kappa_{\eta\eta}^2) + \nu(\kappa_{\xi\xi}+\kappa_{\eta\eta})^2\right] \end{aligned} \tag{4.52}$$

where $E_B = Eh^3/[12(1-\nu^2)]$ is the plate bending rigidity, $E$ is the volume modulus of elasticity, and $\nu$ is the Poisson ratio (e.g. Fung 1965, p. 461).

In terms of the strain energy function, the stress resultants and bending moments are given by

$$\begin{aligned} \tau_{\xi\xi} &= \frac{\partial W^S}{\partial \epsilon_{\xi\xi}}, \qquad \tau_{\xi\eta} = \tau_{\eta\xi} = \frac{1}{2}\frac{\partial W^S}{\partial \epsilon_{\xi\eta}}, \qquad \tau_{\eta\eta} = \frac{\partial W^S}{\partial \epsilon_{\eta\eta}}, \\ m_{\xi\xi} &= \frac{\partial W^S}{\partial \kappa_{\xi\xi}}, \qquad m_{\xi\eta} = m_{\eta\xi} = \frac{1}{2}\frac{\partial W^S}{\partial \kappa_{\xi\eta}}, \qquad m_{\eta\eta} = \frac{\partial W^S}{\partial \kappa_{\eta\eta}}. \end{aligned} \tag{4.53}$$

For the strain energy function expressed by (4.52),

$$\begin{aligned} \tau_{\xi\xi} &= \frac{Eh}{1-\nu^2}(\epsilon_{\xi\xi} + \nu\epsilon_{\eta\eta}), \qquad \tau_{\xi\eta} = \tau_{\eta\xi} = \frac{Eh}{1+\nu}\epsilon_{\xi\eta}, \\ \tau_{\eta\eta} &= \frac{Eh}{1-\nu^2}(\epsilon_{\eta\eta} + \nu\epsilon_{\xi\xi}), \\ m_{\xi\xi} &= \frac{1}{12}\frac{Eh^3}{1-\nu^2}(\kappa_{\xi\xi} + \nu\kappa_{\eta\eta}), \qquad m_{\xi\eta} = m_{\eta\xi} = \frac{1}{12}\frac{Eh^3}{1+\nu}\kappa_{\xi\eta}, \\ m_{\eta\eta} &= \frac{1}{12}\frac{Eh^3}{1-\nu^2}(\kappa_{\eta\eta} + \nu\kappa_{\xi\xi}), \end{aligned} \tag{4.54}$$

It should be emphasized that the preceding formulas apply only for small membrane deformations. The recent work of Steigmann & Ogden (1997, 1999) establishes a framework for computing bending moments at finite deformations. In this theory, the tension tensor $\tau$ is decomposed into a symmetric part and a skew-symmetric part, where the latter is defined in equation (4.5). Constitutive equations for the symmetric part of the tension tensor and for the bending moments tensor are then developed in terms of a strain energy function dependent on strain and bending measures.

**5. Axisymmetric membranes.** We turn our attention now to axisymmetric membranes generated by rotating a line around the $x$ axis, as depicted in Figure 5. Kinematics and dynamics are described in polar cylindrical coordinates with axial position $x$, distance from the $x$ axis, and meridional angle measured around the $x$ axis with origin in the $xy$ plane forming the triplet $(x, \sigma, \varphi)$. The flow inside and outside the capsule, and the membrane tensions and bending moments developing due to the deformation are assumed to be axisymmetric. Following the formalism of thin-shell theory, we consider the mid-surface of the membrane and introduce: (a) the azimuthal and meridional tensions $\tau_s$ and $\tau_\varphi$, which are the principal tensions of the in-plane stress resultants, (b) the transverse shear tension $q$, and (c) the azimuthal and meridional bending moments $m_s$ and $m_\varphi$, as illustrated in Figure 5.

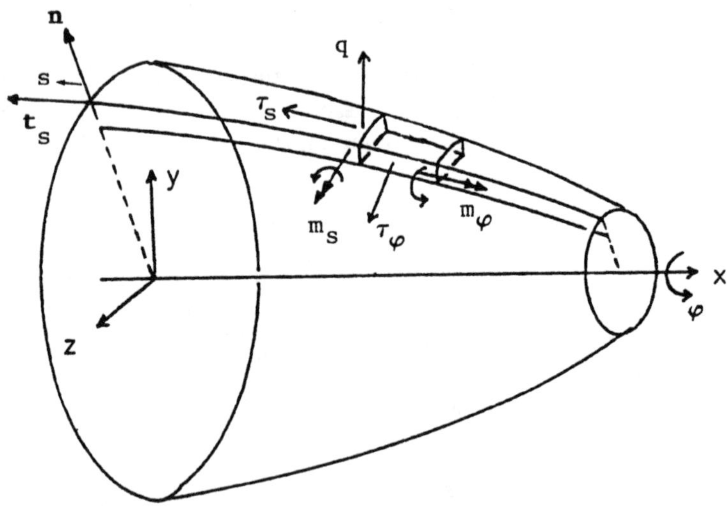


FIG. 5. *Elastic tensions and bending moments developing around the edges of a patch on an axisymmetric membrane.*

As a preliminary, we introduce the arc length along the trace of the membrane in a meridional plane denoted by $s$, and the unit vector that is tangential to the membrane and lies in a meridional plane corresponding to a certain value of the azimuthal angle $\varphi$, denoted by $\mathbf{t}_s$. The unit vector normal to the interface, $\mathbf{n}$, is directed into the ambient suspending fluid, as illustrated in Figure 5. The principal curvatures of the membrane in a meridional plane and its conjugate plane are denoted by $\kappa_s$ and $\kappa_\varphi$.

Using fundamental relations from differential geometry, we find that if the radial position of the membrane is described by the equation

$$\sigma = f(s) = g(x), \tag{5.1}$$

then the principal curvatures are given by

$$\kappa_s = -\frac{f''}{\sqrt{1-f'^2}} = -\frac{g''}{(1+g'^2)^{3/2}}, \tag{5.2}$$

and

$$\kappa_\varphi = \frac{1}{\sigma}\frac{\partial x}{\partial s} = \frac{1}{\sigma}\sqrt{1-f'^2} = \frac{1}{\sigma}\frac{1}{\sqrt{1+g'^2}} \tag{5.3}$$

(e.g. Pozrikidis 1997, p. 162). These expressions are consistent with Godazzi's formula

$$\kappa_s = \frac{\partial}{\partial \sigma}(\sigma \kappa_\varphi), \tag{5.4}$$

which allows us to compute one of the principal curvatures in terms of the other.

**5.1. Force and torque balances.** To compute the jump in the hydrodynamic traction across the membrane, we consider a small section of the membrane that is confined between: (a) two meridional planes passing through the $x$ axis, and (b) two parallel planes that are perpendicular to the $x$ axis and enclose a small section of the interface in a meridional plane with arc length $\Delta s$. Performing a force balance over this section, we find

$$\Delta \mathbf{f} = (\sigma^{(s)} - \sigma^{(c)}) \cdot \mathbf{n} = \Delta f^n\, \mathbf{n} + \Delta f^s\, \mathbf{t}_s \tag{5.5}$$

where the normal jump is given by

$$\Delta f^n = \kappa_s\, \tau_s + \kappa_\varphi\, \tau_\varphi - \frac{1}{\sigma}\frac{\partial}{\partial s}(\sigma q), \tag{5.6}$$

and the tangential jump is given by

$$\Delta f^s = -\frac{\partial \tau_s}{\partial s} - \frac{1}{\sigma}\frac{\partial f}{\partial s}(\tau_s - \tau_\varphi) - \kappa_s q. \tag{5.7}$$

The function $f(s)$, describing the shape of the membrane, was defined in (5.1). An analogous torque balance shows that the transverse shear tension is related to the bending moments by

$$q = \frac{1}{\sigma}\frac{\partial}{\partial s}(\sigma m_s) - \frac{1}{\sigma}\frac{\partial f}{\partial s}\, m_\varphi = \frac{1}{\sigma}\frac{\partial f}{\partial s}\left[\frac{\partial}{\partial \sigma}(\sigma m_s) - m_\varphi\right] \tag{5.8}$$

(e.g. Møllmann 1981, p. 33). Far from the axis of symmetry, equations (5.5)–(5.8) reduce to equations (3.3) and (3.4) for a two-dimensional membrane. Substituting the right-hand side of (5.8) in place of the shear tension in (5.6) and (5.7), we obtain relations in terms of the in-plane stress resultants and bending moments alone.

It is reassuring to confirm that expressions (5.6)–(5.8) are consistent with the more general expressions for three-dimensional membranes discussed in Section 4. For this purpose, we identify the surface curvilinear coordinate $\xi$ with the arc length measured along the trace of the membrane in a meridional plane denoted by $s$, and the curvilinear coordinate $\eta$ with the meridional angle $\varphi$, whereupon the arc length metric coefficients are given by $h_\xi = 1$ and $h_\eta = \sigma$. Since all tensions and moments have been assumed axisymmetric, the principal axes coincide with the chosen curvilinear axes, and equations (4.15)–(4.20) reproduce equations (5.6)–(5.8).

**5.2. Hydrostatics.** It is instructive to compute the distribution of the stress resultants in hydrostatics on the basis of the equilibrium condition (5.5) (Green & Adkins 1970, pp. 159–160). If gravitational effects are neglected, the jump in hydrodynamic traction across the membrane is given by

$$\Delta\mathbf{f} = (\sigma^{(s)} - \sigma^{(c)}) \cdot \mathbf{n} = (P_c - P_s)\,\mathbf{n} \tag{5.9}$$

where $P_c$ and $P_s$ are the uniform pressures inside and outside the capsule. In the absence of significant bending moments, the normal and tangential components of the interface force balance provide us with the equilibrium equations

$$\kappa_s\,\tau_s + \kappa_\varphi\,\tau_\varphi = \Delta P \equiv P_c - P_s, \tag{5.10}$$

and

$$\frac{\partial}{\partial s}(\sigma\tau_s) = \tau_\varphi\,\frac{\partial f}{\partial s}. \tag{5.11}$$

Solving (5.10) for $\tau_\varphi$, substituting the result into (5.11), and rearranging, we derive the following ordinary differential equation for $\tau_s$ with respect to the azimuthal arc length $s$,

$$\sigma\,\kappa_\varphi\,\frac{\partial}{\partial s}(\sigma\tau_s) + (\sigma\tau_s)\,\kappa_s\,\frac{\partial f}{\partial s} = \Delta P\,\sigma\,\frac{\partial f}{\partial s}. \tag{5.12}$$

Substituting Godazzi's relation (5.4) into the right-hand side of (5.12), we find

$$\frac{\partial}{\partial s}(\sigma^2\,\kappa_\varphi\,\tau_s) = \Delta P\,\sigma\,\frac{\partial f}{\partial s} \tag{5.13}$$

or

$$\frac{1}{\sigma}\,\frac{\partial}{\partial\sigma}(\sigma\,\frac{\partial x}{\partial s}\,\tau_s) = \Delta P. \tag{5.14}$$

Integrating (5.14) with respect to $s$ from a point $s = 0$ up to an arbitrary point $\sigma_0$, we find

$$\tau_s(\sigma) = \frac{1}{\sigma \frac{\partial x}{\partial s}} \left[\sigma_0 \left(\frac{\partial x}{\partial s}\right)_{\sigma_0} \tau_s(\sigma_0) + \frac{1}{2} \Delta P \left(\sigma^2 - \sigma_0^2\right)\right] \tag{5.15}$$

which shows that an unphysical singularity occurs develops when $\partial x/\partial s = 0$ and $\sigma \neq 0$.

This result reveals that stress resultants alone are not able to support shapes where the $x$ position reaches a local maximum along the trace of the membrane in a meridional plane, including the biconcave disk assumed by red blood cells in the resting condition. In the absence of bending moments, the membrane must be unstressed: $\tau_s = 0$ and $\Delta P = 0$.

**5.3. Constitutive equations for elastic tensions.** To evaluate the right-hand side of equations (5.6)–(5.8), we require constitutive relations for the elastic stress resultants and bending moments. To derive relations for the elastic tensions, we introduce the principal extension ratios

$$\lambda_s = \frac{\partial s}{\partial s^R}, \qquad \lambda_\varphi = \frac{\sigma}{\sigma^R}, \tag{5.16}$$

where the superscript $R$ denotes a reference state. If the area of the membrane is locally and thus globally conserved,

$$\lambda_s \lambda_\varphi = 1. \tag{5.17}$$

To this end, we have two main choices reflecting the assumed nature of the membrane.

First, we may regard the membrane as a distinct two-dimensional hyperelastic medium, and express the principal stress resultants in terms of the surface strain energy function $W^S$ using equations (4.35), where $\tau_1^P = \tau_s$, $\tau_2^P = \tau_\varphi$, $\lambda_1 = \lambda_s$, and $\lambda_2 = \lambda_\varphi$. Alternatively, we may regard the membrane as a thin sheet of a three-dimensional incompressible material and work with the strain invariants shown in (4.46); the principal elastic tensions derive from the volume strain energy function $W^V$ using relations (4.47).

**5.4. Constitutive equations for bending moments.** To compute the bending moments developing in a hyperelastic membrane, we introduce the bending measures of strain

$$K_s = \lambda_s \, \kappa_s - \kappa_s^R, \qquad K_\varphi = \lambda_\varphi \, \kappa_\varphi - \kappa_\varphi^R, \tag{5.18}$$

where the superscript $R$ denotes a reference configuration corresponding to the unstressed shape where the bending moments vanish (Reissner 1949, 1950, 1963, 1969). Zarda et al. (1977) expressed the bending moments in terms of the surface bending energy function $W_B$ in a form that is analogous to that shown in equations (4.35), as

$$m_s = \frac{1}{\lambda_\varphi} \frac{\partial W_B}{\partial K_s}, \qquad m_\varphi = \frac{1}{\lambda_s} \frac{\partial W_B}{\partial K_\varphi}, \tag{5.19}$$

Love's first approximation expressed by the expression on the right-hand side of (4.52) yields the bending energy function

$$W_B = \frac{E_B}{2} \left(K_s^2 + 2\,\nu\, K_s K_\varphi + K_s^2\right), \tag{5.20}$$

where $E_B$ and $\nu$ are physical constants expressing membrane material properties.

**5.5. Deformed capsules in hydrostatics.** Zarda et al. (1977) computed osmotically swelled capsules with specified unstressed shapes, regarded as models of deformed red blood cells. In hydrostatics, and in the absence of gravitational effects, the normal and tangential components of the equilibrium equation (5.5) reduce to

$$\kappa_s\, \tau_s + \kappa_\varphi\, \tau_\varphi - \frac{1}{\sigma} \frac{\partial}{\partial s}(\sigma q) = \Delta P, \tag{5.21}$$

and

$$\frac{\partial \tau_s}{\partial s} + \frac{1}{\sigma} \frac{\partial f}{\partial s} \left(\tau_s - \tau_\varphi\right) + \kappa_s\, q = 0, \tag{5.22}$$

The transverse shear tension $q$ is related to the bending moments by equation (5.8). For a specified surface area corresponding to an assumed resting shape, the pressure difference $\Delta P$ is a function of the cell volume, and vice versa.

In one series of computations, Zarda et al. (1977) accounted for the incompressibility of the membrane by adding to the right-hand sides of equations (4.35) a position-dependent isotropic tension, and requiring the incompressibility constraint (5.17), as will be discussed in Section 6. The elastic tensions and bending moments were computed using the strain energy functions (4.40) and (5.20). In the numerical implementation, the equilibrium shapes were computed based on a variational principle. An alternative involves marking the trace of the membrane in a meridional plane with a group of marker points, requiring the satisfaction the equilibrium equations at the marker points, approximating the various derivatives with finite differences, specifying either the pressure difference $\Delta P$ or the cell volume, and then solving the resulting system of nonlinear algebraic equations for the position of the marker points using an iterative method.

**6. Discussion.** We have reviewed and derived expressions for the jump in the hydrodynamic traction across interfaces with a membrane-like constitution. Inserting theses expressions into the boundary-integral equation allows us to compute the interface velocity by solving an integral equation, and thereby simulate the evolution of the capsules from a specified initial shape (Pozrikidis 2000b).

Several researchers have conducted numerical simulations taking into account the in-plane elastic tensions, as reviewed in the Introduction. Transverse shear tensions and bending moments were included by Zarda et al. (1977) for axisymmetric membranes in hydrostatics, as discussed in Section 5, and by Pozrikidis (2000a) and Kwak & Pozrikidis (2000) for three-dimensional and axisymmetric membranes in hydrodynamics.

**6.1. Incompressible interfaces.** Biological membranes consisting of lipid bilayers have a large modulus of dilatation, that is, they behave like two-dimensional nearly incompressible fluids. To account for the membrane incompressibility, a position-dependent isotropic tension playing the role of surface pressure may be added to the elastic in-plane stress resultants. The introduction of an additional surface function provides us with an additional degree of freedom that allows the satisfaction of the incompressibility constraint at every point over the membrane.

In global Cartesian coordinates, the incompressibility constraint is expressed by the equation

$$\theta = \frac{1}{h_S}\frac{Dh_S}{Dt} = (\mathbf{P}\cdot\nabla)\cdot\mathbf{u} = 0, \tag{6.1}$$

where $\theta$ is the rate of surface dilatation, $D/Dt$ is the material derivative, $h_s$ is the surface metric associated with the convected surface curvilinear coordinates, and $\mathbf{u}$ is the fluid velocity (Secomb & Skalak 1982).

In surface curvilinear coordinates $(\xi, \eta)$, the incompressibility constraint (6.1) takes the form

$$\begin{aligned}\theta &= \frac{1}{h_S}\frac{Dh_S}{Dt} = \frac{1}{h_S}\mathbf{n}\cdot(\frac{\partial \mathbf{u}}{\partial \xi}\times\frac{\partial \mathbf{x}}{\partial \eta}+\frac{\partial \mathbf{x}}{\partial \xi}\times\frac{\partial \mathbf{u}}{\partial \eta}) \\ &= (\mathbf{P}\cdot\nabla)\cdot(\mathbf{P}\cdot\mathbf{u}) + 2\kappa_m\,\mathbf{u}\cdot\mathbf{n} = 0 \\ &= u^\xi|_\xi + u^\eta|_\eta + 2\kappa_m\,\mathbf{u}\cdot\mathbf{n} = 0\end{aligned} \tag{6.2}$$

where $\kappa_m$ is the interface mean curvature (Waxman 1984, eq. 3.2b; Zhou & Pozrikidis 1995; Pozrikidis 1997, p. 21), and a vertical bar signifies the covariant derivative taken with respect to the subscribed variable, defined in terms of the Christoffel symbols (e.g. Aris 1962). The numerical implementation of (6.2) for capsules deforming under the influence of simple shear flow was discussed by Zhou & Pozrikidis (1995).

Considering axisymmetric flow, we express the azimuthal and meridional tensions in terms of an unknown mean and a deviatoric component, where the latter derives from a strain energy function. Pozrikidis

(1990) simulated the transient deformation of axisymmetric capsules enclosed by incompressible elastic membranes evolving under the influence of an elongational flow. The distribution of the isotropic membrane tension was computed using the incompressibility constraint (5.17), and the deviatoric elastic tension was computed using a constitutive equation. Requiring $D(\lambda_s \lambda_\varphi)/Dt = 0$, where $D/Dt$ is the material derivative, we obtain a scalar constraint on the distribution of the membrane velocity $\mathbf{u}$,

$$\mathbf{t}_s \cdot \frac{\partial \mathbf{u}}{\partial s} + \frac{1}{\sigma}\,\mathbf{u}\cdot\mathbf{e}_\sigma = 0 \tag{6.3}$$

where $\mathbf{t}_s$ is the unit vector tangential to the membrane in a meridional plane, and $\mathbf{e}_\sigma$ is the unit vector normal to the $x$ axis. It can be shown by straightforward rearrangement that (6.3) is consistent with equations (6.1) and (6.2).

In the case of two-dimensional flow, the incompressibility constraint simplifies to

$$\mathbf{t}\cdot\frac{\partial \mathbf{u}}{\partial l} = \frac{\partial(\mathbf{u}\cdot\mathbf{t})}{\partial l} + \kappa\,\mathbf{u}\cdot\mathbf{n} = 0, \tag{6.4}$$

where $\mathbf{t}$ is the unit vector tangential to the membrane, $l$ is the arc length measured in the direction of $\mathbf{t}$, and $\kappa$ is the curvature of the membrane in the $xy$ plane. The numerical implementation of (6.4) is discussed by Zhou & Pozrikidis (1995).

**6.2. Viscous interfaces.** In addition to surface elasticity, membranes exhibit surface viscosity, that is, they behave like two-dimensional Newtonian or viscoelastic Boussinesq fluids (e.g. Scriven 1960, Waxman 1984, Barthès-Biesel & Sgaier 1985, Slattery 1990, Edwards et al. 1991). Scriven (1960) proposed a constitutive equation for the Newtonian interfacial tensions in surface curvilinear coordinates. Secomb & Skalak (1982) observed that the coupling of the interfacial dynamics to the hydrodynamics on either side of the interface is facilitated by working in global Cartesian coordinates, and expressed the Newtonian surface tension tensor in the form

$$\tau = \gamma\,\mathbf{P} + (\kappa - \epsilon)\,\theta\,\mathbf{P} + 2\,\epsilon\,\mathbf{E}^S, \tag{6.5}$$

where $\theta$ is the rate of dilatation given in equations (6.1) and (6.2), $\mathbf{P} = \mathbf{I} - \mathbf{nn}$ is the tangential projection operator, $\epsilon$ and $\kappa$ are two physical constants expressing the interface shear and dilatational viscosity, and $\mathbf{E}^S$ is the Cartesian surface rate of deformation tensor given by

$$E^S_{ij} = \frac{1}{2}\,P_{ik}\,P_{jl}\,(\frac{\partial u_k}{\partial x_l} + \frac{\partial u_l}{\partial x_k}). \tag{6.6}$$

The two projections on the right-hand side of (6.6) remove derivatives of the velocity in the direction normal to the interface, as well derivatives of the normal component of the velocity in directions that are tangential to the interface. Thus, the right-hand side of (6.6) may be computed from knowledge of the velocity distribution over the interface.

Performing a force balance over a small interfacial patch, and taking the limit as the size of the patch becomes infinitesimal, we find that the jump in the hydrodynamic traction is given by

$$\Delta \mathbf{f} = -(\mathbf{P} \cdot \nabla) \cdot \tau. \tag{6.7}$$

Substituting this expression into the boundary-integral equation, we obtain an integral equation of a non-standard kind for the interfacial velocity (Pozrikidis 1994).

**6.3. Alternative formulations.** We have discussed membrane tensions and bending moments in the context of continuum mechanics and under the auspices of the theory of thin shells. The macromolecular structure of certain membranes suggests that an alternative formulation that views a membrane as a network of viscoelastic links defined by computational nodes might be more appropriate.

For example, Hansen et al. (1996, 1997a,b) set up a molecular network model based on random Delaunay triangulation representing the red blood cell membrane cytoskeleton, and computed the macroscopic elastic shear and modulus of areal expansion. Eggleton & Popel (1998) discretized the membrane of a capsule into a collection of flat triangles representing elastic plates, and computed the elastic tensions at the nodes under the assumption that strains and tensions exist only in the individual plane of each triangle. Boey et al. (1998) and Discher et al. (1998) developed a coarse-grained molecular model that permits the direct coupling of classical hydrodynamics to the dynamics of the molecular layers and networks comprising the membrane, in a manner that circumvents the use of a macroscopic constitutive equation (see also Seifert 1998).

An alternative method of computing the jump in the hydrodynamic traction across a membrane hinges on the concept of configurational membrane energy playing the role of an effective Hamiltonian (e.g. Cai & Lubensky 1995, Anderson et al. 1998, Navot 1998). As an example, the instantaneous energy of a membrane consisting of a symmetric lipid bilayer with a flat resting shape may be expressed in terms of surface integrals in the form

$$E = \int_{Membrane} \tau \, dS + 2\,\kappa_B \int_{Membrane} \kappa_M^2 \, dS, \tag{6.8}$$

where $\kappa_B$ is a physical constant expressing the bending stiffness, $\kappa_m$ is the mean curvature of the membrane, and $\tau$ is a position dependent in-plane tension developing to ensure membrane incompressibility (Seifert 1997). Let the instantaneous shape of the membrane be described by the equation $G(\mathbf{x}) = 0$, where $G$ is a suitable function, and express the interface energy in the form $E = E(G(\mathbf{x}))$ evaluated at $G(\mathbf{x}) = 0$, where $E(\cdot)$ is a nonlinear integro-differential functional defined over all possible membrane configurations. The jump in the hydrodynamic traction across the membrane may be found using the principle of virtual displacements (e.g. Navot 1998) [101].

Kraus et al. (1996) discretized the membrane of a vesicle into a collection of flat triangles defined by computational nodes, and represented the flow by a superposition of elementary flows induced by point forces located at the vertices. The strength of the point force located at the $j$th node, denoted by $\mathbf{x}^{(j)}$, was computed by the equation

$$F_i^{(j)} = -\frac{\partial E}{\partial x_i^{(j)}}. \tag{6.9}$$

Although computationally convenient, the discrete models are sensitive to the method of surface discretization - flat versus curved triangulation.

In conclusion, we point out the need for developing efficient computational methods that couple low-Reynolds-number hydrodynamics to the mechanics of biological membranes regarded as thin-shells. The ultimate objective is the availability of an integrated program that allows the accurate simulation of the deformation of solitary capsules and suspensions of capsules encountered in physiology and bioengineering.

**7. Acknowledgement.** I express by appreciation to the organizers of this workshop for extending me an invitation to participate and contribute to this volume. This research has been supported by a grant provided by the National Science Foundation.

## REFERENCES

ANDERSON, D.M., MCFADDEN, G.B. & WHEELER, A.A., 1998, Diffuse-interface methods in fluid mechanics. *Annu. Rev. Fluid Mech.* **30**, 139–165.

ARIS, R., 1962, *Vectors, Tensors, and the Basic Equations of Fluid Mechanics.* Prentice-Hall.

BARTHÈS-BIESEL, D., 1980, Motion of a spherical microcapsule freely suspended in a linear shear flow. *J. Fluid Mech.* **100**, 831–853.

BARTHÈS-BIESEL, D., 1991, Role of interfacial properties on the motion and deformation of capsules in shear flow. *Physica A* **172**, 103–124.

BARTHÈS-BIESEL, D. & RALLISON, J.M., 1981, The time-dependent deformation of a capsule freely suspended in a linear shear flow. *J. Fluid Mech.* **113**, 251–267.

BARTHÈS-BIESEL, D. & SGAIER, H., 1985 Role of membrane viscosity in the orientation and deformation of a spherical capsule suspended in shear flow. *J. Fluid Mech.* **160**, 119–135.

BEATTY, M.F., 1987, Topics of finite elasticity: Hyperelasticity of rubber, elastomers, and biological tissues - with examples. *Appl. Mech. Rev.* **40**, 1699–1734.

BOEY, S.K., BOAL, D.H. & DISCHER, D.E., 1998, Simulations of the erythrocyte cytoskeleton at large deformation. I. Microscopic models. *Biophys. J.* **75**, 1573–1583.

BREYIANNIS, G. & POZRIKIDIS, C., 2000, Simple shear flow of suspensions of elastic capsules. *Theor. Comp. Fluid Dyn.* **13**, 327–347.

BUDIANSKY, B. & SANDERS, J.L. JR., 1963, On the "Best" first-order linear shell theory. In *Progress in Applied Mechanics*, Prager Anniversary Volume, McMillan, pp. 129–140.

BUDIANSKY, B., 1968, Notes on nonlinear shell theory. *J. Applied Mech.* **35**, 393–401.

CAI, W. & LUBENSKY, T.C., 1995, Hydrodynamics and dynamic fluctuations of fluid membranes. *Phys. Rev. E* **52**, 4252–4266.

CHARLES, R. & POZRIKIDIS, C., 1998, Significance of the dispersed-phase viscosity on the simple shear flow of suspensions of two-dimensional liquid drops. *J. Fluid Mech.* **365**, 205–234.

CORNELIUSSEN, A.H. & SHIELD, R.T., 1961, Finite deformation of elastic membranes with application to the stability of an inflated and extended tube. *Arch. Rational Mech. Anal.* **7**, 273–304.

DIAZ, A., PELEKASIS, N. & BARTHÈS-BIESEL, D., 2000, Transient response of a capsule subjected to varying flow conditions: Effect of internal fluid viscosity and membrane elasticity. *Phys. Fluids* **12**, 948–957.

DISCHER, D.E., BOAL, D.H. & BOEY, S.K., 1998, Simulations of the erythrocyte cytoskeleton at large deformation. II. Micropipette aspiration. *Biophys. J.* **75**, 1584–1597.

EDWARDS, D.A., BRENNER, H. & WASAN, D.T., 1991, *Interfacial Transport Processes and Rheology.* Butterworth Heinemann, Massachusetts.

EGGLETON, C.D. & POPEL, A.S., 1998, Large deformation of red blood cell ghosts in a simple shear flow. *Phys. Fluids* **10**, 1834–1845.

EVANS, E.A. & SKALAK, R., 1980, *Mechanics and Thermodynamics of Biomembranes.* CRC Press.

FLÜGGE, W., 1973, *Stresses in Shells.* Springer-Verlag.

FUNG, Y.C., 1965, *Foundations of Solid Mechanics.* Prentice-Hall.

GREEN, A.E. & ADKINS, J.E., 1970, *Large Elastic Deformations.* Clarendon Press, Oxford.

GREEN, A.E. & ZERNA, W., 1968, *Theoretical Elasticity.* Dover.

GURTIN, M.E. & MURDOCH, A.I., 1975, A continuum theory of elastic material surfaces. *Arch. Ration. Mech. Anal.* **57**, 291–323,

HANSEN, J.C., SKALAK, R., CHIEN, S. & HOGER, A., 1996, An elastic network model based on the structure of the red blood cell membrane skeleton. *Biophys. J.* **70**, 146–166.

HANSEN, J.C., SKALAK, R., CHIEN, S. & HOGER, A., 1997a, Influence of network topology on the elasticity of the red blood cell membrane skeleton. *Biophys. J.* **72**, 2369–2381.

HANSEN, J.C., SKALAK, R., CHIEN, S. & HOGER, A., 1997b, Spectrin properties and the elasticity of the red blood cell membrane skeleton. *Biorheology* **34**, 327–348.

KRAUS, M., WINTZ, W., SEIFERT, U. & LIPOWSKY, R., 1996, Fluid vesicles in shear flow. *Phys. Rev. Lett.* **77**, 3685–3688.

KWAK, S. & POZRIKIDIS, C., 2000. Effect of membrane bending stiffness on the deformation of capsules in uniaxial extensional flow. *Submitted for publication.*

LE DRET, H. & RAOULT, A., 1995, The nonlinear membrane model as variational limit of nonlinear three-dimensional elasticity. *J. Math. Pures Appl.* **74**, 549–578.

LEYRAT-MAURIN, A. & BARTHÈS-BIESEL, D., 1994, Motion of a deformable capsule through a hyperbolic constriction. *J. Fluid Mech.* **279**, 135–163.

LI, X.Z., BARTHÈS-BIESEL, D. & HELMY, A., 1988, Large deformations and burst of a capsule freely suspended in an elongational flow. *J. Fluid Mech.* **187**, 179–196.

LI, X., CHARLES, R. & POZRIKIDIS, C., 1996, Simple shear flow of suspensions of liquid drops. *J. Fluid Mech.* **320**, 395–416.

LI, X. & POZRIKIDIS, C., 2000, Wall-bounded and channel flow of suspensions of liquid drops. *Int. J. Multiphase Flow* **26**, 1247–1279.

LIBAI, A. & SIMMONDS, J.G., 1998, *The Nonlinear Theory of Elastic Shells*, Cambridge University Press.

LIPOWSKY, R., 1991, The conformation of membranes. *Nature* **349**, 475–481.

LIPOWSKY, R. & SACKMANN, E., 1995, *Structure and Dynamics of Membranes*, Vol. 1A, 1B. Elsevier.

MCDONALD, P., 1996, *Continuum Mechanics.* PWS.

MOHANDAS, N. & EVANS, E., 1994, Mechanical properties of the red cell membrane in relation to molecular structure and genetic defects. *Annu. Rev. Biophys. Biomol. Struct.* **23**, 787–818.

MØLLMANN, H., 1981, *Introduction to the Theory of Thin Shells.* Wiley.

NAGHDI, P.M., 1972, *Theory of shells and plates.* In Handbuch der Physik, vol VIa/2, C. Truesdell (Edt.), pp. 435–640.

NAVOT, Y., 1998, Elastic membranes in viscous shear flow. *Phys. Fluids* **10**, 1819–1833.

OGDEN, R.W., 1984, *Non-linear Elastic Deformations.* Dover.

POZRIKIDIS, C., 1990, The axisymmetric deformation of a red blood cell in uniaxial straining flow. *J. Fluid Mech.* **216**, 231–254.

POZRIKIDIS, C., 1992, *Boundary Integral and Singularity Methods for Linearized Viscous Flow.* Cambridge University Press.

POZRIKIDIS, C., 1994, Effects of surface viscosity on the finite deformation of a liquid drop and the rheology of dilute emulsions in simple shearing flow. *J. Non-Newt. Fluid Mech.* **51**, 161–178.

POZRIKIDIS, C., 1995, Finite deformation of liquid capsules enclosed by elastic membranes in simple shear flow. *J. Fluid Mech.* **297**, 123–152.

POZRIKIDIS, C., 1997, *Introduction to Theoretical and Computational Fluid Dynamics.* Oxford University Press.

POZRIKIDIS, C., 2000a, Effect of membrane bending stiffness on the deformation of capsules in simple shear flow. *Submitted for publication.*

POZRIKIDIS, C., 2000b. Interfacial dynamics for Stokes flow. *J. Comp. Phys.* To appear.

RAMANUJAN, S. & POZRIKIDIS, C., 1998, Deformation of liquid capsules enclosed by elastic membranes in simple shear flow: large deformations and the effect of fluid viscosities. *J. Fluid Mech.* **361**, 117–143.

REISSNER, E., 1949, On the theory of thin elastic shells. In *Contributions to Applied Mechanics*, H. Reissner Anniversary Volume, J.W. Edwards (Edt.), Ann Arbor, pp. 231–247.

REISSNER, E., 1950, On axisymmetrical deformations of thin shells of revolution. *Proceedings, Third Symposium in Applied Mathematics*, pp. 27–52.

REISSNER, E., 1963, On the equations for finite symmetrical deflections of thin shells of revolution. In *Progress in Applied Mechanics*, Prager Anniversary Volume, McMillan, pp. 171–178.

REISSNER, E., 1969, On finite symmetrical deflections of thin shells of revolution. *J. Appl. Mech.* **36**, Trans. ASME 91, Series E, 267–270.

SANDERS, J.L., 1963, Nonlinear theories of thin shells. *Quart. Appl. Math.* **21**, 21–36.

SCRIVEN, L.E., 1960, Dynamics of a fluid interface. *Chem. Eng. Sci.* **12**, 98–108.

SECOMB, T.W. & SKALAK, R., 1982, Surface flow of viscoelastic membranes in viscous fluids. *Q. J. Mech. Appl. Maths* **35**, 233–247.

SECOMB, T.W., SKALAK, R., ÖZKAYA, N. & GROSS, J.F., 1986, Flow of axisymmetric red blood cells in narrow capillaries. *J. Fluid Mech.* **163**, 405–423.

SEIFERT, U., 1997, Configurations of fluid membranes and vesicles. *Adv. Phys.* **46**, 13–137.

SEIFERT, U., 1998, Modelling nonlinear red cell elasticity. *Biophys J.* **75** 1141–1142.

SIMMONDS, J.G. & DANIELSON, D.A., 1972, Nonlinear shell theory with finite rotation and stress-function vectors. *J. Applied Mech.* **39**, 1098–1090.

SKALAK, R., ÖZKAYA, N. & SKALAK, T.C., 1989, Biofluid mechanics. *Annu. Rev. Fluid Mech.* **21**, 167–204.

SKALAK, R. TÖZEREN, A., ZARDA, P.R. & CHIEN, S., 1973, Strain energy function of red blood cell membranes. *Biophys. J.* **13**, 245–264.

SLATTERY, J.C., 1990, *Interfacial Transport Phenomena.* Springer-Verlag.

STEIGMANN, D.J. & OGDEN, R.W., 1997, Plane deformations of elastic solids with intrinsic boundary elasticity. *Proc. R. Soc. London A* **453**, 853–877.

STEIGMANN, D.J. & OGDEN, R.W., 1999, Elastic surface-substrate interactions. *Proc. R. Soc. London A* **455**, 437–474.

VALID, R., 1995, *The Nonlinear Theory of Shells through Variational Principles.* Wiley.

WAXMAN, A.M., 1984, Dynamics of a couple-stress fluid membrane. *Stud. Appl. Math.* **70**, 63–86.

ZARDA, P.R., CHIEN, S. & SKALAK, S., 1977, Elastic deformations of red blood cells. *J. Biomechanics.* **10**, 211–221.

ZINEMANAS, D. & NIR, A., 1988, On the viscous deformation of biological cells under anisotropic surface tension. *J. Fluid Mech.* **193**, 217–241.

ZHOU, H. & POZRIKIDIS, C., 1995, Deformation of liquid capsules with incompressible interfaces in simple shear flow. *J. Fluid Mech.* **283**, 175–200.

# UNSTEADY AERODYNAMICS OF TWO DIMENSIONAL INSECT FLIGHT

Z. JANE WANG*

**Abstract.** Motivated by our interest in unsteady aerodynamics of insect flight, we compute the Navier-Stokes equation around a two dimensional flapping wing. The analysis of unsteady flows in forward flight reveals a mechanism of frequency selection, which results from the two intrinsic time scales associated with the shedding of leading and trailing edge vortices. The predicted preferred frequency scales inversely with the size of the wing, which is consistent with the zoological observation. The computation of hovering flight uncovers an intrinsic mechanism of generating lift by creating a downward jet of counter-rotating vortex pairs. The average computed forces in a generic hovering flight are shown to be sufficient to support typical insect weight.

**1. Introduction.** The myth 'bumble-bees can not fly according to conventional aerodynamics' simply reflects our poor understanding of unsteady viscous fluid dynamics. In particular, we lack a theory of vorticity shedding due to dynamic boundaries at the intermediate Reynolds numbers relevant to insect flight, typically between $10^2$ and $10^4$, where both viscous and inertial effects are important. The complicated interactions between unsteady vortical flows and dynamic wings are keys to insects' aerial versatility, but they are inconvenient for constructing a simple theory.

In the early studies of insect flight, it was hoped that unsteady aerodynamics effects might be relatively unimportant, as argued by Weis-Fogh and Jensen [1]. Later, several workers have suggested that unsteady effects might play a significant role, especially during hovering flight – see reviews by Maxworthy [2], Ellington [3], and Spedding [4]. Most recently, exciting experiments by Ellington's group at Cambridge and Dickinson's group at Berkeley have quantified a number of unsteady effects by studying flows and forces around robotic wings that are dynamically similar to insect wings [5, 6]. In particular, Ellington et al. observed a coherent three-dimensional leading-edge vortex and argued that three-dimensional effects are important for hovering. In addition, Dickinson and collaborators systematically studied the interactions between translational and rotational motions of the wing, and measured time-dependent forces on a robot fruitfly wing, thereby demonstrating the importance of wing rotation.

What appears to be missing are the resolved unsteady flows. To complement the experiments, we build computational tools to solve unsteady flows around a flapping wing which mimics insect motion. Analysis of these computations allows us to decipher the underlying mechanism of vortex shedding and its implication in flapping flight. In this article, we focus on two dimensional computations, which are first steps in our efforts. In

*Theoretical and Applied Mechanics, Cornell University, Ithaca, NY 14853; Email: jane.wang@cornell.edu.

particular, we illustrate that the unsteady aerodynamics can select an optimal frequency range in forward flight, and we also show the formation of a dipole jet and instantaneous forces in two dimensional hovering flight. The details of these two results are reported in references [7] and [8].

**2. Two dimensional computations.** Two-dimensional computation is an instructive step and indispensable for several reasons. Firstly, from a practical point of view, it is possible to resolve two dimensional flows at a large range of Reynolds numbers relevant to insect flight; secondly, it helps us to build intuition about unsteady effects, such as time scales of vortex growth, without being hampered by technical difficulties such as 3D visualization; and finally, by comparing with experiments and future three dimensional computation, we will be able to determine the relevant importance of two and three dimensional effects.

In one of the handful of computational studies, Gustafson et. al. [9] computed two-dimensional hovering flight in two generic strokes. These authors showed the snapshots of vorticity field which are in qualitative agreement with experiments near the wing, but they did not report the corresponding instantaneous forces. More recently Liu et. al. [10] computed three-dimensional hovering flight of a hawkmoth and confirmed experimentally observed axial flows, but did not report the vorticity field, and their force measurements were inconclusive. Therefore, it is worthwhile to improve upon these works to resolve both the vorticity field and instantaneous forces.

To solve the flows around a flapping wing, we employ a fourth order essentially compact finite difference scheme (EC4) for the incompressible Navier-Stokes equation [11]. An advantage of the scheme is that at each time step, only two Poisson solvers are required to achieve a fourth order spatial accuracy. The scheme uses the vorticity-stream function formulation, and the vorticity boundary condition is explicitly enforced to satisfy the no-slip boundary condition.

To efficiently resolve the two dimensional flow between $Re = 10^2$ and $Re = 10^4$, we adapt the scheme to the elliptic coordinates and treat the far field boundary condition on the stream function almost exactly in the Poisson solvers [12]. The elliptic coordinates $(\mu, \theta)$ are shown in Figure 1, where constant $\mu's$ and $\theta's$ correspond to the confocal ellipses and hyperbolae.

An uniform mesh in $(\mu, \theta)$ is naturally clustered near the tips of the wing and is coarse at the far field in $(x, y)$ space. Such a grid is suitable for our problem since the vorticity is strongest near the tip of the ellipse and is weaker away from the body. Furthermore, the elliptic coordinates can be mapped onto a Cartesian coordinates via the conformal transformation:

$$x + iy = \cosh(\mu + i\theta), \tag{1}$$

which preserves the Laplacian up to a local scaling function $S(\mu, \theta)$:

$$\Delta_{x,y} = \Delta_{\mu,\theta}/S(\mu, \theta), \qquad \text{with} \qquad S(\mu, \theta) = \cosh^2 \mu - \cos^2 \theta. \tag{2}$$

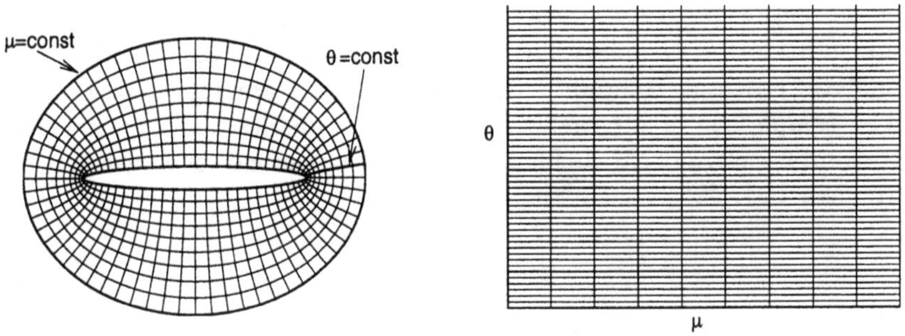


FIG. 1. *The left figure shows the elliptic coordinates* $(\mu, \theta)$ *fitted to a two-dimensional wing. Confocal ellipses and hyperbolae correspond to constant* $\mu$*'s, and* $\theta$*'s, respectively. The right figure shows the Cartesian coordinates, with periodic boundary conditions in the* $\theta$ *direction, conformally mapped from the elliptic coordinates.*

The dimensionless Navier-Stokes equation for the two-dimensional vorticity field in the elliptic coordinates becomes:

$$\frac{\partial(S\omega)}{\partial t} + (\sqrt{S}\mathbf{u} \cdot \nabla)\omega = \frac{1}{Re}\Delta\omega, \tag{3}$$

where $\mathbf{u}$ is the velocity field, $\omega$ the vorticity field, and $Re$ the Reynolds number. For incompressible flows, $\mathbf{u}$ and $\omega$ can be expressed in terms of the stream function $\Psi$:

$$S\omega = \Delta\Psi, \tag{4}$$

$$\sqrt{S}\mathbf{u} = -\nabla \times \Psi, \tag{5}$$

where the derivatives are taken with respect to $(\mu, \theta)$. The conformal transformation results in a constant coefficient Poisson equation for $\Psi$, which can be solved via FFT efficiently.

We solve the Navier-Stokes equation in the body-frame. For translational motions considered here, no fictitious force appears in the vorticity equation. The no-slip boundary condition at the wing is enforced explicitly through the vorticity boundary condition. The radius of the computational boundary is chosen to be 5 to 10 times the half chord length. This large radius is affordable due to the stretched mesh in the elliptic coordinates.

To resolve the flow, we keep at least 10 grid points along the radial direction in the boundary layer, and at least 30 points in the azimuthal direction around each tip, whose length scale is estimated by its radius of curvature. The typical resolution for $Re \sim 1000$ is $128 \times 256$ for single strokes, and $256 \times 256$ for the flapping motion.

To test the code, we compute flow past a cylinder, a limiting case of an ellipse, and compare the results with experiments [13, 14]. Good agreements are found in the separation angle, as well as the velocity field as a function of time and space. As an additional check, the computed flow outside the boundary layer in early time is also compared with the analytic potential solution. These comparisons are shown in Figure 2. The comparison of lift coefficient between numerics and experiments of Dickinson and Götz can be found in [7], and numerical convergence studies are done in [15].

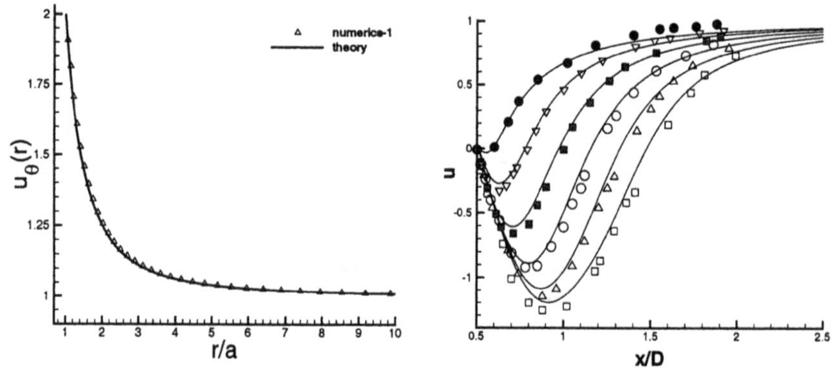


FIG. 2. *Left figure: numerical result on azimuthal velocity $u_\theta(r)$ vs. $r$ compared with theory. Right figure: Velocity (u) at different locations along the symmetric axis $(x/D)$ in the wake, where $D$ is the diameter of the cylinder. Different symbols correspond to time of measurement (rescaled by $D/u_0$) in the increment of 0.5, starting from $t_s = 0.5$ (solid circle). Lines are from the numerics, and points are from experiments (Bouard and Coutanceau (1980)).*

The forces on the ellipse can be computed from the vorticity field. In particular, the pressure force $F_p$ is determined by the the vorticity flux, and the viscous force $F_\nu$ by the vorticity along the ellipse. More specifically,

$$F_p = \rho\nu \int \frac{\partial \omega}{\partial \mu}(\sinh \mu_0 \sin\theta \ \hat{\mathrm{x}} + \cosh \mu_0 \cos\theta \hat{\mathrm{y}}) d\theta \tag{6}$$

$$F_\nu = \rho\nu \int \omega(-\cosh \mu_0 \sin\theta \ \hat{\mathrm{x}} + \sinh \mu_0 \cos\theta \ \hat{\mathrm{y}}) d\theta \tag{7}$$

where $\rho$ is the fluid density. Or equivalently, the forces can be evaluated by the rate of change of total momentum in the fluid. It is conventional to define the lift and drag to be orthogonal and parallel to the velocity at infinity $u_\infty$. The dimensionless forces are the lift and drag coefficients:

$$C_d = \frac{F_\perp}{\rho u_\infty^2 c}, \qquad C_l = \frac{F_\parallel}{\rho u_\infty^2 c} \tag{8}$$

However, we are interested in the average forces. It is convenient to decompose the forces in the direction with respect to the mean horizontal flow velocity, $F_x$ and $F_y$. We denote the corresponding dimensionless forces to be the thrust ($C_x$) and lift ($C_y$) coefficients,

$$C_x = \frac{F_x}{\rho u_1^2 c}, \qquad C_y = \frac{F_y}{\rho u_1^2 c}. \tag{9}$$

The two sets of definitions are simply choices of reference frames, and have no physical consequences apart from the convenience in discussions.

**3. Forward flight.** In this section, we explore a consequence of the unsteady effects in forward flapping flight, namely how they can select an optimal frequency range.

To illustrate the idea, let us consider a simple model describing the forward flapping motions as drawn schematically in Figure 3. The ellipse represents a wing element in chord direction. The motion of the wing consists of the superposition of the mean forward motion, denoted by $u_0$ and a sinusoidal flapping motion $u_1(t)$, in the direction of a line inclined at angle $\beta$ to the horizontal, with an amplitude $A$ and frequency $f$, i.e., $u_1(t) = 2\pi f A \sin(2\pi f t)$.

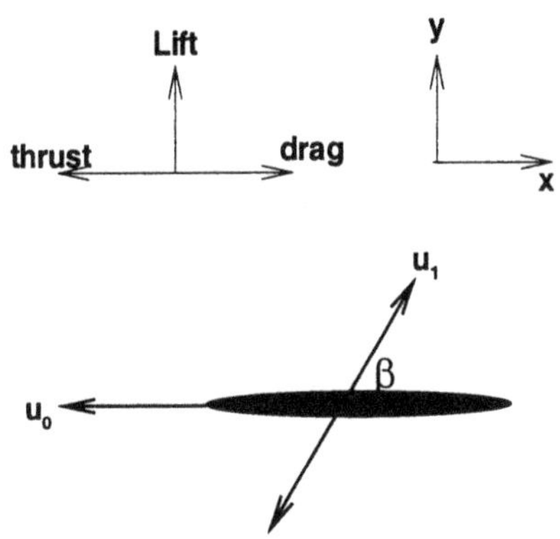


FIG. 3. *Flapping model. The ellipse represents a cross section of a flapping wing in the chord direction. The thickness ratio is chosen to be 0.125 for most of our computations unless otherwise specified.* $u_0$ *is the mean flight velocity,* $u_1$ *the flapping velocity, and* $\beta$ *the angle between the stroke plane and x-axis. Mean thrust and lift are defined with respect to the direction of* $u_0$.

For simplicity, we choose the mean angle of attack to be zero, and $\beta$ to be $\pi/2$. The averaged resultant force in this case has only a horizontal component. In reality, insect and bird flight generates both lift and thrust by tilting the wing relative to the direction of flight.

The parameters, $u_0$, $f$, and $A$, together with kinematic viscosity $\nu$ and wing chord $c$, form three dimensionless quantities: the Reynolds number, $Re$, and the two Strouhal numbers, $St_a$ and $St_c$, as defined in the following:

$$Re = \frac{u_0 c}{\nu} \tag{10}$$

$$St_a = \frac{fA}{u_0} \tag{11}$$

$$St_c = \frac{fc}{u_0} \tag{12}$$

Note that $St_a$ measures the ratio of the maximum flapping velocity and the forward velocity, and $St_c$ is the dimensionless flapping frequency. To study the optimal flapping frequency, we define the optimal flapping to be the one which produces the maximum thrust per unit power input. We study the dependence of average thrust and efficiency as a function of frequency. Because the parameter $St_a$ was shown experimentally to be a scaling variable [16], we will first fix $St_a$ and study the dependence of the thrust on various combinations of frequency and amplitude, and later vary $St_a$.

**3.1. Optimal flapping: Observations and qualitative features.** In this set of numerical experiments, we set $Re = 1000$, and $St_a = 0.16$, a number based on forward dragonfly flight with $f = 40Hz$, $A = 2cm$ and $u = 5m/s$ [17]. Fig. 4 shows the thrust and lift coefficients vs. the dimensionless time $t_s$, defined as $t_s = tu_0/c$, for four different frequencies, $St_c = 0.25, 0.5, 1.0$, and 2.0. Note we have set $u_0 = 1$ and $c = 2$ to fix our units.

The lift coefficients $C_y$'s in the bottom plot vary sinusoidally about the zero mean, thus the average vertical force is zero as expected from the symmetric flapping.

On the other hand, the thrust coefficients $C_x$'s are asymmetric about the zero axis, because the fore and after symmetry is broken by the mean forward velocity field. A negative value of $C_x$ corresponds to a thrust in the mean forward direction. In Table 1, we tabulate the average thrust coefficient and the efficiency $Q$. Note that $\langle C_x \rangle$ and $Q$ reach maxima at $St_c \sim 1$.

Next we examine the qualitative features of the associated vortex dynamics in these four cases. During each period, the trailing edge vortex begins to shed and grow at the turn of a flap, but the shedding of the leading edge vortex depends on the angle of attack and the flapping frequency. In Figure 5, we compare the vorticity contours for different flapping frequencies at 1/8th of a period. We sketch the basic vortex features next to these figures. In the case of the highest $St_c$, the trailing edge vortex is relatively small, as is the net circulation around the wing and the forces. As $St_c$ decreases, the trailing edge vortex grows and the forces increase. Finally, as $St_c$ further increases, the leading edge vortex, which has the opposite sign of the trailing edge vortex starts to shed and reduce the net circulation and the forces. The middle value of $St_c$ corresponds to the observed optimal thrust and efficiency.

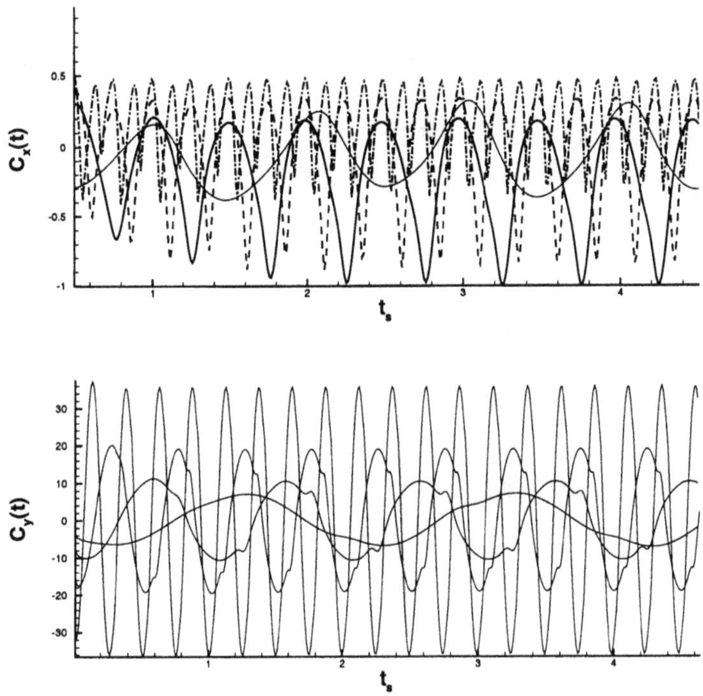


FIG. 4. *Time dependent thrust and lift coefficients for fixed* $St_a = 0.16$, *and different frequencies,* $f = 0.25Hz, 0.5Hz, 1Hz$, *and* $2Hz$, *or equivalently,* $St_c = 0.5, 1, 2$, *and* 4. *The frequency of each curve can be read directly from the graph by counting peaks.*

TABLE 1
*Thrust coefficient* $\langle C_x \rangle$ *and its two contributions from pressure force* $(C_{xp})$ *and the viscous force* $(C_{x\nu})$ *as a function of* $St_c$.

| $St_c$ | 4 | 2 | 1 | 0.5 |
|---|---|---|---|---|
| $\langle C_x \rangle$ | 0.18 | -0.06 | -0.24 | -0.08 |
| $\langle C_{xp} \rangle$ | -0.02 | -0.18 | -0.27 | -0.12 |
| $\langle C_{x\nu} \rangle$ | 0.20 | 0.12 | 0.03 | 0.04 |
| $Q(\%)$ | 1.2 | 1.2 | 13.5 | 11.5 |

**3.2. Optimal flapping: explanations.** The previous pictures suggest that the following two processes may play significant roles in selecting a frequency: 1) the unsteady forces due to the trailing edge vortex growth, similar to the Wagner effect [18], and 2) the shedding of the leading edge vortex as a function of $St_c$. We can investigate these two effects quantita-

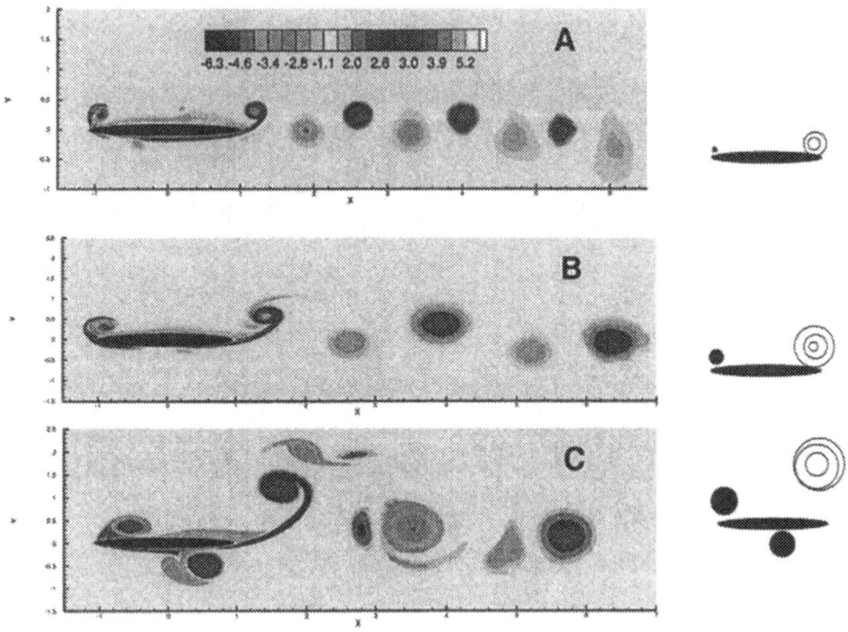


FIG. 5. *Contour plot of the wake vorticity at 1/8th of a period for different frequencies: A) $f = 1.0(St_c = 2.0)$, B) $f = 0.5(St_c = 1.0)$, C) $f = 0.25(St_c = 0.5)$. The grey scale indicates the strength of the vorticity. Other parameters are: $Re = 1000$ and $St_a = 0.16$. The sketches on the right-hand side summarize the basic structure of shed-vortices. The leading and trailing edge vortices have different signs. In case A, the trailing edge vortex is small and leading edge vortex is bound to the ellipse; in case B, the trailing edge vortex is larger compared to case A while the leading edge is still bound; in case C, both the leading and trailing edge vortices are shed (the vortex at the bottom of the ellipse is the leading edge vortex shed in the previous cycle, which is best viewed in the animation of the simulation).*

tively in the case of a single stroke moving with velocity $U$ at an angle of attack $\alpha$, which serves as a snapshot of a flapping wing, with $U$ the sum of $u_0$ and $u_1$, $\alpha = \tan^{-1}[u_1(t)/u_0]$.

Fig. 6 shows the lift coefficient as a function of angle of attack. The lift peaks at $45^0 - 60^0$ for $Re \sim 1000$. Fig. 6 shows the thrust coefficient. The thrust exists only at or above a minimum angle of attack, $\alpha_{min} = 20^0$, as seen in the left of Figure 6. More interestingly, both lift and drag are time-dependent, and the thrust only occurs in a time window $t_s \in (0.2, 1.5)$, or equivalently, $t \in (0.4, 3)$. The lower bound can be associated with the time scale governing the growth of the trailing edge vortex hence the lift. The upper bound can be attributed to the time scale governing the shedding of the leading edge vortex which reduces the lift.

Now we can imagine that during a flapping motion, the effective angle of attack increases from $0^0$ to a maximum angle of attack over time $T/4$. In

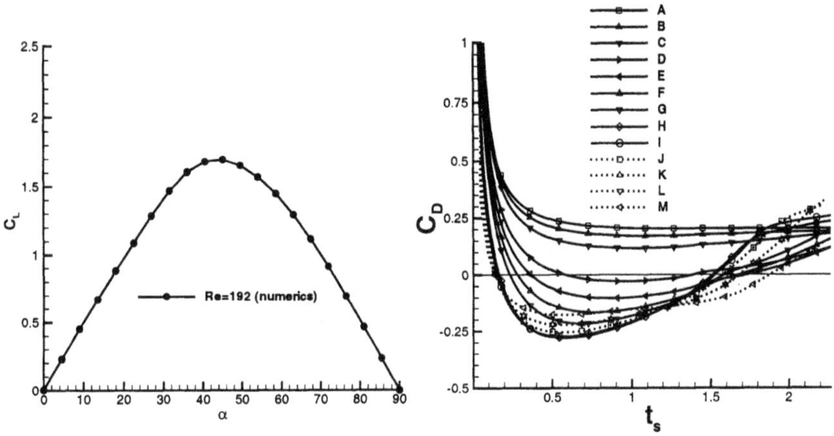


FIG. 6. *Left figure: Lift coefficients as a function of angle of attack. The data are measured at* $t_s = 2.0$. *Right figure: Thrust window for different angles of attack:* $\alpha = 5^0(A)$, $10^0(B)$, $15^0(C)$, $25^0(D)$, $30^0(E)$, $35^0(F)$, $40^0(G)$, $45^0(H)$, $55^0(I)$, $60^0(J)$, $65^0(K)$, $70^0(L)$, *and* $75^0(M)$.

order to achieve a high thrust, there are two conditions. First, as suggested by Fig. 6, the maximum angle of attack should be in the range of $45^0 - 60^0$, which sets $St_a$ to be in the range of $0.16 - 0.27$, Secondly, for each fixed $St_a$, the wing stroke should be just long enough to stay inside the thrust window. Among the four different frequencies shown in Fig. 5, $f = 2Hz$ falls outside the thrust window, and indeed we find a drag instead of a thrust; $f = 0.5Hz$ corresponds to a situation where the flap just reaches its $\max(C_x)$ at the end of the half cycle to benefit from the whole thrust window to produce a large thrust. The other two cases correspond to the intermediate situation.

To confirm this picture, we repeat similar experiments for different $St_a$'s. The results are summarized in Figure 7, where the absolute values of the thrust coefficient and efficiency are plotted against $St_c$. Note that for each fixed $St_c$, $|C_x|$ increases with $St_a$ as expected. And finally, the maxima occur at $St_c \sim 0.7$ independent of $St_a$ at higher $St_a$. This is consistent with Fig. 6, where we observe that maximum thrust occurs at a time which is insensitive to $St_a$ for sufficiently high $St_a$.

**3.3. Discussion.** Insect flight is certainly far more complex and diverse than the simple model considered here. Therefore it is difficult to compare directly the exact numerical values of the observed optimal frequencies with the biological data. Nevertheless, the scaling result shown in Fig. 7, i.e., $St_c \sim 0.7$, suggests that the optimal frequencies are inversely proportional to the dimension of the wing. This is consistent with the data compiled by Greenewalt [19] in his Fig. 10 and Fig. 12 for birds and insect

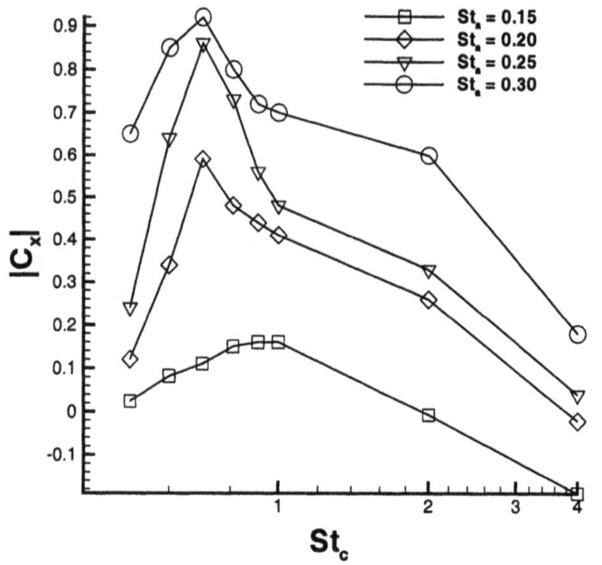


FIG. 7. *Thrust coefficient as a function of $St_c$ for a given $St_a$. Different $St_a$'s are represented by different symbols.*

flight, covering scales over almost three decades, despite scatters in the data.

**4. Hovering flight.** Hovering is an extreme mode of flight where the forward velocity is zero. Because the mean forward velocity is zero, insects have to draw clean air from the ambient flow and to get rid of the "messy vortices" they have created so to obtain a periodic lift. So what do they do and why does it work?

A special class of hovering insects, e.g., a small wasp (encasia formosa), use a cleaver "clap and fling" mechanism, discovered by Weis-Fogh, later analyzed by Lighthill [20], and experimentally investigated by Maxworthy [2]. Such a strategy can create unusually large lift by shutting and opening a pair of wings, each of which gains a circulation that serves as the starting vortex for the other wing, thereby sidestepping the classical Wagner effect of delay in force generation [18]. Lighthill also argued that in three dimensional case, pairs of counter rotating tip-vortices are shed into a wake [20].

However, a large class of insects do not use "clap and fling". For example, dragonflies, hawkmoth, and fruitflies employ "figure of eight" strokes; each wing sweeps back and forth while rotating about the spanwise axis. The required large lift must result from other unsteady effects. These are emphasized in the recent experiments of Ellington et al. and Dickinson et al. [5, 6].

The goal of this work is to quantify the vortex dynamics that is essential for hovering and identify a minimal two dimensional model that produces sufficient lift. We compute flows around a two dimensional hovering wing which follows a simplified motion based on dragonfly sequence, and find that the right combination of flapping and rotation can create an organized dipole wake jetted downward thus generates a lift on the wing [8]. This dipole wake is consistent with the early experimental observations [9]. A snapshot of one dipole is shown in Fig. 8.

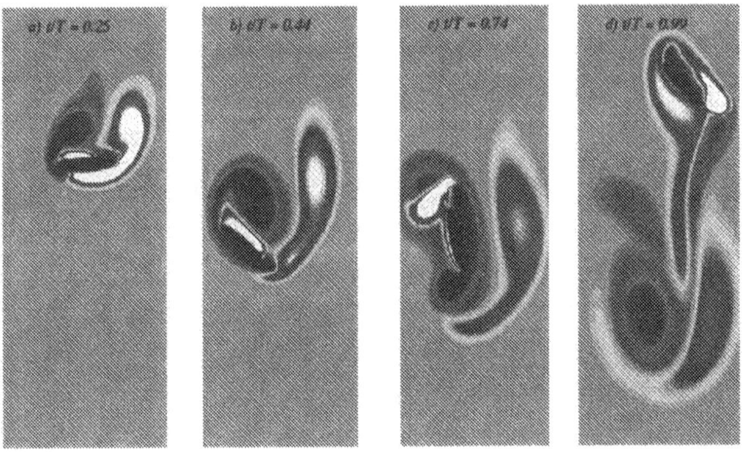

FIG. 8. *Snapshots of the vorticity field illustrate the formation and dynamics of a dipole jet during hovering. During the down-stroke, as shown in a) and b), a pair of counter-rotating vortices is generated by the translational motion, and during the upstroke, as shown in c) and d), the rotation fuses the leading and trailing edge vortices to form a co-moving dipole pair. The dipole and wing then separate in the upstroke phase. At the end of the upstroke, shown in d), the wing is sufficiently far away from the strong vortices generated in the previous cycle, and is ready to repeat the whole process without interfering with the previous vortices. The colors indicate the strength of the local vorticity, with blue indicating counter clockwise rotation and red clockwise rotation. The parameters are* $A = 2.5cm$, $T = 0.025s$, $\beta = \pi/3$, $\phi = 0$, $\nu = 2.0cm^2/s$, *thus* $Re = 157$.

The vortex dynamics suggests that three dimensionality is not essential for hovering. Previously it was assumed the dipole vortices were formed from the shedding of tip-vortices [21, 22]– a three dimensional effect. But apparently the figure-eight motion alone can generate a downward flow.

Despite the differences, we observe similarities between the dynamics of vortices in two and three dimensions. In particular, the dominant leading-edge vortex is attached to the wing during the downward phase, which is consistent with experiments [5]. The "dynamic stall" is present in two-dimensional hovering partly because the vortex separation time scale is larger than half of the flapping period and partly because the wing rotates.

To quantify the lift and compare it with the weight of a typical insect, we plot lift (vertical force) and drag (horizontal force) as a function of time in Fig. 9.

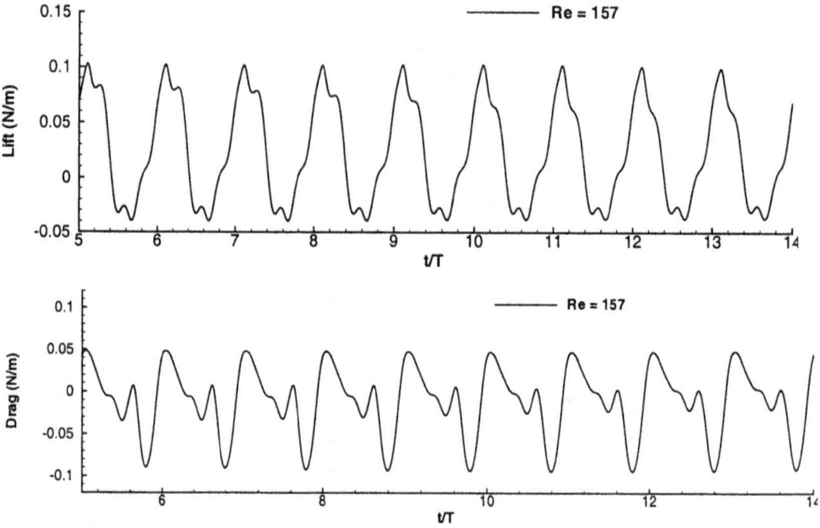


FIG. 9. *Time dependent two dimensional lift and drag. The time is normalized by its period,* $0.025s$.

The forces quickly reach a periodic state after one period, because of negligible interference from vortices generated in previous cycles as seen in Fig. 8. The double peaks in up and down strokes resemble typical experimental measurements [6, 23]. The averaged two-dimensional lift and thrust over 50 periods are $2.8 \times 10^{-2}$ N/m and $5.2 \times 10^{-3}$ N/m, respectively. Repeated computation with twice the spatial resolution gives the same answer in the significant digits.

Based on these values, we estimate the forces on a three dimensional wing using a standard elliptical loading along the wing span [24]. For a wing 5 cm long and 1 cm wide at mid-span, we find the total lift supported by one wing is $1.1 \times 10^{-3}$N. It follows that two pairs of wing can support $4.4 \times 10^{-3}$N, which is more than the weight of a typical dragonfly ranging between $1 \times 10^{-3}$N to $2.5 \times 10^{-3}$N [25].

To show that these results are not fortuitous, we vary separately the phase, the viscosity, and the amplitude, to study their effects on the forces. The average force is insensitive to viscosity at sufficiently small viscosities, because the lift is dominated by the dynamic pressure rather than by viscous force. Finally lift increases with amplitude almost quadratically and the lift coefficient, normalized by maximum translational velocity, is close to one. To show that these results are not fortuitous, we vary separately the phase, the viscosity, and the amplitude, to study their effects on the forces. The average force is insensitive to viscosity at sufficiently small viscosities, because the lift is dominated by the dynamic pressure rather than by viscous force. Finally lift increases with amplitude almost quadratically and the lift coefficient, normalized by maximum translational velocity, is close to one [8].

We may speculate what will happen to these vortices in three dimensions. Ellington et al. [5] pointed out that the axial flow along the leading edge, an intrinsic three dimensional effect, in principal can stabilize the leading edge vortex and prevent it from shedding. On the other hand, early experiments by Freymuth et al. [9] have shown a dipole wake around a mechanic wing undergoing generic sinusoidal motion. Our computations suggest that the dipoles are formed by leading and trailing edges vortex along the wing span. We suspect that in three dimensions, the two spanwise vortex lines along the centers of leading and trailing edge vortex are connected by the tip vortex, and they will shed as a whole at the end of each stroke. This scenario is one of many possibilities, and future experiments and three dimensional computation will offer more definitive answers.

**5. Summary.** In summary, we implemented and tested a computational tool to resolve the time-dependent vorticity and forces in unsteady viscous flows around a two dimensional moving wing. The analysis of unsteady flows in forward flight revealed a mechanism of frequency selection, which resulted from the two intrinsic time scales associated with the shedding of leading and trailing edge vortices. The predicted preferred frequency scales inversely with the size of the wing, which is consistent with the zoological observation. The computation of hovering flight uncovered the origin of an organized dipole jet in hovering flight and suggested a fundamental leading order two dimensional mechanism for lift generation in generic hovering flight employing heaving and pitching motions.

**Acknowledgments.** I am grateful to Steve Childress and Mike Shelley for many useful discussions, Jianguo Liu for the ec4 code on cavity flows, and NERSC for providing the supercomputer CPU time. The work is supported in part by NSF grant No. DMS-9400912 and DMS-9510356 and DOE grant No. DE-FG02-88ER25053 through Courant Institute of Mathematical Sciences.

## REFERENCES

[1] T. Weis-Fogh and M. Jensen, Proc. Roy. Soc. B. 239 **239**, 415 (1956).
[2] T. Maxworthy, Ann. Rev. Fluid Mech. **13**, 329 (1981).
[3] C. P. Ellington, Phil. Trans. R. Soc. Lond. **B 305**, 1 (1984).
[4] G. R. Spedding, Contemporary Mathematics **141**, 401 (1993).
[5] C. P. Ellington, C. van den Berg, A. P. Willmott, and A. L. R. Thomas, Nature **384**, 626 (1996).
[6] M. H. Dickinson, F. O. Lehmann, and S. P. Sane, Science **284**, 1954 (1999).
[7] Z. J. Wang, J. Fluid Mech. **410**, 323 (2000).

[8] Z. J. Wang, Phys. Rev. Lett. **85**, 10, 2035 (2000).
[9] P. Freymuth, K. Gustavson, and R. Leben, in *Votex method and vortex motion*, edited by K. Gustavson and J. Sethian (SIAM, Philadelphia, 1991), p. 143.
[10] H. Liu *et al.*, J. Exp. Biol. **201**, 461 (1998).
[11] W. E and J. Liu, J. Comp. Phys. **126**, 122 (1996).
[12] Z. J. Wang, J. of Comp. Phys. **153**, 666 (1999).
[13] M. Coutanceau and R. Bouard, J. Fluid Mech. **79**, 231 (1977).
[14] R. Bouard and M. Coutanceau, J. Fluid Mech. **101**, 583 (1980).
[15] Z. J. Wang, J.-G. Liu, and S. Childress, Phys. Fluids. **11**, 9 (1999).
[16] M. S. Triantafyllou, G. S. Triantafyllou, and R. Gopalkrishnan, Phys. Fluids A **3**, 12 (1991).
[17] R. A. Norberg, In Swimming and Flying in Nature(ed. T. Y. Wu, C. J. Brokaw and C. Brennen) .
[18] H. Wagner, Z. angew. Math. Mech. **5**, 17 (1925).
[19] C. H. Greenewalt, *Dimensional Relationships for Flying Animals, publication 4477* (Smithsonian Miscellaneous Collections, Washington D.C., 1962).
[20] S. J. Lighthill, J. Fluid Mech. **60**, (1973).
[21] J. Rayner, J. Fluid Mech. **91**, 697 (1979).
[22] M. J. Lighthill, *Mathematical Biofluiddynamics*, SIAM, Philadelphia, 1975.
[23] M. Cloupeau, J. F. Devillers, and D. Devezeaux, J. Exp. Biol. **80**, 1(1979).
[24] A. M. Kuethe and C.-Y. Chow, *Foundations of Aerodynamics* John Wiley & Sons. New York, 1976.
[25] J. M. Wakeling, and C. P. Ellington, J. Exp. Biol. **200**, 557(1997), S. Savage, B. G. Newman, and D. T.-M. Wong, J. exp. Biol. **83**, 59(1979).

## LIST OF WORKSHOP PARTICIPANTS

- Christian Alscher, FSP Mathematisierung, University of Bielefeld
- Kevin Anderson, Institute for Mathematics and its Applications
- Kayne Arthurs, Department of Mathematics, Duke University
- Bruce Ayati, Institute for Mathematics and its Applications
- J.R. Blake, School of Mathematics and Statistics, University of Birmingham
- Jacob J. Blum, Department of Cell Biology, Duke University Medical Center
- Danielle Carr, Department of Mathematics, Bryn Mawr College
- Jun Choe, Department of Mathematics, Korea Advanced Institute of Science and Technology
- Dan Coombs, Program in Applied Mathematics, University of Arizona
- Ricard Cortez, Department of Mathematics, Tulane University
- Robert Dillon, Department of Mathematics, Washington State University
- Fred Dulles, Institute for Mathematics and its Applications
- Lisa J. Fauci, Department of Mathematics, Tulane University
- Aaron Fogelson, Department of Mathematics, University of Utah
- Don Gaver, Department of Biomedical Engineering, Tulane University
- Stuart Goldstein, GCB Department, University of Minnesota
- Joseph F. Grcar, Sandia National Laboratory
- Shay Gueron, Department of Mathematics, Technion - Israel Institute of Technology
- Mansoor Haider, Department of Mathematics, Duke University
- Dave Halpern, Department of Mathematics, University of Alabama
- Cushing Hamlen, Medtronic, Inc.
- Matt Hopkins, Department of Mathematics, Rose-Hulman Institute
- Chris Jordan, Department of Environmental Population Organismic Biology, University of Colorado
- Mimi Koehl, Department of Integrative Biology, University of California-Berkeley
- Harold Layton Department of Mathematics Duke University
- Randall Leveque, Department of Applied Mathematics, University of Washington

- Charles Lindemann, Department of Biological Sciences, Oakland University
- Nadov Liron, Mathematics Department, Technion, Israel Institute of Technology
- Sharon Lubkin, Biomathematics Program, North Carolina State University
- David McQueen, Courant Institute, New York University
- Willard Miller, Institute for Mathematics and its Applications
- Peter Moore, Department of Mathematics, Tulane University
- Patrick Nelson, Department of Mathematics, Duke University
- Mette Olufsen, Mathematics, Center for Biodynamics, Boston University
- Tim Pedley, Department of Applied Mathematics and Theoretical Physics, University of Cambridge
- C. Pozrikidis, AMES, University of California-San Diego
- Zvi Priel, Department of Chemistry, Ben Gurion University
- Robert Root, Department of Mathematics, Lafayette College
- Peter Satir, Jack and Peal Resnick Campus, Albert Einstein College of Medicine
- Zachariah Sinkala, Middle Tennessee State University
- Michael Sleigh, School of Biological Sciences, University of Southampton
- Carl S. Smith, Urology Division, Hennepin County Medical Center
- John M. Stockie, Department of Mathematics and Statistics, Simon Fraser University
- Z. Jane Wang, Theoretical and Applied Mechanics, Cornell University
- Harvey Williams, Mathematics and Biomedical Engineering, Tulane University
- Ralf Wittenberg, Department of Mathematics, University of Michigan

## IMA SUMMER PROGRAMS

1987 Robotics
1988 Signal Processing
1989 Robust Statistics and Diagnostics
1990 Radar and Sonar (June 18–29)
New Directions in Time Series Analysis (July 2–27)
1991 Semiconductors
1992 Environmental Studies: Mathematical, Computational, and Statistical Analysis
1993 Modeling, Mesh Generation, and Adaptive Numerical Methods for Partial Differential Equations
1994 Molecular Biology
1995 Large Scale Optimizations with Applications to Inverse Problems, Optimal Control and Design, and Molecular and Structural Optimization
1996 Emerging Applications of Number Theory (July 15–26)
Theory of Random Sets (August 22–24)
1997 Statistics in the Health Sciences
1998 Coding and Cryptography (July 6–18)
Mathematical Modeling in Industry (July 22–31)
1999 Codes, Systems, and Graphical Models (August 2–13, 1999)
2000 Mathematical Modeling in Industry: A Workshop for Graduate Students (July 19–28)
2001 Geometric Methods in Inverse Problems and PDE Control (July 16–27)
2002 Special Functions in the Digital Age (July 22–August 2)

## IMA "HOT TOPICS" WORKSHOPS

- Challenges and Opportunities in Genomics: Production, Storage, Mining and Use, April 24–27, 1999
- Decision Making Under Uncertainty: Energy and Environmental Models, July 20–24, 1999
- Analysis and Modeling of Optical Devices, September 9–10, 1999
- Decision Making under Uncertainty: Assessment of the Reliability of Mathematical Models, September 16–17, 1999
- Scaling Phenomena in Communication Networks, October 22–24, 1999
- Text Mining, April 17–18, 2000
- Mathematical Challenges in Global Positioning Systems (GPS), August 16–18, 2000
- Modeling and Analysis of Noise in Integrated Circuits and Systems, August 29–30, 2000
- Mathematics of the Internet: E-Auction and Markets, December 3–5, 2000
- Analysis and Modeling of Industrial Jetting Processes, January 10–13, 2001

- Wireless Networks, August 6–10 2001

## SPRINGER LECTURE NOTES FROM THE IMA:

*The Mathematics and Physics of Disordered Media*
Editors: Barry Hughes and Barry Ninham
(Lecture Notes in Math., Volume 1035, 1983)

*Orienting Polymers*
Editor: J.L. Ericksen
(Lecture Notes in Math., Volume 1063, 1984)

*New Perspectives in Thermodynamics*
Editor: James Serrin
(Springer-Verlag, 1986)

*Models of Economic Dynamics*
Editor: Hugo Sonnenschein
(Lecture Notes in Econ., Volume 264, 1986)

*Current Volumes:*

1 **Homogenization and Effective Moduli of Materials and Media** J. Ericksen, D. Kinderlehrer, R. Kohn, and J.-L. Lions (eds.)
2 **Oscillation Theory, Computation, and Methods of Compensated Compactness** C. Dafermos, J. Ericksen, D. Kinderlehrer, and M. Slemrod (eds.)
3 **Metastability and Incompletely Posed Problems** S. Antman, J. Ericksen, D. Kinderlehrer, and I. Muller (eds.)
4 **Dynamical Problems in Continuum Physics** J. Bona, C. Dafermos, J. Ericksen, and D. Kinderlehrer (eds.)
5 **Theory and Applications of Liquid Crystals** J. Ericksen and D. Kinderlehrer (eds.)
6 **Amorphous Polymers and Non-Newtonian Fluids** C. Dafermos, J. Ericksen, and D. Kinderlehrer (eds.)
7 **Random Media** G. Papanicolaou (ed.)
8 **Percolation Theory and Ergodic Theory of Infinite Particle Systems** H. Kesten (ed.)
9 **Hydrodynamic Behavior and Interacting Particle Systems** G. Papanicolaou (ed.)
10 **Stochastic Differential Systems, Stochastic Control Theory, and Applications** W. Fleming and P.-L. Lions (eds.)
11 **Numerical Simulation in Oil Recovery** M.F. Wheeler (ed.)
12 **Computational Fluid Dynamics and Reacting Gas Flows** B. Engquist, M. Luskin, and A. Majda (eds.)
13 **Numerical Algorithms for Parallel Computer Architectures** M.H. Schultz (ed.)
14 **Mathematical Aspects of Scientific Software** J.R. Rice (ed.)
15 **Mathematical Frontiers in Computational Chemical Physics** D. Truhlar (ed.)
16 **Mathematics in Industrial Problems** A. Friedman
17 **Applications of Combinatorics and Graph Theory to the Biological and Social Sciences** F. Roberts (ed.)
18 ***q*-Series and Partitions** D. Stanton (ed.)
19 **Invariant Theory and Tableaux** D. Stanton (ed.)
20 **Coding Theory and Design Theory Part I: Coding Theory** D. Ray-Chaudhuri (ed.)
21 **Coding Theory and Design Theory Part II: Design Theory** D. Ray-Chaudhuri (ed.)
22 **Signal Processing Part I: Signal Processing Theory** L. Auslander, F.A. Grünbaum, J.W. Helton, T. Kailath, P. Khargonekar, and S. Mitter (eds.)

23 **Signal Processing Part II: Control Theory and Applications of Signal Processing** L. Auslander, F.A. Grünbaum, J.W. Helton, T. Kailath, P. Khargonekar, and S. Mitter (eds.)
24 **Mathematics in Industrial Problems, Part 2** A. Friedman
25 **Solitons in Physics, Mathematics, and Nonlinear Optics** P.J. Olver and D.H. Sattinger (eds.)
26 **Two Phase Flows and Waves** D.D. Joseph and D.G. Schaeffer (eds.)
27 **Nonlinear Evolution Equations that Change Type** B.L. Keyfitz and M. Shearer (eds.)
28 **Computer Aided Proofs in Analysis** K. Meyer and D. Schmidt (eds.)
29 **Multidimensional Hyperbolic Problems and Computations** A. Majda and J. Glimm (eds.)
30 **Microlocal Analysis and Nonlinear Waves** M. Beals, R. Melrose, and J. Rauch (eds.)
31 **Mathematics in Industrial Problems, Part 3** A. Friedman
32 **Radar and Sonar, Part I** R. Blahut, W. Miller, Jr., and C. Wilcox
33 **Directions in Robust Statistics and Diagnostics: Part I** W.A. Stahel and S. Weisberg (eds.)
34 **Directions in Robust Statistics and Diagnostics: Part II** W.A. Stahel and S. Weisberg (eds.)
35 **Dynamical Issues in Combustion Theory** P. Fife, A. Liñán, and F.A. Williams (eds.)
36 **Computing and Graphics in Statistics** A. Buja and P. Tukey (eds.)
37 **Patterns and Dynamics in Reactive Media** H. Swinney, G. Aris, and D. Aronson (eds.)
38 **Mathematics in Industrial Problems, Part 4** A. Friedman
39 **Radar and Sonar, Part II** F.A. Grünbaum, M. Bernfeld, and R.E. Blahut (eds.)
40 **Nonlinear Phenomena in Atmospheric and Oceanic Sciences** G.F. Carnevale and R.T. Pierrehumbert (eds.)
41 **Chaotic Processes in the Geological Sciences** D.A. Yuen (ed.)
42 **Partial Differential Equations with Minimal Smoothness and Applications** B. Dahlberg, E. Fabes, R. Fefferman, D. Jerison, C. Kenig, and J. Pipher (eds.)
43 **On the Evolution of Phase Boundaries** M.E. Gurtin and G.B. McFadden
44 **Twist Mappings and Their Applications** R. McGehee and K.R. Meyer (eds.)
45 **New Directions in Time Series Analysis, Part I** D. Brillinger, P. Caines, J. Geweke, E. Parzen, M. Rosenblatt, and M.S. Taqqu (eds.)

46 **New Directions in Time Series Analysis, Part II**
D. Brillinger, P. Caines, J. Geweke, E. Parzen, M. Rosenblatt, and M.S. Taqqu (eds.)

47 **Degenerate Diffusions**
W.-M. Ni, L.A. Peletier, and J.-L. Vazquez (eds.)

48 **Linear Algebra, Markov Chains, and Queueing Models**
C.D. Meyer and R.J. Plemmons (eds.)

49 **Mathematics in Industrial Problems, Part 5** A. Friedman

50 **Combinatorial and Graph-Theoretic Problems in Linear Algebra**
R.A. Brualdi, S. Friedland, and V. Klee (eds.)

51 **Statistical Thermodynamics and Differential Geometry of Microstructured Materials**
H.T. Davis and J.C.C. Nitsche (eds.)

52 **Shock Induced Transitions and Phase Structures in General Media** J.E. Dunn, R. Fosdick, and M. Slemrod (eds.)

53 **Variational and Free Boundary Problems**
A. Friedman and J. Spruck (eds.)

54 **Microstructure and Phase Transitions**
D. Kinderlehrer, R. James, M. Luskin, and J.L. Ericksen (eds.)

55 **Turbulence in Fluid Flows: A Dynamical Systems Approach**
G.R. Sell, C. Foias, and R. Temam (eds.)

56 **Graph Theory and Sparse Matrix Computation**
A. George, J.R. Gilbert, and J.W.H. Liu (eds.)

57 **Mathematics in Industrial Problems, Part 6** A. Friedman

58 **Semiconductors, Part I**
W.M. Coughran, Jr., J. Cole, P. Lloyd, and J. White (eds.)

59 **Semiconductors, Part II**
W.M. Coughran, Jr., J. Cole, P. Lloyd, and J. White (eds.)

60 **Recent Advances in Iterative Methods**
G. Golub, A. Greenbaum, and M. Luskin (eds.)

61 **Free Boundaries in Viscous Flows**
R.A. Brown and S.H. Davis (eds.)

62 **Linear Algebra for Control Theory**
P. Van Dooren and B. Wyman (eds.)

63 **Hamiltonian Dynamical Systems: History, Theory, and Applications**
H.S. Dumas, K.R. Meyer, and D.S. Schmidt (eds.)

64 **Systems and Control Theory for Power Systems**
J.H. Chow, P.V. Kokotovic, R.J. Thomas (eds.)

65 **Mathematical Finance**
M.H.A. Davis, D. Duffie, W.H. Fleming, and S.E. Shreve (eds.)

66 **Robust Control Theory** B.A. Francis and P.P. Khargonekar (eds.)

67 **Mathematics in Industrial Problems, Part 7** A. Friedman

68 **Flow Control** M.D. Gunzburger (ed.)

69 **Linear Algebra for Signal Processing**
A. Bojanczyk and G. Cybenko (eds.)

70 **Control and Optimal Design of Distributed Parameter Systems**
J.E. Lagnese, D.L. Russell, and L.W. White (eds.)

71 **Stochastic Networks** F.P. Kelly and R.J. Williams (eds.)

72 **Discrete Probability and Algorithms**
D. Aldous, P. Diaconis, J. Spencer, and J.M. Steele (eds.)

73 **Discrete Event Systems, Manufacturing Systems, and Communication Networks**
P.R. Kumar and P.P. Varaiya (eds.)

74 **Adaptive Control, Filtering, and Signal Processing**
K.J. Åström, G.C. Goodwin, and P.R. Kumar (eds.)

75 **Modeling, Mesh Generation, and Adaptive Numerical Methods for Partial Differential Equations** I. Babuska, J.E. Flaherty, W.D. Henshaw, J.E. Hopcroft, J.E. Oliger, and T. Tezduyar (eds.)

76 **Random Discrete Structures** D. Aldous and R. Pemantle (eds.)

77 **Nonlinear Stochastic PDEs: Hydrodynamic Limit and Burgers' Turbulence** T. Funaki and W.A. Woyczynski (eds.)

78 **Nonsmooth Analysis and Geometric Methods in Deterministic Optimal Control** B.S. Mordukhovich and H.J. Sussmann (eds.)

79 **Environmental Studies: Mathematical, Computational, and Statistical Analysis** M.F. Wheeler (ed.)

80 **Image Models (and their Speech Model Cousins)**
S.E. Levinson and L. Shepp (eds.)

81 **Genetic Mapping and DNA Sequencing**
T. Speed and M.S. Waterman (eds.)

82 **Mathematical Approaches to Biomolecular Structure and Dynamics**
J.P. Mesirov, K. Schulten, and D. Sumners (eds.)

83 **Mathematics in Industrial Problems, Part 8** A. Friedman

84 **Classical and Modern Branching Processes**
K.B. Athreya and P. Jagers (eds.)

85 **Stochastic Models in Geosystems**
S.A. Molchanov and W.A. Woyczynski (eds.)

86 **Computational Wave Propagation**
B. Engquist and G.A. Kriegsmann (eds.)

87 **Progress in Population Genetics and Human Evolution**
P. Donnelly and S. Tavaré (eds.)

88 **Mathematics in Industrial Problems, Part 9** A. Friedman

89 **Multiparticle Quantum Scattering With Applications to Nuclear, Atomic and Molecular Physics** D.G. Truhlar and B. Simon (eds.)

90 **Inverse Problems in Wave Propagation** G. Chavent, G. Papanicolau, P. Sacks, and W.W. Symes (eds.)

91 **Singularities and Oscillations** J. Rauch and M. Taylor (eds.)

92 **Large-Scale Optimization with Applications, Part I: Optimization in Inverse Problems and Design**
L.T. Biegler, T.F. Coleman, A.R. Conn, and F. Santosa (eds.)

93 **Large-Scale Optimization with Applications, Part II: Optimal Design and Control**
L.T. Biegler, T.F. Coleman, A.R. Conn, and F. Santosa (eds.)

94 **Large-Scale Optimization with Applications, Part III: Molecular Structure and Optimization**
L.T. Biegler, T.F. Coleman, A.R. Conn, and F. Santosa (eds.)

95 **Quasiclassical Methods**
J. Rauch and B. Simon (eds.)

96 **Wave Propagation in Complex Media**
G. Papanicolaou (ed.)

97 **Random Sets: Theory and Applications**
J. Goutsias, R.P.S. Mahler, and H.T. Nguyen (eds.)

98 **Particulate Flows: Processing and Rheology**
D.A. Drew, D.D. Joseph, and S.L. Passman (eds.)

99 **Mathematics of Multiscale Materials** K.M. Golden, G.R. Grimmett, R.D. James, G.W. Milton, and P.N. Sen (eds.)

100 **Mathematics in Industrial Problems, Part 10** A. Friedman

101 **Nonlinear Optical Materials** J.V. Moloney (ed.)

102 **Numerical Methods for Polymeric Systems** S.G. Whittington (ed.)

103 **Topology and Geometry in Polymer Science** S.G. Whittington, D. Sumners, and T. Lodge (eds.)

104 **Essays on Mathematical Robotics** J. Baillieul, S.S. Sastry, and H.J. Sussmann (eds.)

105 **Algorithms For Parallel Processing** M.T. Heath, A. Ranade, and R.S. Schreiber (eds.)

106 **Parallel Processing of Discrete Problems** P.M. Pardalos (ed.)

107 **The Mathematics of Information Coding, Extraction, and Distribution** G. Cybenko, D.P. O'Leary, and J. Rissanen (eds.)

108 **Rational Drug Design** D.G. Truhlar, W. Howe, A.J. Hopfinger, J. Blaney, and R.A. Dammkoehler (eds.)

109 **Emerging Applications of Number Theory** D.A. Hejhal, J. Friedman, M.C. Gutzwiller, and A.M. Odlyzko (eds.)

110 **Computational Radiology and Imaging: Therapy and Diagnostics**
C. Börgers and F. Natterer (eds.)

111 **Evolutionary Algorithms** L.D. Davis, K. De Jong, M.D. Vose, and L.D. Whitley (eds.)

112 **Statistics in Genetics** M.E. Halloran and S. Geisser (eds.)

113 **Grid Generation and Adaptive Algorithms** M.W. Bern, J.E. Flaherty, and M. Luskin (eds.)

114 **Diagnosis and Prediction** S. Geisser (ed.)

115 **Pattern Formation in Continuous and Coupled Systems: A Survey Volume** M. Golubitsky, D. Luss, and S.H. Strogatz (eds.)

116 **Statistical Models in Epidemiology, the Environment, and Clinical Trials** M.E. Halloran and D. Berry (eds.)

117 **Structured Adaptive Mesh Refinement (SAMR) Grid Methods**
S.B. Baden, N.P. Chrisochoides, D.B. Gannon, and M.L. Norman (eds.)

118 **Dynamics of Algorithms**
R. de la Llave, L.R. Petzold, and J. Lorenz (eds.)

119 **Numerical Methods for Bifurcation Problems and Large-Scale Dynamical Systems**
E. Doedel and L.S. Tuckerman (eds.)

120 **Parallel Solution of Partial Differential Equations**
P. Bjørstad and M. Luskin (eds.)

121 **Mathematical Models for Biological Pattern Formation**
P.K. Maini and H.G. Othmer (eds.)

122 **Multiple-Time-Scale Dynamical Systems**
C.K.R.T. Jones and A. Khibnik (eds.)

123 **Codes, Systems, and Graphical Models**
B. Marcus and J. Rosenthal (eds.)

124 **Computational Modeling in Biological Fluid Dynamics**
L.J. Fauci and S. Gueron (eds.)

**Forthcoming Volumes:**

1998–1999: Mathematics in Biology

Membrane Transport and Renal Physiology

Mathematical Approaches for Emerging and Reemerging Infectious Diseases

Decision Making under Uncertainty: Energy and Environmental Models

1999–2000: Reactive Flow and Transport Phenomena

Fire

Confinement and Remediation of Environmental Hazards and Resource Recovery

Dispersive Corrections to Transport Equations, Simulation of Transport in Transition Regimes, Multiscale Models for Surface Evolution and Reacting Flows

Mathematical Foundations of Speech Processing and Recognition, Mathematical Foundations of Natural Language Modeling

Image Processing and Low Level Vision, Image Analysis and High Level Vision, Mathematical Methods in Speech and Image Analysis

Mathematics of the Internet: E-auction and Markets

www.ingramcontent.com/pod-product-compliance
Lightning Source LLC
LaVergne TN
LVHW012304141025
823511LV00003B/194
*9780387952338*